Carbon-Neutral Infrastructure

Carbon-Neutral Infrastructure

Guest Editors

Wentong Wang
Dongdong Yuan
Augusto Cannone Falchetto
Fucheng Guo

MDPI

Basel • Beijing • Wuhan • Barcelona • Belgrade • Novi Sad • Cluj • Manchester

Guest Editors
Wentong Wang
Faculty of Transportation
Shandong University of
Science and Technology
Qingdao
China

Dongdong Yuan
Key Laboratory for Special
Area Highway Engineering of
Ministry of Education
Chang'an University
Xi'an
China

Augusto Cannone Falchetto
Department of Civil,
Environmental, and
Architectural Engineering
University of Padua
Padua
Italy

Fucheng Guo
School of Civil Engineering
Lanzhou Jiaotong University
Lanzhou
China

Editorial Office
MDPI AG
Grosspeteranlage 5
4052 Basel, Switzerland

This is a reprint of the Special Issue, published open access by the journal *Buildings* (ISSN 2075-5309), freely accessible at: https://www.mdpi.com/journal/buildings/special_issues/BJSHW8S61B.

For citation purposes, cite each article independently as indicated on the article page online and as indicated below:

Lastname, A.A.; Lastname, B.B. Article Title. *Journal Name* **Year**, *Volume Number*, Page Range.

ISBN 978-3-7258-5791-3 (Hbk)
ISBN 978-3-7258-5792-0 (PDF)
https://doi.org/10.3390/books978-3-7258-5792-0

Contents

buildings

MDPI

Editorial

Carbon-Neutral Infrastructure

Wentong Wang [1,*], Dongdong Yuan [2], Augusto Cannone Falchetto [3] and Fucheng Guo [4]

1 Faculty of Transportation, Shandong University of Science and Technology, Qingdao 266590, China
2 School of Highway, Chang'an University, Middle-Section of Nan'er Huan Road, Xi'an 710064, China; ddy@chd.edu.cn
3 Department of Civil Engineering, Aalto University, 02150 Espoo, Finland; augusto.cannonefalchetto@aalto.fi
4 School of Civil Engineering, Lanzhou Jiaotong University, Lanzhou 730070, China; fcguo@lzjtu.edu.cn
* Correspondence: wwt@sdust.edu.cn

1. Preamble

As global awareness of climate change intensifies, the significance of carbon neutrality has become increasingly evident and a pivotal sustainable development target for human society to pursue. As a fundamental element of economic and social advancement, carbon emissions during infrastructure construction and operation must be given due consider. Therefore, comprehensive assessments of the impact of the entire life cycle of infrastructure on carbon emissions are imperative.

The life cycle of infrastructure comprises a series of phases, including material selection, construction processes, and maintenance management. To achieve carbon neutrality, it is essential to conduct comprehensive research and pursue innovative solutions in all of these areas. A substantial corpus of research has been conducted across various academic disciplines, including the exploration of low-carbon materials, utilization of recycled materials, promotion of green construction technology, total life cycle management, and intelligent monitoring management and control. Considering the prevailing trends in research and development, this journal is committed to collecting high-quality scientific papers, disseminating advanced scientific and technological knowledge within the aforementioned fields, and discussing the technical feasibility of infrastructure construction in the context of intelligent, carbon-neutral infrastructure development, and technological innovation.

In this context, this Special Issue focuses on carbon -neutral and collecting in-depth and detailed studies from various aspects, aiming at exploring the effective application and great potential of various engineering materials and advanced technologies in the ambitious process of gradually realizing the low-carbon development and sustainable development of infrastructure. Through the analysis of the characteristics of different engineering materials and in-depth studies on innovative technologies, we intend to provide practical solutions for the low-carbon development of infrastructure.

2. Innovative Technologies or Methods Used in Infrastructure

From the perspective of the resource utilization of industrial wastes, research has focused on the treatment and application of solid wastes, including red mud, dredging sludge, waste steel slag, and construction waste. From an environmental standpoint, the vast quantities of these wastes represent a significant burden. Conversely, their application in the construction of infrastructure, such as roads, can address the issue of waste disposal, reduce reliance on natural resources, lower project costs, and facilitate resource recycling, thereby contributing to the attainment of carbon neutrality. From this perspective, a strength prediction neural network has been proposed for modified red mud, which is used as a road material in engineering applications, and the network is used to optimize the material solution so that it can meet the strength requirements of the project while reducing the cost [1]. This study was predicated on experimental data accumulated over the course of years of engineering practice, as well as artificial neural networks, utilizing a backpropagation (BP) neural network. The results show that the BP neural network model exhibits high prediction accuracy in predicting the performance of modified red mud pavement materials. Another study related to the resource utilization of waste used scrap steel slag and composite coagulants to improve dredged mud and sludge, exploring the changes in its physical and chemical properties and its potential for application in engineering [2]. This study investigated the influence of the particle size and dosage of waste steel slag on physical improvement, as well as the interaction effect of curing agent components on unconfined compressive strength. The optimal dosage of waste steel slag and the particle size ratio were determined. Micro-mechanism analysis demonstrated that the waste steel slag enhanced the pore structure of dredged silt, thereby providing a skeletal framework. In addition, the hydration products produced by the curing agent fulfilled the roles of filling and cementing. Another study related to road base materials conducted an experiment on the mechanical properties of cement-stabilized crushed stone road base materials using recycled aggregates from construction waste [3]. The feasibility of using these raw materials to prepare cement-stabilized macadam roadbeds was determined through experimental verification. Subsequently, an analysis was conducted to determine the effects of varying cement dosages and curing periods on the strength of the materials. This analysis was performed using the unconfined compressive strength test. The results show that the recycled aggregates were characterized in accordance with the specifications and presented satisfactory mechanical properties. Furthermore, the curing agent was observed to enhance the early strength of the recycled aggregate base material.

Following the resolution of the issue pertaining to the sourcing of industrial waste, the subsequent challenge concerns performance enhancement. The exploration of innovative materials, including microencapsulated self-healing cementitious materials, rubber-modified asphalt, and g-C_3N_4/TiO_2 coatings, has yielded novel concepts and methodologies for the optimization of infrastructure performance. These materials offer unique advantages in improving the durability, functionality, and energy utilization efficiency of engineered structures. For example, a recent study related to microencapsulated self-healing cementitious materials explored the repair effect of urea–formaldehyde resin/epoxy resin-microencapsulated self-healing cementitious materials at the microscopic level [4]. This study explored the feasibility of self-healing from the perspective of molecular mechanisms and evaluated the repair performance of microencapsulated self-healing cementitious mortars with different microcapsule concentrations, curing agent types, and curing ages. The results of the study demonstrate that the adhesion and mechanical properties of the cement matrix were enhanced following the incorporation of microcapsules. The repair agent released from the microcapsule core exhibited higher strength after curing compared to its strength in the uncured state. Another study related to red mud modification explored the possibility of applying red mud as a solid waste material in road construction to alleviate the problem of red mud accumulation [5]. Titanium gypsum and phosphogypsum were used as curing agents to modify red mud. The findings indicate that the optimal moisture content of titanium-gypsum-cement-stabilized red mud (RTC) surpassed that of phosphorus-gypsum-cement-stabilized red mud (RPC), thereby affirming the superiority of RPC as a modified red mud material. Rubber-modified asphalt is a carbon-neutral asphalt-based material that facilitates the recycling of waste rubber materials. Therefore, it is necessary to evaluate the interfacial interaction of the rubber–asphalt system and its effect on the viscosity properties of rubber-modified asphalt and to analyze the contact properties of rubber particles in asphalt at the microscopic level [6]. The rubber swelling test and solvent elution test were carried out for the rubber–bitumen system under different preparation conditions. The swelling ratio, degradation ratio, and swelling–degradation ratio were proposed to evaluate interfacial interactions. The results demonstrate that following the establishment of a stable three-dimensional network structure at the interface of the rubber–asphalt system, the greater the effective interfacial interaction, the higher the viscosity. Additionally, the gel film on the surface of the rubber particles, during the dissolution and effective degradation processes, can contribute to enhanced viscosity. Another study related to $g\text{-}C_3N_4/TiO_2$ coating materials explored the use of $SrAl_2O_4{:}Eu^{2+},Dy^{3+}$ long afterglow materials doped in $g\text{-}C_3N_4/TiO_2$ coatings for photodegradation [7]. Automobile exhaust gas purification tests were carried out on the prepared samples, and the optimum mass ratio of $g\text{-}C_3N_4/TiO_2$ to $SrAl_2O_4{:}Eu^{2+},Dy^{3+}$ was determined. The microstructures and performances of the samples were evaluated with characterization tests, such as XRD, FT-IR, XPS, and TG-DSC. The experimental results show that the incorporation of long afterglow materials greatly improved the efficiency of the photocatalysts in purifying exhaust gas under a low-light environment.

In addition, a novel fatigue testing method for semi-rigid subgrade considering the spatial stress state was proposed in a paper related to the fatigue performance assessment of semi-rigid subgrades for a key technical issue in infrastructure construction [8]. Based on this method, the fatigue performance of three reinforced semi-rigid subgrade materials (steel wire mesh, plastic geogrid, and fiberglass) was tested and compared with that of unreinforced materials. The fatigue strain evolution laws of these materials were analyzed, and the fatigue strain eigenvalues in the limit state were proposed as the evaluation index of the fatigue failure of subgrade materials. The results show that the fatigue performance is can be ordered as follows: plastic geogrid > wire mesh > glass fiber > unreinforced specimen. Another study related to building thermal demand assessment developed an intelligent assessment model for building thermal demand [9]. The model is an interpretable machine learning model developed using Bayesian-optimized LightGBM and SHAP (SHapley Additive exPlanations) methods for assessing the thermal comfort design of buildings in different regions and for different uses. The results show that the optimized LightGBM model exhibits excellent evaluation performance on the test set, outperforming other machine learning models such as XGBoost and SVR (Support Vector Regression). An accurate building thermal comfort design based on this evaluation model will facilitate the implementation of carbon neutral strategies.

In the contemporary era, the pursuit of carbon neutrality has emerged as a global imperative. The development of low-carbon materials and associated technologies has been identified as a pivotal factor in attaining sustainable development and carbon neutrality in the transport sector. The extant research in this area is predominantly focused on the concept of carbon-neutral infrastructure, aiming to attract articles on novel materials or innovative technologies for carbon-neutral transport infrastructure. These contributions offer a substantial theoretical foundation and practical references, thereby promoting the development of infrastructure construction in the direction of low-carbon, green and sustainable development. In the future, there is potential to encourage the deep interdisciplinary cross-fertilization of materials science, civil engineering, environmental science, computer science, and other disciplines to optimize materials design and structural performance analysis using artificial intelligence technology and to study the environmental friendliness and sustainability of materials in conjunction with environmental science.

Author Contributions: Conceptualization, W.W.; methodology, D.Y.; software, A.C.F.; validation, D.Y. and F.G.; formal analysis, A.C.F.; investigation, D.Y.; resources, F.G.; data curation, D.Y.; writing—original draft preparation, W.W.; writing—review and editing, W.W.; visualization, A.C.F.; supervision, F.G.; project administration, A.C.F.; funding acquisition, W.W. All authors have read and agreed to the published version of the manuscript.

Acknowledgments: We would like to thank all the authors, reviewers, and staff of the *Buildings* Editorial Office for their great support during the preparation of this Special Issue.

Conflicts of Interest: The authors declare no conflicts of interest.

References

1. Ji, Q.; Jia, X.; Wang, Y.; Cheng, Y. Study on Strength Prediction and Material Scheme Optimization for Modified Red Mud Based on Artificial Neural. *Buildings* **2024**, *14*, 3544. [CrossRef]
2. Ji, Q.; Wang, Y.; Jia, X.; Cheng, Y. Experimental Study on Mechanical Properties and Stability of Marine Dredged Mud with Improvement by Waste Steel Slag. *Buildings* **2024**, *14*, 3472. [CrossRef]
3. Yuan, Y.; Hu, X.; Wang, K.; Liu, Z.; Zhong, M.; Meng, K. Study on Mechanical Properties of Road Cement-Stabilized Macadam Base Material Prepared with Construction Waste Recycled Aggregate. *Buildings* **2024**, *14*, 2605. [CrossRef]
4. Mao, H.; Cao, X.; Guo, M.; Jiang, C.; Chen, D. Study on the Repair Effect of Self-Healing Cementitious Material with Urea-Formaldehyde Resin/Epoxy Resin Microcapsule. *Buildings* **2024**, *14*, 2201. [CrossRef]
5. Yan, S.; Cheng, Y.; Wang, W.; Jin, L.; Ding, Z. Gypsum-Enhanced Red Mud Composites: A Study on Strength, Durability, and Leaching Characteristics. *Buildings* **2024**, *14*, 1979. [CrossRef]
6. Su, J.; Li, P.; Zhu, G.; Wang, X.; Dong, S. Interface Interaction of Waste Rubber-Asphalt System. *Buildings* **2024**, *14*, 1868. [CrossRef]
7. Zhou, B.; Li, H.; Cui, A.; Wang, D.; Guo, F.; Wang, C. Implementing All-Weather Photocatalysis of Exhaust Fumes Based on the g-C_3N_4/TiO_2/$SrAl_2O_4$: Eu^{2+}, Dy^{3+} Ternary Composite Coating. *Buildings* **2024**, *14*, 1743. [CrossRef]
8. Wang, L.; Zhao, Z.; Liang, H.; He, Y.; Kang, X.; Xu, M. Research on Laboratory Testing Method of Fatigue Performance of Semi-Rigid Base Considering Spatial Stress State. *Buildings* **2024**, *14*, 365. [CrossRef]
9. Liu, H.; Ma, E. An Explainable Evaluation Model for Building Thermal Comfort in China. *Buildings* **2023**, *13*, 3107. [CrossRef]

buildings

MDPI

Article

Study on Strength Prediction and Material Scheme Optimization for Modified Red Mud Based on Artificial Neural Networks

Qiaoling Ji *, Xiuru Jia, Yingjian Wang and Yu Cheng

College of Transportation, Shandong University of Science and Technology, Qingdao 266590, China; jiaxiuru@sdust.edu.cn (X.J.); skd996718@sdust.edu.cn (Y.C.)
* Correspondence: qiaolingji@sdust.edu.cn

Abstract: Focusing on the complex nonlinear problems of strength prediction and the material scheme design of modified red mud for use as a road material in engineering applications, a strength prediction neural network is established and utilized to optimize the material scheme, including the compound-solidifying agent ratio, water content, and curing age, based on experimental data accumulated during years of engineering practice and an artificial neural network. In this study, a backpropagation (BP) neural network is adopted, and 114 sets of experimental data are used to train the parameters of the unconfined compressive strength prediction model. Then, using the BP strength prediction model, the material scheme optimization process is carried out, with the strength and material costs as the objectives. The results show that the BP neural network model has a high prediction accuracy, the relative prediction error is basically within 10%, the root-mean-squared error is less than 0.04, and the correlation coefficient is more than 0.99. According to the strength requirements of modified red mud in different road projects and the constraints of each property, an optimal material scheme with a lower cost and higher 7 d target strength is obtained using a mix of polymer agent–fly-ash–cement–speed-cement in a ratio of 0.02%:1.96%:4.78%:0%, with a 33.93% water content of raw red mud, so that the target strength and material cost are 2.987 MPa and 17.099 CNY/T. This study creates an optimal material scheme, incorporating the compound-solidifying agent ratio, curing age, and water content of the modified red mud road material according to the strength requirements of different projects, thereby promoting the popularization of the utilization of red mud with better engineering practicability and economy.

Keywords: modified red mud; optimization; strength prediction; BP neural network; solidifying-agent ratio; water content

1. Introduction

Red mud is the industrial solid waste discharged from alumina extraction in the aluminum industry. It is alkaline and can pollute surface water and groundwater. With the development of the aluminum industry and the reduction in bauxite grade, the amount of red mud is increasing, and so the comprehensive utilization of red mud is becoming an increasingly important research topic [1–3]. The modified red mud is a kind of resource utilization with wide application prospects. Roadbed filling, stabilizing roadbed gravel, improving roadbed soil, prefabricating road construction components, and filling hollow areas are among its applications [4–6]. These engineering uses have specific technical requirements for the strength of modified red mud road material. However, the strength of modified red mud road material is affected by many things, such as the amount and ratio of compound-solidifying agents, the amount of water in the material, the curing age [7,8], and more. Some of these factors have an interaction effect on the strength. Addressing these issues will incur significant costs in terms of manpower. Therefore, it is vital to reasonably optimize the material scheme of modified red mud road material.

In previous studies, researchers used linear and nonlinear regression models to predict the unconfined compressive strength of modified red mud road materials based on a substantial number of experiments [9]. Liu [10] utilized reliable and mature multivariate linear equations to fit the long-term compressive strength test data for steel slag fine-aggregate red-mud-based concrete, including five influence factors affecting red mud admixtures, curing age, curing temperature, curing humidity, and 28 d unconfined compressive strength. Fu et al. [11] used a Gauss–Markov linear model to create a binary prediction model that took into account the curing age and the amount of coal system metakaolin. The model was 5% accurate for predicting the compressive strength of red mud that had been changed by the coal metakaolin. The studies, within the confines of the research scope, establish a regression prediction model for the strength of a specific modified red mud material, utilizing a moderate amount of experimental data. However, the strength requirements for using modified red mud as a road material vary due to the different sources of raw red mud, engineering projects, and road types. When it comes to predicting the strength of modified red mud road materials based on different application conditions, the existing prediction models struggle to meet all requirements.

The excellent adaptive learning ability, high parallelism, and fault tolerance of artificial neural networks (ANN) have made them a popular choice for predicting the strength of engineering materials in recent years. ANN, inspired by biological neural systems, are computational models designed to mimic the human brain's mechanisms for information processing and learning. These networks consist of numerous interconnected artificial neurons, forming network structures through weighted connections. The learning process of an ANN typically employs the backpropagation algorithm, which minimizes the error between outputs and expected values by adjusting the connection weights. Compared to traditional engineering design methods, ANN excel in tasks involving complex nonlinear relationships, pattern recognition, classification, and prediction. Due to their superior adaptability and efficient information-processing capabilities, ANN are considered essential tools for solving complex problems, especially in the context of big data analysis and machine learning. ANN have shown a lot of promise for use, especially in predicting the buckling capacity and compressive strength of solidified soil, checking the performance of structures, and finding the best materials for a job.

For example, Abarkan et al. [12] used an ANN to find the best design of circular hollow-section stainless steel short columns and then compared their findings to Eurocode 3. This showed that their model accurately predicted how different design parameters would affect the columns' ability to hold weight. Rabi et al. [13–15] applied ANN to predict the web-post-buckling capacity of high-strength steel beams with elliptical openings, the sectional load-bearing capacity of cold-formed circular hollow sections (CHS), and the buckling capacity of hot-formed CHS beam-columns. Their results showed that ANN models provided accurate predictions in terms of large-scale data analysis and performed excellently when addressing nonlinear relationships between structural response and material properties. Moreover, they also validated the effectiveness of ANN models with experimental data and existing design standards, emphasizing their potential in complex structural analysis. A study by Jweihan et al. [16] created an empirical model for the rutting-susceptibility index (RSI) of asphalt mixtures. They used a hybrid machine-learning approach that combined the strengths of artificial neural networks (ANN) and gene expression programming (GEP). Their study demonstrated the potential of ANN to improve asphalt pavement design and performance evaluation under various environmental conditions. Kewalramani et al. [17] used ANN for the non-destructive testing of concrete compressive strength via ultrasonic pulse velocity, showing that ANN were more effective than other methods in terms of predicting concrete compressive strength. Yoon et al. [18] made a prediction model that was more accurate than most regression models by looking at the complicated and nonlinear relationships between the parts of lightweight aggregate concrete and its compressive strength and elastic modulus. Ridho et al. [19] constructed three highly accurate ANN models using the Levenberg–Marquardt algorithm, with recycled aggregate content as input and compressive strength as output. These studies indicate that ANN exhibits significant advantages and application potential in improving mixture-design efficiency and accuracy and predicting the mechanical properties of materials.

The relationship between the strength of solidified soil and the ratio is a complex, non-linear one. The BP neural network, an artificial neural network, is capable of powerful self-learning and nonlinear mapping, which makes it an effective tool for predicting the relationship between the strength of solidified soil and the ratio. Consequently, the BP neural network is frequently employed for the purpose of predicting the properties of solidified soil. Li et al. [20] constructed a compressive strength prediction model for coral concrete using BP neural network and optimized the model parameters and found that the optimized model produced reliable results. Han et al. [21] conducted an analysis on the impact of various fly ash dosages on the strength of concrete, leading to the establishment of a BP-neural-network-based compressive strength prediction model for fly ash concrete. Ma [22] created a BP neural network to predict the compressive strength of fiber-reinforced-polymer-confined concrete and compared it with the traditional regression model. The neural network model is more accurate and stable in its predictions. Li et al. [23] used a BP neural network to establish a prediction model for the compressive strength and flexural strength of hybrid fiber-reinforced concrete. These studies have achieved favorable results, which can replace part of the test screening work, reduce the test workload, and save human and financial resources. However, because these prediction models rely on the mapping relationship between parameters established by concrete or cemented gravel test data, they cannot accurately predict the strength of red mud road materials due to insufficient red mud test data support. Currently, there are few studies on the strength prediction of modified red mud road materials using neural network models. Zhang [24] employed a feed-forward neural network to develop a predictive model for the strength of red-mud-based grouting materials. The model was trained using two inputs: the amount of raw red mud and the ratio of solidifying agents. The objective of the study was to develop a prediction model for the strength of red-mud-based grouting materials utilized in tunnel construction and civil-engineering-disaster prevention and control projects. However, the model was not applicable to the strength prediction of modified red mud road materials due to the differing performance requirements for fluidity and strength.

When it comes to optimizing the road material scheme, most scholars choose the material scheme that exhibits the highest strength and best performance based on experimental tests. An et al. [25] conducted an experimental study on the road performance, durability, and environmental performance of red-mud–steel slag–cement pavement base material. They ultimately determined two sets of optimal ratios for this material. Wu et al. [26] employed an orthogonal test method and a linear regression model in MATLAB to derive a strength-prediction equation for modified red mud and identify the optimal ratio. A review of the literature reveals a lack of studies investigating the use of artificial neural networks (ANN) for the optimization of solidifying-agent ratios. Li et al. [27] established a BP model for optimizing SMA material schemes, taking into account the stability, fluidity, volumetric parameters of the Marshall test, and the properties of both asphalt and minerals. Compared to the traditional method, the artificial neural networks (ANN) can significantly reduce time and test costs while also offering a new approach to optimize mix ratios.

To solve the problems of finding the best red mud road materials for different engineering projects, this study uses the BP neural network as a deep-learning algorithm to predict the strength of modified red mud and suggests a way to use the trained BP model to find the best material scheme. First, we establish an unconfined compressive strength prediction model for modified red mud road materials, using the ratio of compound-solidifying agent (polymer agent, fly ash, cement, speed cement), water content, and curing age as input. We then train this model using accumulated experimental data from multiple regions over several years. The model's prediction performance is validated using 10 sets of experimental data, with evaluation metrics including correlation coefficient (R), maximum and minimum relative error, standard deviation of errors (*SD*), mean absolute error (*MAE*), mean-squared error (*MSE*), and root-mean-squared error (*RMSE*). Next, we use the BP strength-prediction model to estimate a mapping set of strength with various material schemes, taking into account the strength requirements of various road-engineering projects and the constraints on the material components. We then calculate the material cost of each material scheme based on the engineering market. Finally, we utilize a double-objective decision method to optimize material schemes, aiming to achieve both strength and cost objectives. This study offers a method for predicting the unconfined compressive strength of modified red mud road materials within a specific range, as well as a technique for optimizing the material scheme. Its use will reduce the amount of experimental work and offer valuable insights for the strength prediction and material scheme optimization of modified red mud.

2. Experimental Dataset on the Strength of Modified Red Mud Road Material

2.1. Experimental Data Analysis and Selection

The experimental data on the unconfined compressive strength of modified red mud road materials used in this study were from the experiments carried out by our team in a couple of engineering projects located at Longkou in Shandong province, Taiyuan in Shanxi province, and Beihai in Guangxi province of China. We experimentally investigated the effects of the amount and material scheme of the compound-solidifying agent, water content, and curing age on the unconfined compressive strength of the modified red mud road materials in these tests. Table 1 displays the value ranges of each property, from which we randomly selected 114 experimental groups for analysis in this study (Table 2), and Figure 1 displays the frequency distribution histograms of each influence factor.

Table 1. Characteristics of the experimental dataset.

Property	Minimum Value	Maximum Value	Average Value
Polymer agent (%)	0.05	0.54	0.35
Fly ash (%)	0	2.40	0.80
Cement (%)	0.56	11.46	6.34
Speed cement (%)	0	1.80	0.27
Water content (%)	24.10	33.40	27.25
Curing age (d)	3	11.00	5.19
Unconfined compressive strength (MPa)	0.15	4.99	2.94

Table 2. Representative values of the experimental dataset.

Material Scheme	Polymer Agent (%)	Fly Ash (%)	Cement (%)	Speed Cement (%)	Curing Age (d)	Water Content (%)	Strength (MPa)
S1	0.05	0.1	0.855	0	3	25	0.36
S2	0.36	0.8	6.84	0	7	29.6	3.82
S3	0.36	0	7.64	0	4	29.6	3.31
S4	0.135	0.3	2.565	0	6	25	1.27
S5	0.54	0	9.66	1.8	3	25	3.03
...	...	...	...	...	...	...	...
S51	0.225	0.5	4.275	0	3	25	1.73
S52	0.225	0.5	2.775	1.5	3	25	0.97
S53	0.36	1.6	6.04	0	11	29.6	3.01
S54	0.09	0.2	1.11	0.6	6	25	0.61
S55	0.36	1.2	6.44	0	7	24.1	4.06
...	...	...	...	...	...	...	...
S110	0.36	0	6.44	1.2	7	26.7	3.56
S111	0.54	2.4	9.06	0	11	29.6	4.69
S112	0.36	0	6.44	1.2	7	33.4	3.1
S113	0.045	0.1	0.855	0	3	25	0.46
S114	0.36	1.2	6.44	0	3	26.7	3.39

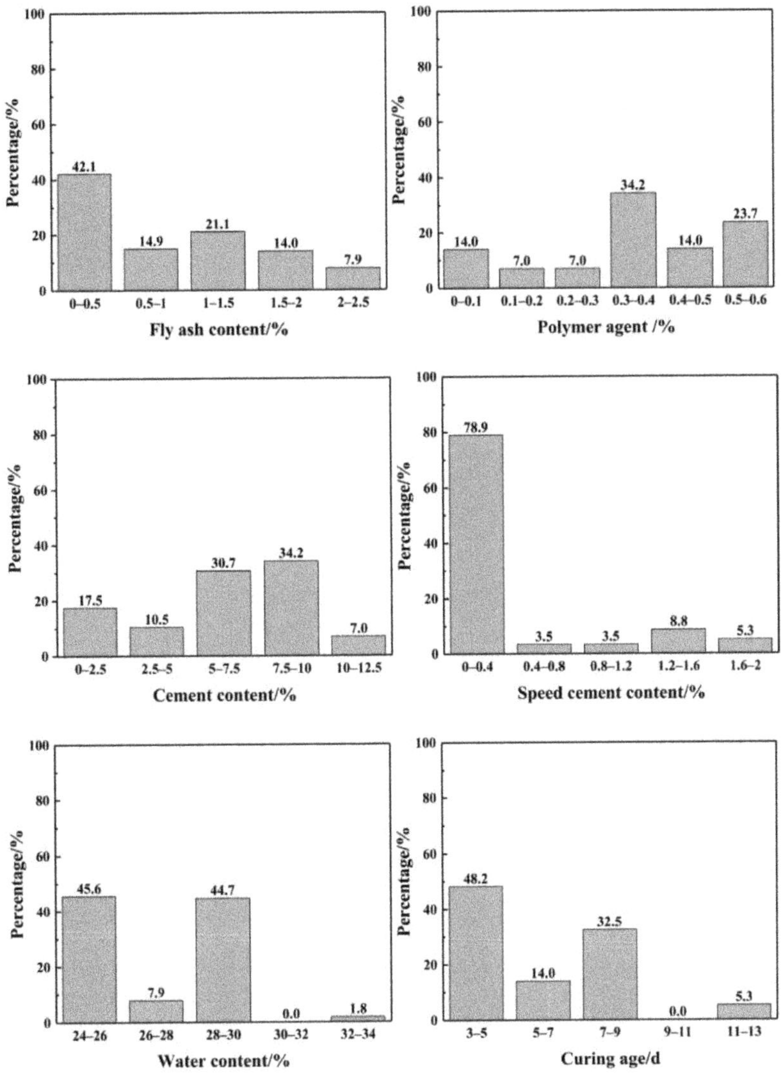

Figure 1. The frequency distributions of each influence factor in the experimental group.

2.2. Influence of the Compound-Solidifying Agent

To give an illustration of the effect of each component in the compound-solidifying agent on the unconfined compressive strength of the modified red mud, four groups of compound-solidifying agent ratio A~D are carefully selected from the 114 data points of experiment tests, as listed in Table 3. Specifically, we prepared A and B to compare the effects of cement and speed cement in the compound-solidifying agent on the unconfined compressive strength of modified red mud road material. The ratio A was polymer agent/fly ash/cement in a ratio of 4.5:10:85.5, while the ratio B was polymer agent/fly ash/cement/speed cement in a ratio of 4.5:10:55.5:30. Ratio B differs from A in that it replaces 30% of the cement with speed cement. We fixed the water content at 25% and set the curing age at 3 and 7 days. Figure 2 presents and compares the unconfined compressive strength results for the four groups, plotting the results of the pure cement for comparison.

Table 3. Four groups of compound-solidifying agent ratio A~D.

Ratio	Polymer Agent (%)	Fly Ash (%)	Cement (%)	Speed Cement (%)
A	4.5	10	85.5	0
B	4.5	10	55.5	30
C	4.5	15	80.5	0
D	4.5	0	80.5	15

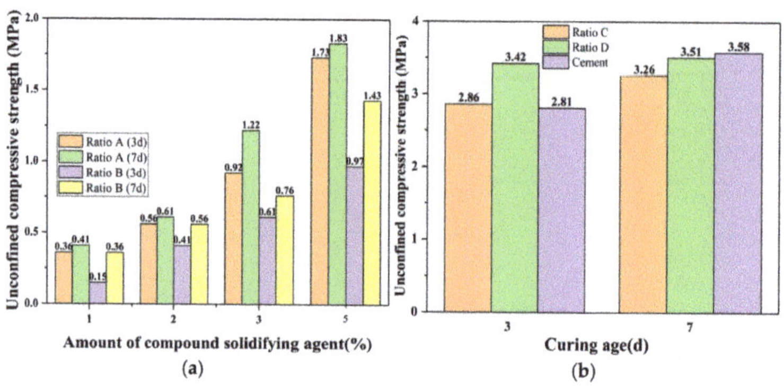

Figure 2. Comparison of four compound-solidifying agent ratio A~D and the pure cement with two curing ages: (**a**) comparison of ratios A and B; (**b**) comparison of ratio C, D, and pure cement.

Figure 2a demonstrates that the unconfined compressive strength of the modified red mud road material increases proportionally with the amount of the compound-solidifying agent. Additionally, ratio A exhibits a superior solidifying effect compared to ratio B, and the gap between their effectiveness widens as the amount of compound-solidifying agent increases, indicating that speed cement (in ratio B) performs worse than regular cement (in ratio A).

We selected two solidifying agent ratios, C and D, to compare their effects on the unconfined compressive strength of modified red mud road material with pure cement in Figure 2b. Ratio C comprises a polymer agent, fly ash, and cement in a ratio of 4.5:15:80.5, whereas ratio D maintains the same ratio but substitutes 15% of the fly ash with speed cement. We set the water content of the two groups at 26.7% and the amount of the compound-solidifying agent at 8%. Figure 2b displays the comparison results at the curing ages of 3 and 7 days, respectively. It shows that at the early curing age, ratio D exhibits a stronger strength than ratio C, primarily due to the faster solidifying rate of speed cement compared to fly ash. Meanwhile, ratio C shows a little more strength than the pure cement, meaning that a small amount of fly ash can make red mud road material stronger. At the curing ages of 7 days, the difference in the unconfined compressive strength between ratios C and D and C and pure cement is reduced to about 7% and 9%, respectively, suggesting that their strengths sufficiently meet general engineering requirements. Considering both cost and environmental benefits, ratio C, which incorporates a polymer ratio and fly ash as a partial replacement for cement, is an optimal choice.

2.3. Data Normalization

It is notable that each influence property has different units and demonstrates a considerable degree of variation in its range of values. Normalizing the initial input data is necessary to eliminate the impact of these magnitude differences to ensure that all values are scaled in a consistent interval range, between 0 and 1. In this study, the following functions are applied for data preprocessing:

$$\hat{X}_i = \frac{(X_i - X_{min})}{(X_{max} - X_{min})} \quad (1)$$

where X_i is the original value of the sample sequence, X_{min} is the minimum value of the sample sequence, X_{max} is the maximum value of the sample sequence, and $\hat{X}_i$ is the normalized value.

3. BP Prediction Model for Modified Red Mud Road Material Strength

3.1. Architecture of the BP Strength Prediction Model

The BP neural network, a type of multilayer feedforward neural network, consists of an input layer, hidden layers, and an output layer. The input layer serves as the first layer, while the output layer is the final layer, with the hidden layers functioning as intermediate layers. A key feature of the BP neural network is the forward transmission of signals in conjunction with the backward propagation of errors. The network continuously adjusts its weights and thresholds based on the predicted error values when the output layer produces results that deviate from the expected outputs. This iterative process allows the BP neural network to progressively approximate the desired output values [28]. Backpropagation neural networks (BP neural networks) are very good at predicting stabilized soil strength with the compound-solidifying agent. This is primarily due to their capacity for nonlinear modeling, flexibility, and accuracy. The network effectively captures the complex nonlinear relationships inherent in the soil stabilization process, making it suitable for various soil types and solidifying agents. The trained BP model could be used in the optimization of the compound-solidifying agent ratio and thereby reduce the number of physical experiments required and the time-consuming process associated with experiments in engineering implementation.

The input variables consist of six influence properties for modified red mud road materials: polymer agent, fly ash, cement, speed cement, water content, and curing age. The output variable is the unconfined compressive strength of the modified red mud road material. Therefore, a three-layer neural network structure is established by using six neurons in the input layer and one neuron in the output layer in Figure 3. We first determine the number of neurons in the hidden layer by using an empirical formula to define a range, and then we choose the optimal value based on the training and fitting results.

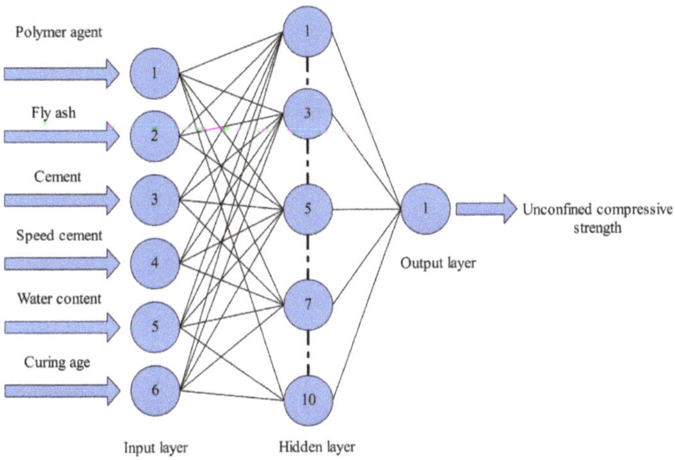

Figure 3. ANN architecture of unconfined compressive strength prediction model for the red mud road material.

The empirical formula for the number of neurons in the hidden layer is as follows:

$$l = \sqrt{m+n} + a \quad (2)$$

where n is the number of neurons in the input layer, l is the number of neurons in the hidden layer, m is the number of neurons in the output layer, and a is a constant between 0 and 10.

According to the empirical formula, the number of neurons in the hidden layer is determined to be a range of 3 to 12, the training set data are used to repeatedly train the neural network, and the number of neurons in hidden layer with the smallest mean-squared error is selected as the optimal number of neurons in hidden layer. Table 4 displays the results of the mean-squared error for different numbers of neurons in the hidden layer.

Table 4. The mean-squared error of different neurons in the hidden layer.

Number of Neurons in the Hidden Layer	Mean Squared Error (*MSE*) of the Training Set
3	0.0327
4	0.0225
5	0.0206
6	0.0754
7	0.0175
8	0.0107
9	0.0117
10	0.0083
11	0.0188
12	0.0296

Table 4 demonstrates that, as the number of neurons in the hidden layer increases, the mean-squared error (*MSE*) exhibits a trend where it first decreases and then increases. When the number of neurons reaches 10, the *MSE* in the training set reaches a minimum value of 0.0083. The increase in neurons may also result in more complex formulas, potentially leading to overtraining of the model. Therefore, we select 10 as the optimal number of neurons in the hidden layer, given its high accuracy with *MSE*. The final ANN architecture of the unconfined compressive strength prediction model for the modified red mud road material is 6-10-1, as shown in Figure 3.

The commonly used BP neural network training algorithms are trainlm, traingd, traindm, trainda, and traindx, among which the one with the optimal iteration accuracy and number of iterations is the trainlm algorithm [29]. Therefore, this study adopts the trainlm algorithm as the B-neural-network-training algorithm.

A hyperbolic tangent sigmoid transfer function is utilized to predict the output-parameter-based normalized values [30], as shown in Equation (3). This transfer function was chosen, as it provided accurate predictions when it was used in other research studies [31,32]. It is one of the most widely used transfer functions when constructing neural networks.

$$y_j = f(net_j) = \left(\frac{2}{1+e^{(-2net_j)}}\right) - 1 \tag{3}$$

where $net_j = \sum_{i=1}^{n} w_{ij}x_i + bias_j$; x_i and y_j are the input and output values, respectively; ω_{ij} and $bias_j$ are the weights and biases, respectively.

The specific principle of choosing the purelin linear function as the transfer function of the neurons in the output layer is shown in Equation (4):

$$y = f\left(\sum_{i}^{n} w_i x_i + b\right) \tag{4}$$

where w_i represents the weight parameter corresponding to the ith input x_i of this neuron, b represents the deviation of this neuron, and $f(x)$ is the activation function of the neuron.

3.2. Accuracy Metrics of Neural Network Output

This study uses a number of accuracy metrics to assess the accuracy of the output predicted by the neural network model. These include the correlation coefficient (*R*), maximum and minimum relative errors, standard deviation of the errors (*SD*), mean absolute error (*MAE*), mean-squared error (*MSE*), and root-mean-squared error (*RMSE*). These metrics provide a comprehensive assessment of the ANN model's accuracy and reliability in predicting outputs.

The correlation coefficient (*R*) measures the degree of correlation between the predicted and experimental values in the sample set. As the value of *R* approaches 1, the accuracy of the neural network prediction increases. The correlation coefficient is calculated using Equation (5):

$$R = 1 - \frac{\sum_{i=1}^{n}(\overline{y}_{\mathrm{i}} - y_{\mathrm{i}})}{\sum_{i=1}^{n}(\overline{y}_{\mathrm{i}} - \overline{y})} \tag{5}$$

where y_{i} is the experimental value of unconfined compressive strength, $\overline{y}_{\mathrm{i}}$ is the predicted value of unconfined compressive strength, $\overline{y}$ is the average of the experimental values, and n is the amount of sample set.

We use the mean-squared error (*MSE*) as an evaluation index to quantify the deviation of the predicted values from the true value in the sample set. A smaller *MSE* indicates higher prediction accuracy for the neural network model [33]. The *MSE* is calculated using the following formula:

$$MSE = \frac{\sum_{i=1}^{n}(y_{\mathrm{i}} - \overline{y}_{\mathrm{i}})^2}{n} \tag{6}$$

Relative error is used as an evaluation index of the deviation rate between the predicted value and the experimental value for a specific sample datum, which is calculated with Equation (7). A lower value indicates greater credibility for the neural network model's predictive capabilities.

$$f_{\mathrm{u}} = \frac{f - f_0}{f_0} = \frac{\overline{y}_{\mathrm{i}} - y_{\mathrm{i}}}{y_{\mathrm{i}}} \times 100\% \tag{7}$$

The standard deviation (*SD*) of the errors is an important statistical measure for assessing the differences between predicted values and true values. A smaller value indicates that the model's predictions are relatively consistent across multiple samples, thereby demonstrating better reliability. The formula for the standard deviation is as follows:

$$SD = \sqrt{\frac{1}{n-1}\sum_{n=1}^{n}(x_{\mathrm{i}} - \overline{x}_{\mathrm{i}})^2} \tag{8}$$

where x_{i} represents the individual error values, and $\overline{x}_{\mathrm{i}}$ denotes the mean of the error values.

The mean absolute error (*MAE*) is the average of the absolute differences between the true values and the predicted values. It reflects the absolute magnitude of the errors, providing a straightforward measure of the prediction accuracy. The formula for *MAE* is as follows:

$$MAE = \frac{1}{n}\sum_{i=1}^{n}|y_{\mathrm{i}} - \overline{y}_{\mathrm{i}}| \tag{9}$$

The root-mean-squared error (*RMSE*) is the square root of the mean-squared error. It shows the average deviation between the predicted value and the actual value. The formula for calculating *RMSE* is as follows:

$$RMSE = \sqrt{MSE} = \sqrt{\frac{\sum_{i=1}^{n}(y_{\mathrm{i}} - \overline{y}_{\mathrm{i}})^2}{n}} \tag{10}$$

3.3. *Training of the BP Strength Prediction Model*

The 114 experimental data are randomly divided into 2 categories: the training sample set with 104 data groups and the evaluating sample set with 10 data groups. The training sample set are divided into 3 categories. The training dataset (70%, 72 groups) is used to train the network and adjust according to the errors; the validation dataset (15%, 16 groups) is used to measure network generalization and stop the training when it stops improving; and the test dataset (15%, 16 groups) refers to data that have not been used in training or validation to independently assess the accuracy of the trained prediction model. We set the number of training samples at 1000, the target error at 10^{-7}, and the learning rate at 0.01. Figure 4 illustrates the BP-neural-network-model-training process.

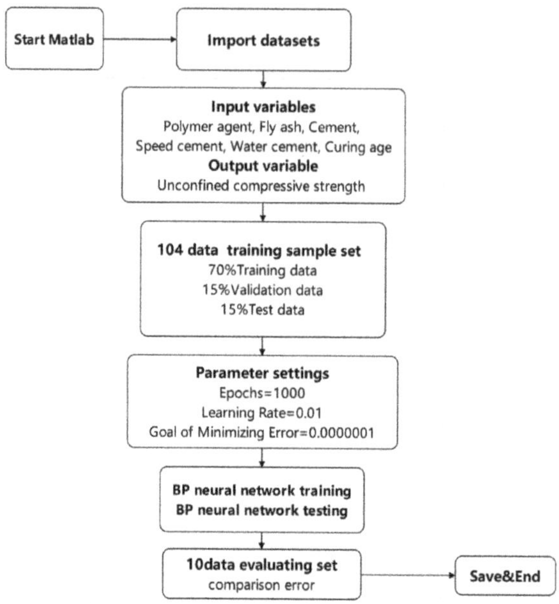

Figure 4. Training process of the BP model.

The predicted and true values of the unconfined compressive strength are shown in Figure 5. These values are compared in the training set, the validation set, the test set, and the total training sample set. Figure 5 reveals a strong correlation between the predicted and true values in the test set and the entire training samples, reaching R = 0.9186 and R = 0.9753, respectively. This strong correlation suggests that the trained BP neural network model performs well.

The mean-squared error (*MSE*) varies with the number of iterations during the training, validation, and testing stages of the BP neural network model as shown in Figure 6. Figure 6 demonstrates that the *MSE* decreases at each stage as the number of iterations increases, reaching its minimum value at 9–10 iterations. At this juncture, the *MSE* reaches a point of equilibrium and exhibits no further decline with the addition of further iterations. Notably, the *MSE* in the validation stage converges to 0.0076, which satisfies the accuracy requirement. This indicates that the neural network model has been well trained and is suitable for practical engineering applications in predicting the unconfined compressive strength of modified red mud road materials.

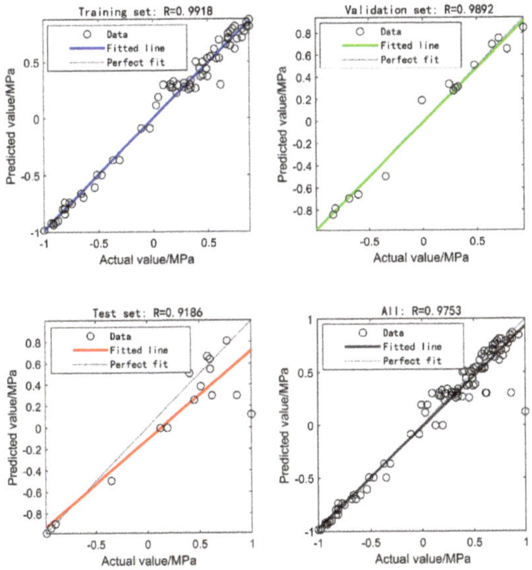

Figure 5. Comparisons between the predicted value and the experimental value in the training process.

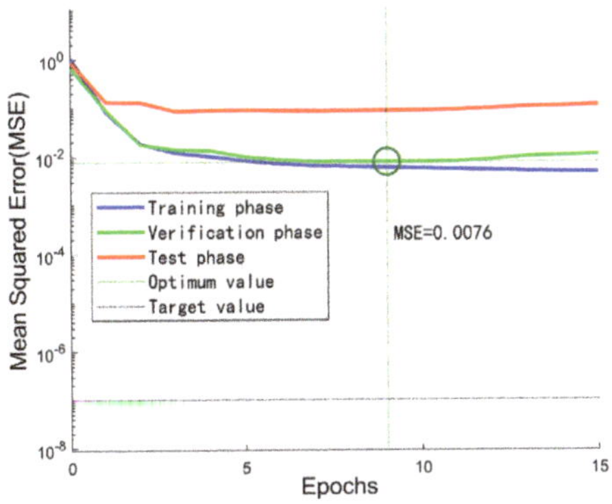

Figure 6. The variation in mean-squared errors in the training, verification, and testing stages with the number of iterations.

3.4. Accuracy Evaluation of BP Strength Prediction Model

In order to evaluate the prediction accuracy of the trained BP strength prediction model, 10 data groups from the evaluating sample set were employed. This process compared the experimental value and the predicted results by the trained BP strength prediction model using the same input in the evaluating sample set. Figure 7 depicts the comparison between the predicted and actual values from the BP neural network model, with the numbers marked on the figure representing the relative errors between the actual and predicted values. Table 5 gives the accuracy-evaluation results, with a maximum relative error of 12.20% and a minimum relative error of 0.21%. The average relative error was found to be 5.58%. The standard deviation of the errors (*SD*) was 0.148, the mean absolute error (*MAE*) was 0.119, the mean-squared error (*MSE*) was 0.022, and the root-mean-squared error (*RMSE*) was 0.149. These accuracy results meet the engineering prediction accuracy requirements [34]. In addition, we conducted a regression analysis to assess the correlation between the predicted and actual values, as shown in Figure 8. The correlation coefficient (R) is 0.9950, indicating high prediction accuracy for the BP neural network strength prediction model.

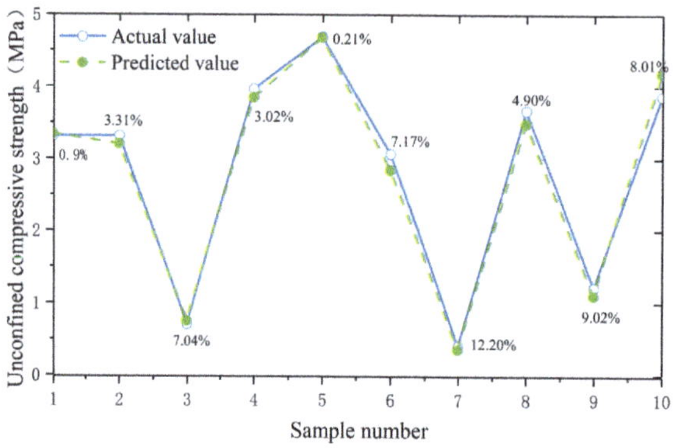

Figure 7. Relative errors of between predictions by the trained BP strength prediction model and the experimental data with evaluating sample set.

Table 5. Prediction accuracy evaluation index.

Maximum RE	*Minimum RE*	*SD*	*RMSE*	*MAE*	*MSE*	*R*
12.20%	0.21%	0.148	0.149	0.119	0.0222	0.995

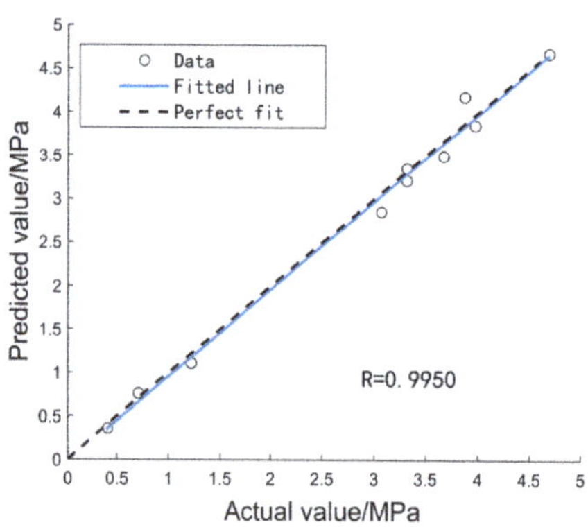

Figure 8. Comparisons between predictions by the trained BP strength prediction model and the experimental data with evaluating sample set.

The total sample set, comprising the training sample set and the evaluating predicted sample set, is employed as the test sample for the purpose of evaluating the prediction error of the model. The trained BP neural network model is then used to predict the unconfined compressive strength of the test sample for comparisons. The errors of the predicted values and the values are analyzed, and the relative error and the distribution of the error range are presented in Table 6.

As demonstrated in Table 6, 51.75% of the BP neural network prediction results exhibit a relative error of less than 5%, while 74.56% demonstrate a relative error of less than 10%. On the whole, it is notable that this BP prediction model for the unconfined compressive strength of modified red mud road material tends to predict with an error of basically less than 10%, indicating that the use of a BP neural network for this purpose is a feasible approach.

Table 6. Relative error distributions by the trained BP prediction model with total sample set.

Relative Error X Distribution Range	BP Sample Size	Relative Error as a Percentage/%
X < 5%	59	51.75
5% ≤ X < 10%	26	22.81
10% ≤ X < 15%	18	15.79
15% ≤ X < 25%	7	6.14
X > 25%	4	3.51
(grand) total	114	100

4. Material Scheme Optimization

The main process of the optimization is divided into four steps. First, the mapping set of strength and the material scheme are calculated using the trained BP model in the constraint conditions of the six variables h_1~h_6. Second, the mapping set of material cost and the material scheme is calculated with Equation (11) for all the material schemes in the strength-mapping set. Hence, we obtain the double-mapping set of strength and material cost from each material scheme. Third, based on the strength needs of red mud road materials for various road projects, the curing age h_5 and its design-unconfined compressive strength are found. Next, the mapping sets of strength and material cost are chosen using the constraints for the curing age h_5 and the strength in Table 7. Fourth, a double-objective decision method is utilized to make an optimization of material schemes with the target objectives of the strength and cost.

Table 7. Unit price of each variable for modified red muds road materials.

Variable	Unit Price (CNY/T)
Polymer agent	850
Fly ash	20
Cement	290
Speed cement	900
1% reduction in red mud water content	2.5

4.1. Objective Material Cost Function

In this optimization, the material cost and the unconfined compressive strength of the modified red mud are considered the objectives. The material scheme is defined as each combination of the six variables, including polymer agent, fly ash, cement, speed cement, curing age, and the water content. Set the material cost as *C*, and set the unconfined compressive strength as *S*. The unconfined compressive strength is calculated using the trained BP strength model in Section 3. The material cost function of modified red mud road material is composed of the six variables in the BP strength model, and the cost calculation formula for 1 ton of modified red mud road material is as follows:

$$C = 850h_1 + 20h_2 + 290h_3 + 900h_4 + 0h_5 + 250(35\% - h_6) \quad (11)$$

where h_1 is the percentage of polymer agent, h_2 is the percentage of fly ash, h_3 is the percentage of cement, h_4 is the percentage of speed cement, h_5 is the curing age, and h_6 is the water content of raw red mud. The material cost is considered for the first four variables h_1~h_4; the curing cost of h_5 is not included. According to the cost accounting in engineering practice in the Jinan-Qingdao Expressway in China [35], the construction cost for reducing water content of raw red mud is utilized as 2.5 CNY/T for every 1% reduction in water content from the natural water content of 35% in average. Table 7 lists all factors and its unit price the considered in the cost function.

4.2. Constraint Conditions for the Mapping Set

The constraint conditions are divided into four types, one is for the content of each component in the compound-solidifying agent, h_1~h_4, as listed in Table 8. The second is the water content of the raw red mud h_1 listed in Table 8, which can be reduced from 34% to 17% in the construction process by special construction technique. The third is for the relative error of the curing age between the prediction value $h_{5-\text{prediction}}$ in the BP model and the design value $h_{5-\text{design}}$ in engineering, which is in the range of 3~14 d in the BP model but usually fixed as 7 d or 14 d in an engineering project. The fourth is for the relative error of the unconfined compressive strength of the modified red mud between the prediction value $y_{\text{prediction}}$ in the BP model and the design value y_{design} in engineering, which has the minimum design standard value in the industry specification.

Table 8. Constraint conditions for the optimization.

Item of Constraint	Constraint Condition
Component of the compound-solidifying agent	$0.0\% \leq h_i \leq 10\%, i = 1, 2, 3, 4$
Red mud water content	$17\% < h_6 < 34\%$
Relative error of curing age	$0\% < \frac{h_{5-\text{prediction}} - h_{5-\text{design}}}{h_{5-\text{design}}} < 5\%$
Relative error of strength	$0\% < \frac{y_{\text{prediction}} - y_{\text{design}}}{y_{\text{design}}} < 5\%$

4.3. Optimization Process

Assuming that the 7-day unconfined compressive strength requirement y of road materials in a project is 2.9~3.0 MPa, about 8028 material schemes are selected from the mapping set predicted by using the trained BP strength model, as shown in the first five columns of Table 9. The material cost of each material scheme is calculated with Equation (11) as listed in the sixth column of Table 9. Then, for the convenience of a double-objective decision analysis, the strength and cost are normalized using Equation (1), as listed in the seventh and eighth columns of Table 9.

Table 9. The mapping set of strength and material cost in a project.

Material Scheme	Polymer Agent (%)	Fly Ash (%)	Cement (%)	Water Content (%)	Strength (MPa)	Cost (CNY/T)	Normalized Strength S	Normalized Cost C	Z
S1	0.02	1.96	4.78	33.93	2.989	17.099	0.893	0.002	442.253
S2	0.02	2.18	4.78	33.93	2.907	17.143	0.067	0.002	292.410
S3	0.08	1.75	4.78	33.93	2.970	17.567	0.699	0.010	68.821
S4	0.08	1.96	4.78	33.93	2.994	17.609	0.938	0.011	64.058
S5	0.08	2.18	4.78	33.93	2.962	17.653	0.621	0.012	59.578
S6	0.08	2.4	4.78	33.93	2.906	17.697	0.056	0.013	55.588
S7	0.02	2.4	4.98	33.93	2.981	17.767	0.808	0.014	50.654
S8	0.18	2.4	4.57	33.93	2.918	17.938	0.181	0.017	41.147
S9	0.13	1.96	4.78	33.93	2.999	18.034	0.987	0.019	37.525
S10	0.13	1.75	4.78	33.93	2.911	17.992	0.106	0.018	38.858
……	……	……	……	……	……	……	……	……	……
S8028	0.08	0.44	5.19	20	2.902	69.519	0.020	0.967	0.730

There are two objectives in this optimization: the strength is a positive objective which is needed to be maximized, and the material cost is a negative objective which is needed to be minimized. A weighted-utility optimization approach is adopted in this study using the following formula:

$$\max Z = w_1 S + \frac{w_2 0.7}{C} \tag{12}$$

where Z is the aggregate utility as the sum of the two objective utility functions, S is the normalized unconfined compressive strength by the BP strength model, C is normalized material cost by Equation (11), and w_1 and w_2 are the weights for the strength and cost, which are set as w_1 = 0.3 and w_2 = 0.7 with the reference to the engineering experience.

With the substitution of normalized material cost C and the normalized unconfined compressive strength I into the Formula (12), the mapping set Z is calculated and sorted from maximum to minimum in Table 7. The maximum of Z is chosen from the mapping set Z listed in the first row of Table 7. Therefore, the optimal material scheme is the one corresponding to the maximum Z, which is 0.02% polymer agent, 1.96% fly ash, 4.78% cement, 0% speed cement for the compound-solidifying agent, and 33.93%water content of raw red mud, generating a modified red mud road material with the 7 d unconfined compressive strength of 2.987 MPa and material cost of 14.424 CNY/T. Looking carefully at the data in Table 9, it can be seen that the aggregate utility Z is dominated by the water content, followed by the polymer agent. To demonstrate this tendency, the water content of the material scheme versus material cost and 7-day unconfined compressive strength in the mapping set are plotted in Figure 9. It can be observed that with the reduction in water content, the material cost in every strength level increases, indicating that the reduction in water content is the most expensive procedure and high water content of 30%~33.93% should be chosen to minimize the material cost. However, it should be pointed out that only the material cost is considered in this study, and the other limitations of construction conditions are neglected, for example, the construction cost of red mud with different water content and construction schedule limit. These costs and limits should be counted in a practical engineering project application.

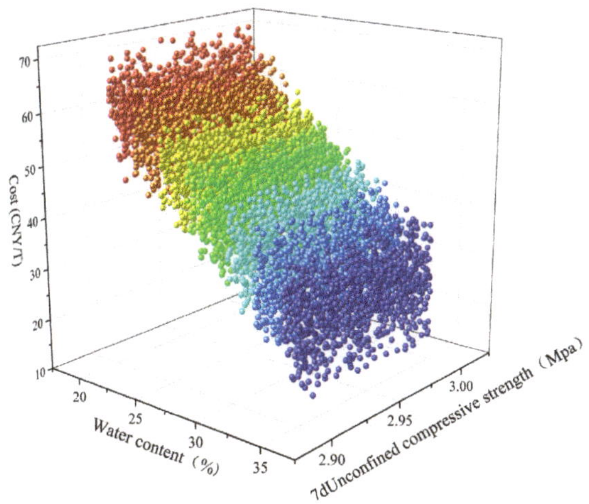

Figure 9. Three-dimensional scatter plot of water content versus material cost and 7-days unconfined compressive strength in a project.

5. Conclusions

In this study, the BP neural network was used to establish the unconfined compressive strength of modified red mud road materials and trained by the experimental data, with the ratio of the compound-solidifying agent (polymer agent, fly ash, cement, speed cement), water content, and curing age as the input, defined as material scheme. Then, the mapping set of strength and cost of each material scheme is calculated using the trained BP model in the constraint conditions of the six variables. Finally, a double-objective decision method is utilized to make an optimization of material schemes with the target objectives of strength and cost. With the established BP strength model, the designer can adjust and optimize the material scheme, including the material scheme of the compound-solidifying agent and water content of the raw red mud according to the target strength and cost, which can reduce the huge experimental work and save the manpower and material resources. The main conclusions are as follows:

(1) The analysis of the influence on the unconfined compressive strength of modified red mud road material shows that the effect of cement is more pronounced than that of quick-hard cement, and its advantage is increasingly evident with an increase in

amount. The addition of 15% fly ash improves the solidification effect, acting as a substitute for cement while offering cost and environmental protection.

(2) The verification results of the trained BP model by the prediction sample set demonstrate that the prediction model exhibits high prediction accuracy with a relative error of less than 10%, a root mean-squared error of less than 0.04, and a correlation coefficient approaching 1.

(3) With the trained BP model, the mapping set of strength and cost of each material scheme is calculated in the constraint conditions of the six variables. A double-objective decision method with a weighted utility optimization approach is utilized to make an optimization of material schemes with the target objectives of the strength and cost.

(4) An optimal material scheme is achieved as polymer composite:fly ash:cement:speed cement = 0.02%:1.96%:4.78%:0%, with a 33.93% water content of raw red mud, resulting in a strength of 2.987 MPa and cost of 17.099 CNY/T, with the aim of achieving a 7-day unconfined compressive strength requirement of 2.9~3.0 MPa for the subbase of extremely heavy and extra heavy traffic roads for second-class highways.

Author Contributions: Conceptualization, Q.J.; methodology, Q.J.; software, Q.J.; validation, Q.J., X.J. and Y.W.; formal analysis, Y.W.; investigation, Q.J. and X.J.; data curation, Y.W.; writing—original draft preparation, Q.J. and X.J.; writing—review and editing, Q.J. and X.J.; visualization, X.J.; supervision, Q.J.; project administration, Q.J. and Y.C.; funding acquisition, Y.C. All authors have read and agreed to the published version of the manuscript.

Funding: This research received no external funding.

Data Availability Statement: The data presented in this study are available on request from the corresponding author on reasonable request. The data are not publicly available due to privacy policies.

Conflicts of Interest: The authors declare no conflicts of interest.

References

1. Liu, X.; Han, Y.X.; He, F.Y.; Li, Y.J.; Gao, P.; Li, W.B. Research status of hazards and comprehensive utilization of red mud. *Met. Mine* **2018**, *11*, 7–12.
2. Geng, C.; Guo, S.H.; Liu, Z.G.; Liu, J.G. Status *Quo* and prospect of comprehensive utilization of red mud resources. *China Nonferrous Metall.* **2022**, *5*, 37–45.
3. Liu, Q.; Wang, Q.; Wu, P.; Wang, J.X.; Lv, X.J. Research progress in application of red mud in cementitious materials. *J. Shandong Univ. Sci. Technol. (Nat. Sci.)* **2022**, *3*, 66–74.
4. Nan, X.L.; Zhang, T.A.; Liu, Y.; Dou, Z.H. Analysis of comprehensive utilization of red mud in China. *J. Process Eng.* **2010**, *A1*, 264–270.
5. Liu, J.R.; Li, H.; Liang, J.L.; Wang, L. Research Progress on Extraction of Valuable Metal Elements from Red Mud. *Compr. Util. Miner.* **2022**, *03*, 107–112.
6. Xue, S.G.; Zhu, M.X.; Yang, X.W.; Guo, X.Y.; Jiang, Y.F.; Huang, S.W.; Zhu, F. Research Progress on Red Mud Excitation Cementitious Materials and Road Use. *Chin. J. Nonferrous Met.* **2023**, *10*, 3421–3439.
7. Zhang, F.K.; Yin, S.Y. Influencing influence factor of modified soil strength of silt based on rough set theory. *Water Transp. Eng.* **2019**, *A2*, 18–21.
8. Shi, M.L.; Tian, X.T.; Wang, H.J.; Yu, C.Y.; Du, X.Y.; Zhang, R.K. Study on the performance of road base layer for cement-lime-phosphogypsum solidification of red mud. *J. Yangtze River Sci. Res. Inst.* **2024**, *1*, 114–120.
9. Song, Z.W. Study on the Properties and Mechanism of Modified Red Mud Combined with Cement Solidified Copper-Contaminated Soil. Doctoral Thesis, Taiyuan University of Technology, Taiyuan, China, 2017.
10. Liu, C. Experimental Study on the Properties of Steel Slag Fine Aggregate Red Mud-Based Concrete. Master's Thesis, Guizhou University, Guizhou, China, 2022.
11. Fu, P.C.; Liu, J.P.; Yuan, C.; Bai, X.H. Strength test study on solidified red mud of coal metakaolin. *Non-Met. Ore* **2019**, *1*, 63–66.
12. Abarkan, I.; Rabi, M.; Ferreira, F.P.V.; Shamass, R.; Limbachiya, V.; Jweihan, Y.S.; Santos, L.F.P. Machine Learning for Optimal Design of Circular Hollow Section Stainless Steel Stub Columns: A Comparative Analysis with Eurocode 3 Predictions. *Eng. Appl. Artif. Intell.* **2024**, *132*, 107952. [CrossRef]
13. Rabi, M.; Jweihan, Y.S.; Abarkan, I.; Ferreira, F.P.V.; Shamass, R.; Limbachiya, V.; Tsavdaridis, K.D.; Santos, L.F.P. Machine Learning-Driven Web-Post Buckling Resistance Prediction for High-Strength Steel Beams with Elliptically-Based Web Openings. *Results Eng.* **2024**, *21*, 101749. [CrossRef]

14. Rabi, M.; Ferreira, F.P.V.; Abarkan, I.; Limbachiya, V.; Shamass, R. Prediction of the Cross-Sectional Capacity of Cold-Formed CHS Using Numerical Modelling and Machine Learning. *Results Eng.* **2023**, *17*, 100902. [CrossRef]
15. Rabi, M.; Abarkan, I.; Rabee, S. Buckling Resistance of Hot-Finished CHS Beam-Columns Using FE Modelling and Machine Learning. *Steel Constr. Des. Res.* **2023**, *17*, 93–103.
16. Jweihan, Y.S.; Al-Kheetan, M.J.; Rabi, M. Empirical Model for the Retained Stability Index of Asphalt Mixtures Using Hybrid Machine Learning Approach. *Appl. Syst. Innov.* **2023**, *6*, 93. [CrossRef]
17. Kewalramani, M.A.; Gupta, R. Concrete compressive strength prediction using ultrasonic pulse velocity through artificial neural networks. *Autom. Constr.* **2006**, *15*, 374–379. [CrossRef]
18. Yoon, J.Y.; Kim, H.; Lee, Y.J.; Sim, S.H. Prediction Model for Mechanical Properties of Lightweight Aggregate Concrete Using Artificial Neural Network. *Materials* **2019**, *12*, 2678. [CrossRef] [PubMed]
19. Ridho, B.K.A.M.A.; Ngamkhanong, C.; Wu, Y.B.; Kaewunruen, S. Recycled Aggregates Concrete Compressive Strength Prediction Using Artificial Neural Networks (ANNs). *Infrastructures* **2021**, *6*, 17. [CrossRef]
20. Li, Z.X.; Wei, Z.B.; Shen, J.L. Compressive strength prediction model of coral concrete based on BP neural network. *Concrete* **2016**, *1*, 64–74.
21. Han, J.J.; Zhao, D.S.; Li, J.P. Compressive strength prediction model of garbage fly ash concrete based on BP neural network. *Concrete* **2022**, *9*, 78–81.
22. Ma, G.; Liu, K. Prediction of compressive strength of CFRP constrained concrete based on BP neural network. *J. Hunan Univ. (Nat. Sci. Ed.)* **2021**, *9*, 88–97.
23. Li, N.; Zhao, J.H.; Wang, J.; Wu, S. Strength prediction of hybrid fiber concrete based on RBF neural network. *Concrete* **2014**, *7*, 23–26.
24. Zhang, J. Composition Design, Chemical Mechanism and Performance Regulation of Red Mud And Multi-Source Solid Waste Pulp Materials. Doctoral Thesis, Shandong University, Shandong, China, 2021.
25. An, Y.C.; Liu, Q.; Tan, B.; Huang, H. Experimental study on preparation of pavement base material by red mud, steel slag and cement. *J. Highw. Transp. Res. Dev.* **2023**, *5*, 35–43.
26. Wu, L.B.; Wang, Y.M.; Chen, W.; Liu, J.M.; Chen, Z.Q. Ratio Optimization of Red Mud-Fly Ash Paste Filling Material Based on Orthogonal Experiment. *Min. Res. Dev.* **2020**, *5*, 45–49.
27. Li, Q.F.; Liang, B.; Liu, C.H. The application of artificial neural networks in SMA mix design. *Henan Sci. Technol.* **2008**, *2*, 208–211.
28. Dong, Z.Z.; Yang, Y. Research on Target Recognition and classification of picking robots based on LM Optimization Algorithm and BP Neural Network. *Agric. Mech. Res.* **2022**, *8*, 25–29.
29. Wei, Y.W.; Zhong, Q.; Wang, D.Y. Prediction of ultimate strength of Type I metal sandwich plate based on BP neural network. *Chin. Ship Res.* **2022**, *2*, 125–134.
30. Gupta, T.; Patel, K.A.; Siddique, S.; Sharma, R.K.; Chaudhary, S. Prediction of mechanical properties of rubberised concrete exposed to elevated temperature using ANN. *Measurement* **2019**, *147*, 106870. [CrossRef]
31. Limbachiya, V.; Shamass, R. Application of artificial neural networks for web-post shear resistance of cellular steel beams. *Thin-Walled Struct.* **2021**, *161*, 107414. [CrossRef]
32. Ferreira, F.P.V.; Shamass, R.; Limbachiya, V. Lateral–torsional buckling resistance prediction model for steel cellular beams generated by Artificial Neural Networks (ANN). *Thin-Walled Struct.* **2022**, *170*, 108592. [CrossRef]
33. Liu, Y.; Zhang, L.Y. Implementation and Performance Comparison of BP and RBF neural networks. *Electron. Meas. Technol.* **2007**, *4*, 77–80.
34. Li, Y.M.; Ding, J.H.; Sun, B.N.; Guan, S. Comparison of BP and RBF neural networks for short-term prediction of sea surface temperature and salinity. *Adv. Mar. Sci.* **2022**, *2*, 220–232.
35. Shandong High-Speed Co., Ltd. *Research on the Key Technology of Large-Scale Recycling of Solid Industrial Waste Materials for Road Use*; Report 2021; Shandong High-Speed Co., Ltd.: Shandong, China, 2021.

buildings

MDPI

Article

Experimental Study on Mechanical Properties and Stability of Marine Dredged Mud with Improvement by Waste Steel Slag

Qiaoling Ji *, Yingjian Wang, Xiuru Jia and Yu Cheng

College of Transportation, Shandong University of Science and Technology, Qingdao 266590, China; 202282160028@sdust.edu.cn (Y.W.); jiaxiuru@sdust.edu.cn (X.J.); skd996718@sdust.edu.cn (Y.C.)
* Correspondence: qiaolingji@sdust.edu.cn

Abstract: As marine-dredged mud and waste steel slag in coastal port cities continue to soar, the traditional treatment method of land stockpiling has caused ecological problems. Thus, it is necessary to find a large-scale resource-comprehensive utilization method for dredged mud and waste steel slag. This study uses waste steel slag and composite solidifying agents (cement, lime, fly ash) to physically and chemically improve marine-dredged mud. The physical improvement effect of the particle size and dosage of waste steel slag was studied by the shear strength test under the effect of freeze–thaw cycle. Then, based on the Box–Behnken design of the response surface method, the interaction effects of the solidifying agent components on the unconfined compressive strength were studied. Then, the water stability under dry–wet cycles and a microscopic mechanism were analyzed by XRD and SEM tests. The results show that the waste steel slag with a dosage of 30% and a particle size of 1.18~2.36 mm has the best improvement. The interaction between cement and lime and lime and fly ash has a significant effect on the linear effect and surface effect of 7d unconfined compressive strength, and the strength increases first and then decreases with the increase in its dosage. For the 14d unconfined compressive strength, only the interaction between cement and lime is still significant. The unconfined compressive strength prediction model is established to optimize the mix ratio of the composite solidifying agent. In the water stability, the water stability coefficients of the 7d and 14d tests are 0.68 and 0.95, respectively, and the volume and mass loss rates are all below 1.5%, showing a good performance in dry–wet resistance and durability. Microscopic mechanism analysis shows that waste steel slag provides an 'anchoring surface' as a skeleton, which improves the pore structure of dredged mud, and the hydration products generated by the solidifying agent play a role in filling and cementation. The results of the study can provide an experimental and technical basis for the resource engineering of marine-dredged mud and waste steel slag, helping the construction of green low-carbon and resource-saving ports.

Keywords: marine-dredged mud; waste steel slag; response surface method; solidifying agent; unconfined compressive strength; shear strength; stability analysis; microscopic

1. Introduction

With the development of coastal port cities, the amount of construction and maintenance of ports, waterways, oceans and coastal projects in China has doubled, and the stock of marine-dredged mud has been increasing annually. The more common ones are offshore dumping, land reclamation, and resource treatment. Most of the existing dumping areas have been used for many years and are in overloaded operation, which can no longer meet the growing demand for dumping. At the same time, due to the import of iron ore, the steelmaking industry is developed in cities usually built on ports. Under the current trend of protection and utilization of various resources, it is urgent to seek large-scale comprehensive utilization of dredged mud and waste steel slag in coastal port cities.

Domestic and foreign scholars have carried out extensive exploration of the resource treatment of marine-dredged mud [1–3]. At present, the resource utilization methods of marine-dredged mud mainly include the utilization of ecological restoration geotechnical materials [1], solidification/stabilization (building materials) [2,3], soil utilization [4], sea sand desalination [5,6] and so on, and among which solidification/stabilization as building materials is a quantitative resource treatment method. Fly ash has good strength development performance, which is equivalent to the creep and shrinkage properties of Portland cement. The formed cementitious material structure has low permeability porosity. Many researchers [7,8] added fly ash to the solidifying agent to improve the mechanical properties of the solidified body. Yoobanpot et al. [9] studied the unconfined compressive strength and other indicators of stable dredged sediments mixed with fly ash. The results showed that the incorporation of fly ash made the unconfined compressive strength of stable dredged sediments increase to a certain extent. Horpibulsuk et al. [10] studied the stabilizing effect of carbide slag and fly ash on soil, and found that the addition of alkaline waste slag can effectively reduce the sensitivity of solidified soil to water. Especially when the dosage of the solidifying agent exceeds a fixed point, the addition of fly ash has an effect on the long-term unconfined compressive strength, and its filling effect can increase the compactness of solidified soil.

As a solidifying agent, cement can effectively improve the mechanical properties of muddy soil. At present, the amount of cement in most studies is generally high, which will lead to large carbon emissions [11]. Therefore, some scholars turn to physical improvement to reduce the amount of cement by using aggregate [12–16]. Wang et al. [12] used recycled fine aggregate to improve the soft soil solidified by industrial waste residue. It concluded that the reinforcement effect can be further improved when the content of recycled fine aggregate is less than 40%, and increases with the increase in the content. Because waste steel slag is scleroid and hard to grind, it can be used as aggregate instead of traditional sand and gravel to physically improve marine-dredged mud. Mozejko et al. [13] studied the influence of the skeleton effect of waste steel slag on the mechanical properties of solidified clay loess. It was found that when the waste steel slag content reached 12%, the 56d unconfined compressive strength of the mixed material could be increased 3~4 times. Huang et al. [14] studied the performance of waste steel slag-miscellaneous fill base material, and found that among the factors affecting the 7d unconfined compressive strength and 28d splitting tensile strength of waste steel slag-miscellaneous fill base material followed the rank: waste steel slag aggregate > cement > concrete debris > solidifying agent, where waste steel slag aggregate has a greatest influence. Wang et al. [15] studied the effect of waste steel slag aggregate content on the road performance of cement-stabilized soil base. It is concluded that the appropriate amount of waste steel slag can form an interlocking effect, which significantly increases the unconfined compressive strength and splitting strength of cement-stabilized soil. However, when the waste steel slag content exceeds 60%, the contact form of waste steel slag particles is changed, resulting in obvious volume expansion and stress concentration damage. Kang et al. [16] investigated the improvement effects of steel slag on soft dredged soil, and found that the strength of soft dredged clay treated with the steel slag exhibited significant improvement over time.

These studies discussed the improvement effect on the mechanical properties of mucky soil from the aspects of the dosage of the solidifying agent, the skeleton effect of fly ash and waste steel slag aggregate. Most of the test methods used in the study are based on orthogonal tests to find the optimal ratio [17,18]. However, the interaction between multiple factors makes the orthogonal test huge. The systematic research on the physical parameters of waste steel slag, such as content and particle size, and the solidifying agent ratio of marine-dredged mud with improvement by waste steel slag is relatively weak. The response surface method is widely used in various fields [19–24]. The function model between the influence factors and the response value can be established by the response surface method, which can help to understand the influence degree of the influence factors on the response and the interaction between them. It can not only study the influence of a single factor on the response, but also explore the interaction between multiple factors. Therefore, there are many studies that use the response surface method to investigate the interaction of multiple factors. Shao et al. [20] carried out a response surface test on the amount of three admixtures for the solidification of lake soft soil, and obtained the best mix ratio of the three admixtures through interaction analyses. Based on the response surface method, Li et al. [22] optimized the mix ratio of alkali-activated slag–fly ash solidified muddy soil and established a strength prediction model with high accuracy. Based on the response surface method, Wang et al. [23] studied the mechanical properties test and durability test of modified slag composite solidified saline soil under different dosages of desulfurization gypsum-sodium hydroxide-calcium carbide slag. Based on the different performance prediction models of unconfined compressive strength, tensile strength and wear resistance established by the response surface method, the researchers [25,26] also used multi-objective optimization analyses to study the best mix ratio parameters of comprehensive performance, and provided a method of multi-performance index optimization design.

Based on this, the solidification test of marine-dredged mud with improvement by waste steel slag was carried out based on the response surface method. The two-step method was used to carry out the research. Firstly, the shear strength test of marine-dredged mud with improvement by waste steel slag under freeze–thaw cycles was carried out. On this basis, the Box–Behnken design model in the response surface method was used to study the influence of the mix ratio of cement, lime and fly ash on the unconfined compressive strength of solidified soil to obtain the optimal mix ratio. The solidification mechanism of marine-dredged mud with improvement by waste steel slag was analyzed by X-ray diffraction and scanning electron microscopy (SEM), which provided a reference for the resource engineering application of waste steel slag and marine-dredged mud.

2. Test Materials and Schemes

2.1. Test Materials and Their Physical and Chemical Properties

The test scenario is the marine-dredged mud yard of Rizhao Port in Shandong Province and the steel-making waste slag of the industrial park in the port area. The stacked marine-dredged mud and waste steel slag with a particle size of 0~4.75 mm are used as raw materials. The basic physical properties and chemical composition contents of marine-dredged mud are measured, shown in Tables 1 and 2, and the particle distribution curve of marine-dredged mud is shown in Figure 1.

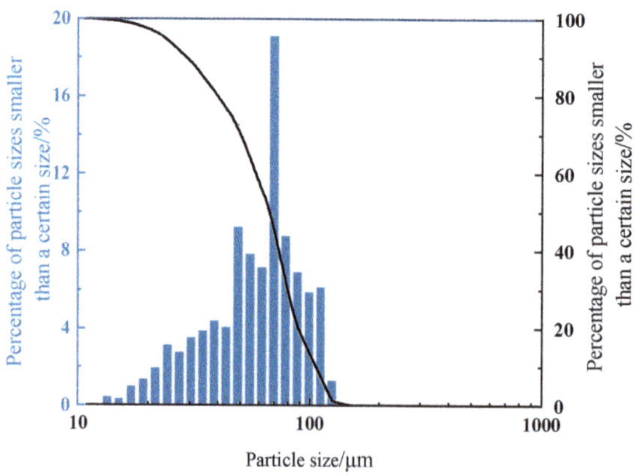

Figure 1. Cumulative distribution curve of marine-dredged mud particle size.

Table 1. Basic physical indexes of marine-dredged mud.

Moisture Content/%	Liquid Limit ω_L/%	Plastic Limit ω_P/%	Fluidity Index I_L	Plasticity Index I_P
54.0	48.2	22.5	1.23	25.7

Table 2. Chemical compositions (%).

Samples	Na_2O	SiO_2	Al_2O_3	K_2O	Fe_2O_3	Br	CaO	MgO	SO_3	TiO_2
Dredge	2.96	60.99	15.26	2.75	5.84	5.14	2.86	1.69	0.49	0.78
Fly ash	0.34	51.36	29.72	1.13	5.26	5.22	3.18	1.39	0.52	1.24

2.2. Test Scheme

(1) Analysis test on particle size and dosage of waste steel slag

In order to reduce the number of variables in the response surface test analyses, before the response surface method test analyses, The direct shear test was used to study the marine-dredged mud with improvement by waste steel slag (hereinafter, MWSS), focusing on the influence of the particle size and dosage of waste steel slag on the shear strength under the condition of 0, 1, 3 and 6 freeze–thaw cycles, so as to determine the optimal particle size and dosage of waste steel slag. According to the 'Test Methods of Soils for Highway Engineering'(JTG 3430-2020) [27], the optimal moisture content of each test group is obtained by a series of standard compaction tests. Referred to some related research [28,29], in order to maximize the use of waste steel slag, the dosage of waste steel slag is set as 30%, 35%, and 40%. The test scheme is shown in Table 3.

Table 3. Test plan for particle size and dosage of waste steel slag of MWSS.

Form	Dosage of Waste Steel Slag/%	Particle Size of Waste Steel Slag/mm	Freeze–Thaw Cycles/Times
Marine dredged mud without waste steel slag	0	0	0, 1, 3, 6
Marine dredged mud with improvement by waste steel slag	30, 35, 40	0~1.18, 1.18~2.36, 2.36~4.75	

(2) Analysis test for mix ratio of solidifying agent

Based on the response surface method, the experimental design uses the Box–Behnken method to test the three factors and three levels of the solidifying agent mix ratio. The response values are set as 7d unconfined compressive strength (Y_7) and 14d unconfined compressive strength (Y_{14}) of marine-dredged mud solidified soil with improvement by waste steel slag (hereinafter, MSWSS). The test design has 3 factors, 3 levels and 5 central points, resulting in a total of 17 groups, as shown in Table 4. Similarly, according to the 4th item of the standard 'Test Methods of Materials Stabilized with Inorganic Binders for Highway Engineering' (T 0843-2009) [30], the optimal moisture content of each test group with the addition of solidifying agent is obtained by a series of standard compaction tests for preparation of unconfined compressive specimens. The samples are made into a diameter of 39.1 mm and a height of 80 mm. Since the solidified soil is usually used in subgrade beds for low strength, the unconfined compressive strength tests are performed according to the 24th item of the standard 'Test Methods of Soils for Highway Engineering'(JTG 3430-2020) [27].

Table 4. A three-factor, three-level Box–Behnken test design scheme for solidifying agent ratios.

Factor \ Number	1	2	3	4	5	6	7	8	9	10	11	12	13	14	15	16	17
Cement: X_A/%	4	8	4	8	4	8	4	8	6	6	6	6	6	6	6	6	6
Lime: X_B/%	6	6	10	10	8	8	8	8	6	10	6	10	8	8	8	8	8
Fly ash: X_C/%	6	6	6	6	4	4	8	8	4	4	8	8	6	6	6	6	6
optimal moisture content w/%	16	18	18	20	17	18	17	22	16	20	18	22	20	20	20	20	20

(3) Analysis test for phase composition and microstructure

In order to study the improvement and solidification mechanism of waste steel slag and optimal mix ratio solidifying agent on marine-dredged mud, SEM test and XRD test were carried out on marine-dredged mud, marine-dredged mud with improvement by waste steel slag and marine-dredged mud solidified soil with improvement by waste steel slag. The SEM test was carried out by Apreo S HiVac field emission scanning electron microscope, and the magnification was set to 2000~8000 times. X-ray diffraction (XRD) experiments were carried out by D8 Advance X-ray diffractometer. The scanning range was $2\theta = 10°\sim80°$, and the rate was 4(°)/min.

3. The Test Results and Analyses of the Influence of Waste Steel Slag Particle Size and Dosage on Shear Strength

3.1. The Influence of Internal Friction Angle

The variation curves of the internal friction angle of marine-dredged mud with improvement by waste steel slag with different particle sizes are shown in Figure 2. Under the condition of different freeze–thaw times and waste steel slag dosage, when the particle size of waste steel slag is 0~1.18 mm and 1.18~2.36 mm, the internal friction angle is relatively large with the improvement of waste steel slag, and when it is 2.36~4.75 mm, the internal friction angle becomes smaller and even lower than the pure marine-dredged mud in some case, revealing that the improvement of the internal friction angle by the particle size has an optimal value. Careful observation of the three particle sizes shows that the particle size of 1.18~2.36 mm has the largest internal friction angle in all studied dosages. The possible reason may be that only a middle-sized steel slag could improve the distribution of marine-dredged mud and thus make the action of interlocking, rubbing and rolling among the particles of marine-dredged mud and steel slag most intense.

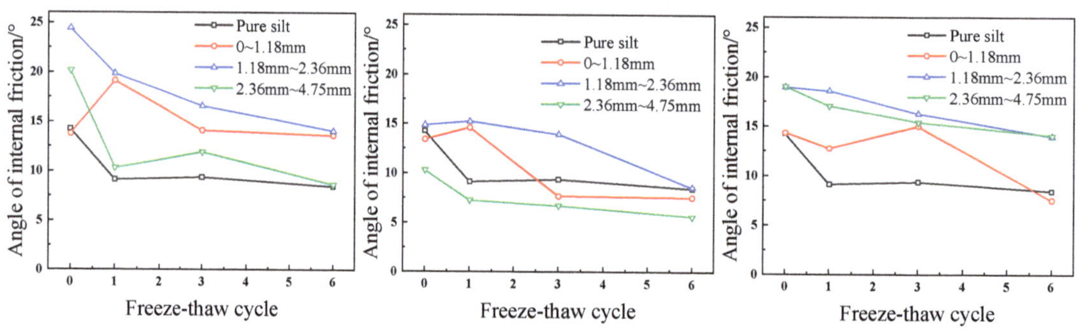

(**a**) Dosage 30% (**b**) Dosage 35% (**c**) Dosage 40%

Figure 2. Variation curves of internal friction angle of marine-dredged mud with improvement by waste steel slag with different waste steel slag particle sizes under freeze–thaw cycle conditions.

The variation curves of the internal friction angle of marine-dredged mud with improvement by waste steel slag with different waste steel slag dosages are shown in Figure 3. It can be seen that under different freeze–thaw times and waste steel slag particle size conditions when the waste steel slag dosage is 35%, its internal friction angle is lower than in other cases. In the case of 3~6 freeze–thaw cycles of 2.36~4.75 mm particle size and 0~1.18 mm particle size, the internal friction angle of marine-dredged mud with improvement by waste steel slag is lower than that of pure marine-dredged mud. For the case of 0~1.18 mm, due to the small particle size of waste steel slag, the improvement effect of pore size and permeability of marine-dredged mud is poor, resulting in a strong capillary effect and high moisture content of the specimens. According to the research of Cai et al. [31], the degree of soil damage caused by freeze–thaw action will increase with the increase in soil moisture content, so the internal friction angle will decrease slightly when the freeze–thaw cycle is 3~6 times. For the case of 2.36~4.75 mm, due to the large particle size of waste steel slag, the structure of the mixed soil may be a dense-suspended type when the content is 35%. Therefore, the waste steel slag aggregates have no contact with each other and cannot form a skeleton, and thus the internal friction resistance is small. When the dosage of waste steel slag with a particle size of 2.36~4.75 mm reaches 40%, the waste steel slag, acting as aggregates, contacts with each other and can form a skeleton. Thus, its structural form may be a framework–pore structure, and the internal friction angle is significantly improved.

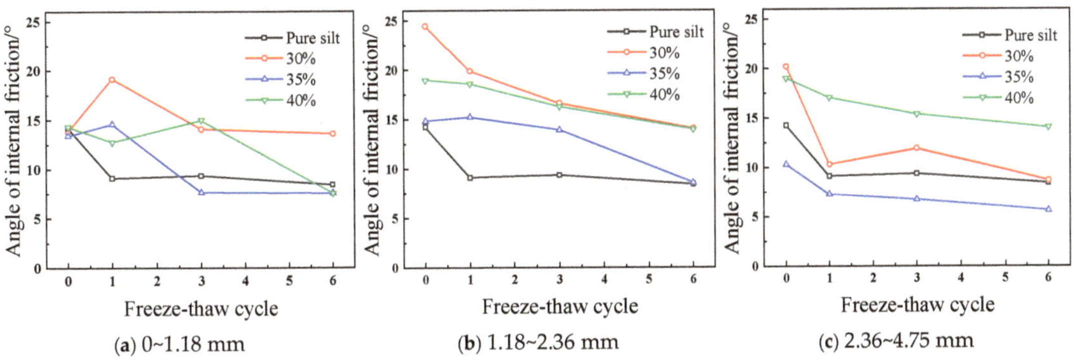

Figure 3. Variation curves of internal friction angle of marine-dredged mud with improvement by waste steel slag with different waste steel slag dosage under freeze–thaw cycle conditions.

3.2. The Influence of Cohesion

The variation curves of cohesion of marine-dredged mud with improvement by waste steel slag with different particle sizes are shown in Figure 4. In general, when the waste steel slag dosage is small, the waste steel slag with different particle sizes has an improving effect on the cohesion of marine-dredged mud. When the waste steel slag dosage is large, the waste steel slag with a particle size of 2.36~4.75 mm has no improvement effect, instead a reduction effect on the cohesion of marine-dredged mud. When the particle size of waste steel slag is 0~1.18 mm and 1.18~2.36 mm, the cohesion of marine-dredged mud with improvement by waste steel slag is higher than that of pure dredged mud and the other two particle sizes. However, when the waste steel slag dosage is 40% (Figure 4c) and the number of freeze–thaw cycles reaches 5~6 times, the cohesion decreases greatly for the case of 0~1.18 mm particle size, while it shows a smaller decrease for the case of 1.18~2.36 mm particle size, and the cohesion is more stable with the variation of freeze–thaw cycles. This may be due to the fact that the electrostatic attraction, van der Waals force and cementation of the soil are weakened by the influence of freeze–thaw cycles. However, because the geometric stacking effect between the waste steel slag and the dredged mud particles with a particle size of 1.18~2.36 mm is better, the apparent cohesion by the interlocking action between particles is stronger, which makes the reduction smaller when the number of freeze–thaw cycles is more.

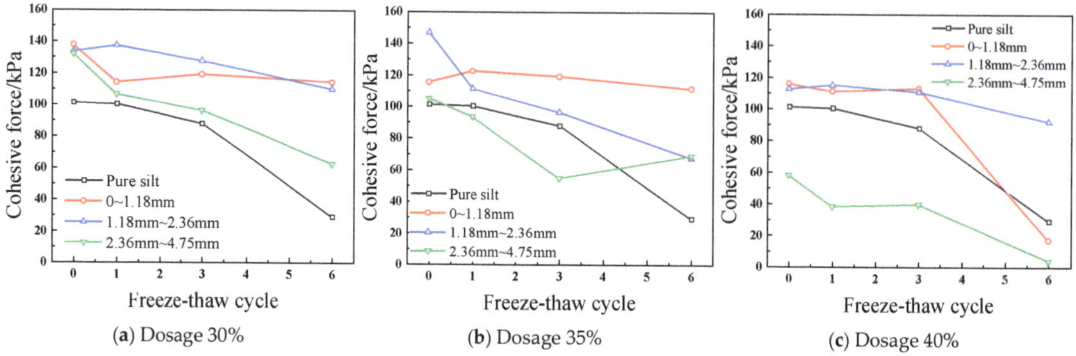

Figure 4. Variation curves of cohesion of marine-dredged mud with improvement by waste steel slag with freeze–thaw cycles for different waste steel slag particle sizes.

The variation curves of cohesion of marine-dredged mud with improvement by waste steel slag with different waste steel slag dosages are shown in Figure 5. It can be clearly seen that no matter what kind of waste steel slag particle size is used to improve the marine-dredged mud, when the waste steel slag dosage is 30%, the cohesion improvement effect is better than in other cases. In general, when the particle size of waste steel slag is small (Figure 5a,b), the cohesion of marine-dredged mud is improved by different dosages of waste steel slag. When the particle size of waste steel slag is large (Figure 5c), the waste steel slag with a dosage of more than 35% has a damaging effect on the cohesion of marine-dredged mud. This is because, with the increase in waste steel slag dosage and particle size, the particle size distribution of the mixture changes greatly, from the dominant position of clay particles to sand particles. Consequently, the electrostatic attraction and molecular attraction between waste steel slags, waste steel slag and soil particles are reduced, and the cementation is weakened, resulting in a decrease in cohesion [32,33].

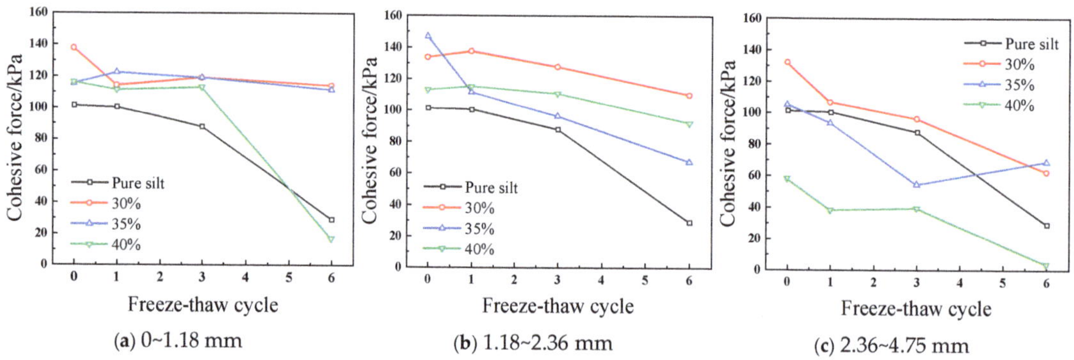

Figure 5. Variation curves of cohesion with freeze–thaw cycles of marine-dredged mud with improvement by waste steel slag with different waste steel slag dosages.

From the above analyses, it can be concluded that the optimal waste steel slag particle size for improving the internal friction angle of marine-dredged mud is 1.18~2.36 mm, and the dosage is 30% and 40%. On the other hand, the optimal waste steel slag particle size for improving the cohesion of marine-dredged mud is 1.18~2.36 mm and the dosage is 30%. Therefore, the follow-up study on the mix ratio of the solidifying agent was carried out with the improved marine-dredged mud by the waste steel slag with a particle size of 1.18~2.36 mm and a dosage of 30%.

4. The Test Results and Analyses of the Influence of Solidifying Agent Mix Ratio on Unconfined Compressive Strength

4.1. Response Surface Model of Unconfined Compressive Strength

The unconfined compressive strength results of marine-dredged mud solidified soil with improvement by waste steel slag (hereinafter, MSWSS) at 7d and 14d ages are shown in Table 5.

Table 5. Unconfined compressive strength results from response surface method tests.

Number	1	2	3	4	5	6	7	8	9	10	11	12	13	14	15	16	17
Y_7/MPa	1.5	0.8	1.4	2.2	1.6	1.0	1.7	1.7	1.5	1.4	1.3	2.4	2.2	2.5	2.7	2.3	2.4
Y_{14}/MPa	1.5	1.4	2.0	2.8	1.8	1.8	1.7	3.2	1.9	2.8	1.3	2.3	2.3	2.7	3.0	2.5	2.9

The regression fitting analyses of the test data in Table 5 was carried out by Design-expert 10 MFC, and the response surface fitting equations Y_7 and Y_{14} for 7d and 14d unconfined compressive strength of MSWSS were obtained, shown in Equations (1) and (2).

$$\begin{aligned} Y_7 = &-7.235 + 0.871x_A + 1.033x_B + 0.671x_C + 0.106x_Ax_B \\ &+0.013x_Bx_C - 0.151x_A^2 - 0.133x_B^2 - 0.126x_C^2 \end{aligned} \tag{1}$$

$$\begin{aligned} Y_{14} = &-10.728 + 1.523x_A + 1.311x_B + 0.804x_C + 0.131x_Ax_B \\ &-0.019x_Ax_C + 0.050x_Bx_C - 0.199x_A^2 - 0.130x_B^2 - 0.093x_C^2 \end{aligned} \tag{2}$$

In order to test the significance of each influencing factor in the response surface equation, the F distribution was used for variance analyses, and the F-value and p-value were calculated and listed in Table 6.

Table 6. Analyses of variance for response surface model equations.

Item	Y_7		Y_{14}	
	F-Value	*p*-Value	*F*-Value	*p*-Value
Model	35.46	<0.0001	32.37	<0.0001
A-cement	0.51	0.4984	5.49	0.0516
B-lime	50.97	0.0002	99.17	<0.0001
C-fly ash	9.57	0.0175	0.34	0.5764
AB	32.73	0.0007	33.63	0.0007
AC	0.45	0.5225	0.69	0.4348
BC	32.73	0.0007	4.88	0.0629
A^2	69.83	<0.0001	81.17	<0.0001
B^2	53.59	0.0002	34.73	0.0006
C^2	0.0002	0.0002	17.58	0.0041
R^2	0.9785		0.9765	
R_a^2	0.9509		0.9464	
C.V./%	8.80		8.43	
Adeq Precision	17.311		16.760	

R^2 is the correlation coefficient, R_a^2 is the adjustment coefficient, *C.V.* is the coefficient of variation.

According to the statistical principle, the significance level (α) was set to 0.05. For the Y_7 response surface model of MSWSS, the $p = 0.5225 > \alpha$ of AC, so the AC item is removed from the regression equation. For the Y_{14} response surface model of MSWSS, the p values of AC and BC are 0.4348 and 0.0629, both greater than α, so the items of AC and BC are dropped from the regression equation. After this correction, the regression equations of each influencing factor based on the coding source (x_i, subscript i indicates each influencing factor) and the 7d and 14d unconfined compressive strengths of MSWSS are obtained. They were coded by Equations (3) and (4):

$$Y_7 = 2.46 - 0.0375x_A + 0.375x_B + 0.1625x_C + 0.425x_Ax_B + 0.425x_Bx_C - 0.605x_A^2 - 0.53x_B^2 - 0.505x_C^2 \quad (3)$$

$$Y_{14} = 2.94 + 0.15x_A + 0.64x_B - 0.038x_C + 0.53x_Ax_B - 0.80x_A^2 - 0.52x_B^2 - 0.375x_C^2 \quad (4)$$

4.2. Interaction Effect of Solidifying Agent on Response Surface

The 3D response surface plots and contour maps of the interaction of different influencing factors (see Figures 6–8) are obtained using the Model Graphs module of the Design-expert software. In these figures, (a) and (b) represent the response surface plot and the contour map, and (c) is the interaction plane plot of two components at 7/14d, where the dotted line is the 95% confidence interval.

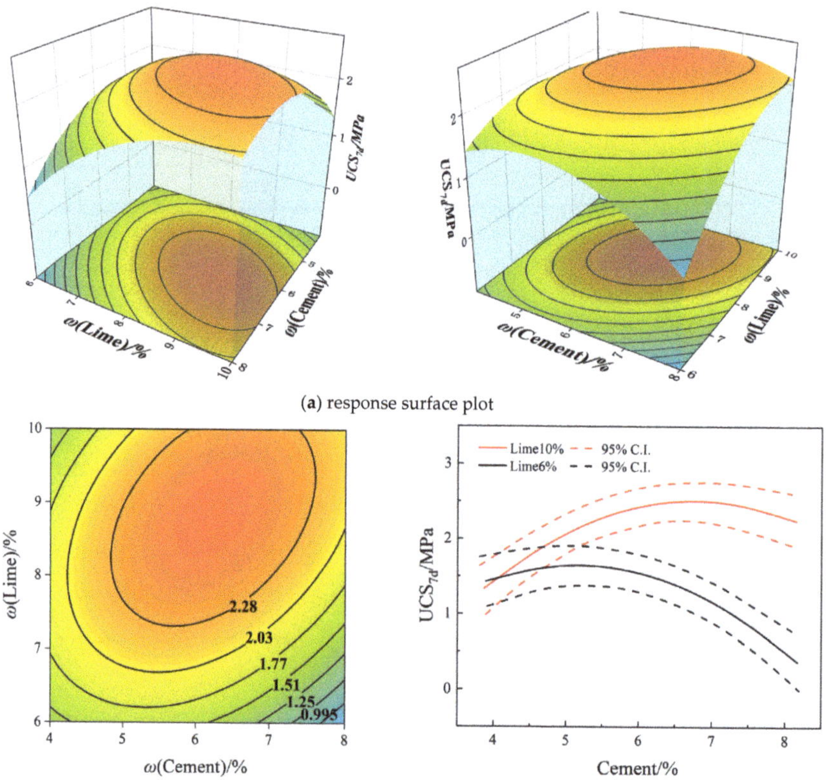

(**a**) response surface plot

(**b**) contour plot

(**c**) 7d cement–lime interaction plan

Figure 6. Response surface, contour and cross-section plots of 7d unconfined compressive strength of MSWSS under cement–lime interaction.

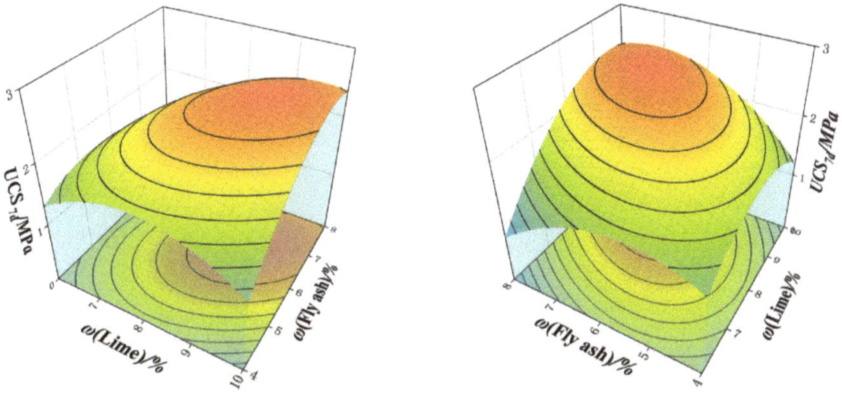

(**a**) response surface plot

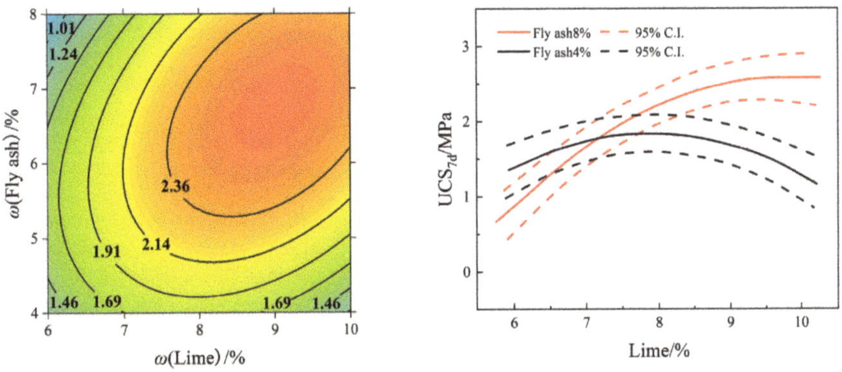

(**b**) contour plot

(**c**) 7d lime fly ash interaction plan

Figure 7. Response surface, contour and cross-section plots of 7d unconfined compressive strength of MSWSS under lime fly ash interaction.

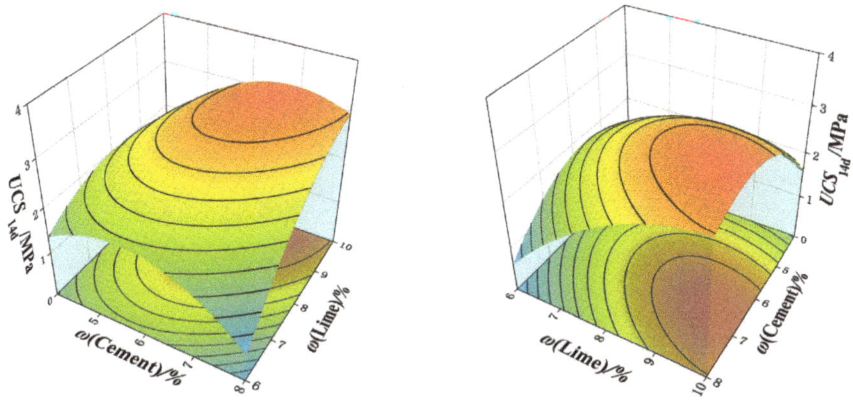

(**a**) response surface plot

Figure 8. *Cont.*

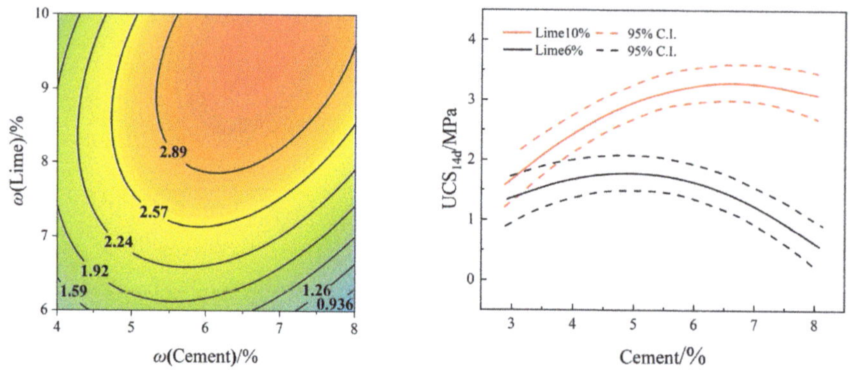

(**b**) contour plot (**c**) 14d cement–lime interaction plan

Figure 8. Response surface, contour and cross-section of 14d unconfined compressive strength of MSWSS under cement–lime interaction.

4.2.1. The Interaction Between Cement and Lime at 7d

From Figure 6a,b, it can be seen that within the range of the designed dosage, the 7d unconfined compressive strength of MSWSS first increases and then decreases with the increase in lime and cement. More specially, in Figure 6c, when the lime dosage is set at 6%, the 7d unconfined compressive strength of MSWSS increases first and then decreases with the increase in cement dosage, and it reaches the highest at 5~6% of the cement dosage. When the lime dosage is set at 10%, the influence trend of cement dosage is similar to that of lime at 6%, but the corresponding to the highest point of the model increases (7%), and the overall strength is higher than that of 6% lime dosage. By analyzing the influence mechanism of this interaction, it can be interpreted that because marine-dredged mud, cement and fly ash contain a large amount of SiO_2 and Al_2O_3, and the lime releases Ca^{2+} and OH^- in the solidification process, which increases the alkalinity in the system [34]. It accelerates the fracture of Si-O and Al-O bonds in marine-dredged mud, cement and fly ash, and promotes the increase in C-S-H (hydrated calcium silicate) and C-A-H (hydrated calcium aluminate). Both products can wrap soil particles and accelerate the granulation between waste steel slag and marine-dredged mud. These formed aggregates are bonded together to form a denser skeleton structure so that the strength shows an increasing trend at low cement dosages. On the other hand, as the cement dosage increases, a high-alkali and high-calcium environment is created, resulting in the rapid formation of C-S-H and C-A-H hydration products and the encapsulation of incompletely reacted aluminosilicate raw materials [35]. Moreover, C-A-H will expand significantly during the hydration process, resulting in the contact surface between dredged mud and waste steel slag not being closed, so that the strength will be reduced.

4.2.2. The Interaction Between Lime and Fly Ash at 7d

From Figure 7a,b, it can be seen that the 7d unconfined compressive strength of MSWSS first increases and then decreases with the increase in lime and fly ash. The reason is similar to the interaction between lime and cement. It is to increase the alkalinity and calcium dosage of the reaction system to increase the formation rate and amount of C-A-H, etc. [35]. Among them, fly ash can produce aluminosilicate and calcareous components, and accelerate the formation of C-A-H in the early stage of the reaction. However, with the increase in lime and fly ash dosage, C-A-H is rapidly formed and wrapped in unreacted cement and dredged mud, leading to incomplete contacts, which has an inhibiting effect on the reaction. As a result, the strength will show a downward trend. It can be concluded from Figure 7c that when the dosage of fly ash is 4%, the 7d unconfined compressive strength of MSWSS first increases and then decreases as the lime increases. When the dosage of fly ash is 8%, the 7d unconfined compressive strength of MSWSS first increases and then tends to be stable with the increase in lime. According to Figure 7c, the 95% confidence interval of the 8% variation curve of fly ash shows a downward trend when the lime dosage is 9%~10%. Therefore, it can be speculated that when the lime dosage is high enough, the unconfined compressive strength will show a trend of first increasing and then decreasing.

4.2.3. The Interaction Between Cement and Lime at 14d

It can be seen from Figure 8a,b that the 14d unconfined compressive strength of MSWSS increases first and then decreases with the increase in cement and lime dosage, which is similar to the trend at 7d. Combined with Figures 6c and 8c, the maximum peak value of the unconfined compressive strength of MSWSS at 14d increases by about 26.8% compared with that at 7d. This is due to the fact that as the age increases, the hydration reaction of the components in the solidified soil is more sufficient, and more C-A-H hydration products are generated, which has a more significant filling effect on the void of MSWSS. At the same time, the highest point of the 14d unconfined compressive strength model with 10% lime dosage appears in the range of 7%~8% cement, which is about 1% higher than the highest point of the 7d unconfined compressive strength model (in the range of 6%~7%). The highest point of the 7d model with 6% lime dosage also has a similar effect to the highest point of the 14d model (from about 5% cement dosage to 5%~6%). It can be inferred that with the increase in curing age, the hydration reaction, carbonation reaction and pozzolanic reaction continue and consume more solidifying agents. Therefore, the dosage of a solidifying agent at the highest point of the unconfined compressive strength model increases.

4.3. Optimization Analyses of Solidifying Agent Mix Ratio

According to the analyses of response surface, cement, lime and fly ash comprehensively affect the strength of the MSWSS, and there is an optimal mix ratio of the three to achieve the best effect of synergistic enhancement of multiple solid wastes. The response surface model is optimized in the Design-expert software. The mix ratio optimization process in the software is shown in Table 7. The first scheme (serial number 1) is the optimal scheme given by the software, that is, the optimal mix ratio of cement, lime and fly ash is 5.60%, 10.00%, and 7.96%.

Table 7. Mixing ratio optimization results.

Number	Cement/%	Lime/%	Fly Ash/%	UCS (7d)/MPa	UCS (14d)/MPa	Usability
1	5.604	10.000	7.963	2.285	2.710	0.829
2	5.608	10.000	7.980	2.283	2.707	0.829
3	5.609	10.000	7.943	2.291	2.718	0.829
4	5.604	10.000	7.984	2.280	2.704	0.829
5	5.587	10.000	7.944	2.283	2.708	0.829
6	5.581	10.000	7.972	2.275	2.697	0.829
7	5.610	10.000	7.999	2.279	2.703	0.829
8	5.628	10.000	7.942	2.297	2.727	0.829
9	5.608	10.000	7.905	2.298	2.728	0.829
10	5.658	10.000	7.900	2.315	2.751	0.828
11	5.455	10.000	7.971	2.230	2.635	0.827

With the optimal mix ratio, the 7d strength of the MSWSS can reach 2.29 MPa, and the 14d strength can reach 2.71 MPa. According to the 'Code for Design of Highway Asphalt Pavement'(JTGD50-2017) [36], the inorganic binder stabilized materials used as the subbase of extremely heavy, extra heavy and heavy traffic roads for expressways and first-class highways should have a minimum 7d unconfined compressive strength of 2.0 MPa. With the optimal mix ratio of the solidifying agent, the 7d unconfined compressive strength of MSWSS can not only meet the engineering requirements of expressways and first-class highways but also significantly improve the resource utilization rate of waste steel slag and marine-dredged mud. Therefore, the practical engineering application of marine-dredged mud could also be further promoted, with significant environmental and economic benefits.

In order to clarify the accuracy and applicability of the optimized response surface model, the experimental response value (measured value) and the model prediction value are compared, as shown in Figure 9. The external studentized residuals are used to detect outliers in regression diagnosis in Figure 10, calculated with the remaining n-1 observations after deleting the *i*th observation, which should be less than 4.81963 in this test. According to Figure 10, the absolute value of the external studentized residuals. It can be seen from Figures 9 and 10 that the maximum error between the measured and predicted values of 7d and 14d unconfined compressive strength of MSWSS is 3.795% and all external studentized residuals are less than the max value, indicating that the optimized response surface model has high accuracy.

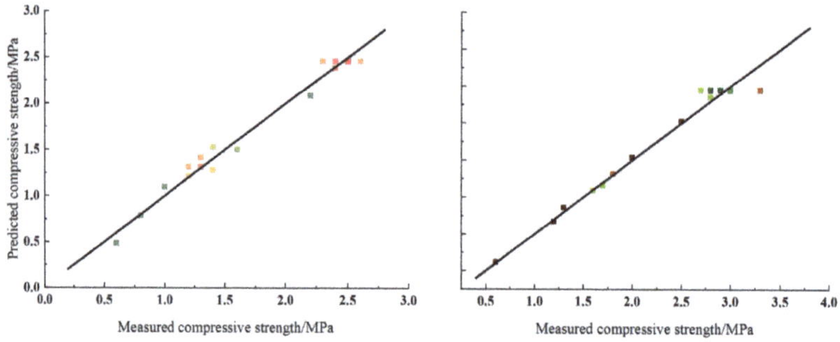

(**a**) 7d (**b**) 14d

Figure 9. Comparison of measured and predicted 7d and 14d unconfined compressive strength of MSWSS.

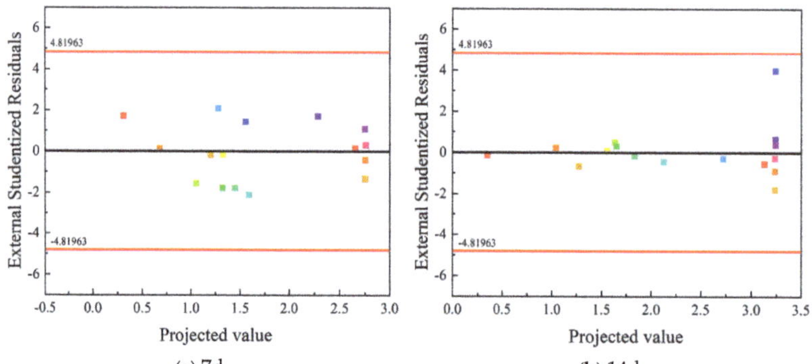

(**a**) 7d (**b**) 14d

Figure 10. Deviation between measured and predicted 7d and 14 unconfined compressive strength of MSWSS.

5. Stability Analyses

Based on the optimal mix ratio of the solidifying agent, the water stability and dry–wet cycle experiments of MSWSS were carried out. The water stability coefficients and the compressive strength characteristics, mass loss rate and volume loss rate under dry–wet cycles were studied.

5.1. Water Stability Analyses

In the 'soil solidified admixture'(CJ/T486-2015), the water stability coefficient is defined as the Equation (5).

$$\gamma_{\omega x} = \frac{R_{\omega x}}{R_{0x}} \times 100 \quad (5)$$

where $\gamma_{\omega x}$ is the water stability coefficient of x d, %. $R_{\omega x}$ is the unconfined compressive strength with normal conditions for $(x-1)$ d and water soaking for 1d. R_{0x} is the unconfined compressive strength of the normal condition x d specimens.

The results of water stability analyses are shown in Figure 11. Among them, the test group number of MWSS without solidification is WS_{NS}, and the test group numbers of MSWSS at the age of 7d and 14d according to the optimal mix ratio of the solidifying agent are M7 and M14, respectively. For WS_{NS}, it is seriously dispersed in the water-soaked state, resulting in its immersion strength of about 0 and poor water stability. The water stability coefficient changes greatly from WS_{NS} to M7. It indicates that the hydration products are formed in MSWSS, such as C-S-H and C-A-H, which can cement the waste steel slag and marine-dredged mud particles, and keep them from collapsing in water. At the same time, the hydration products can fill the pores of MSWSS, and reduce the porosity and permeability of the solidified soil, so that MSWSS has excellent water stability. According to the research of Horpibulsuk et al. [10], the addition of lime to the soil can reduce the water sensitivity of the soil. Under the combined effect of these factors, the MSWSS has a significant change in water stability compared to the MWSS.

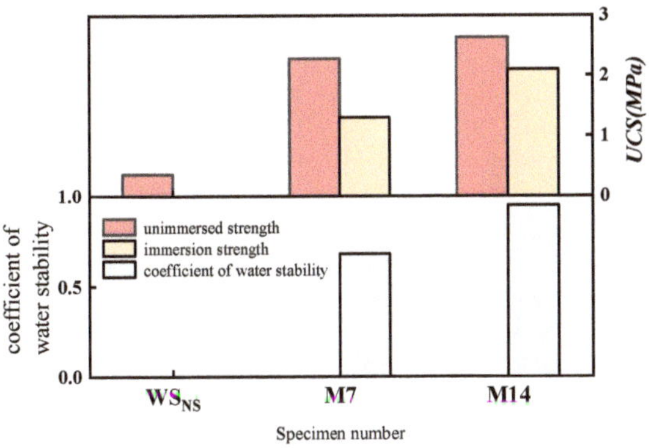

Figure 11. Water stability coefficients of MSWSS.

5.2. Compressive Strength Characteristics Under Dry–Wet Cycle

Under the condition of dry–wet cycle 0~6 times, the variation of 7d and 14d unconfined compressive strength of MSWSS is shown in Figure 12. The loss rate of unconfined compressive strength in dry–wet cycle is defined as the rate of unconfined compressive strength compared to the initial strength without cycle. At the age of 7d, the unconfined compressive strength decreases slowly in the first three times of dry–wet cycle, and decreases sharply at the fourth time, and then decreases slowly. The strength loss rate before 4 dry–wet cycles is less than 0.3, while it increases rapidly and reaches 0.37 at 4 dry–wet cycles. Therefore, the turning point at 7d is in the fourth cycle. At the age of 14d, the strength loss rate of the first cycle is more than 0.3, which is significantly reduced. This is because the first dry–wet cycle has the greatest influence on the specimens, and the cyclic action causes cracks in the specimens [37]. Thereafter, the strength shows a slight increase, which may be due to the fact that as the number of dry–wet cycles increases, the unreacted solidifying agent and the marine-dredged mud in the specimen are fully hydrated, and the cementitious products fill the pores in the specimen, thereby increasing the strength [38]. In addition, the increase in unconfined compressive strength of the solidified soil and the damage caused by dry–wet cycles is a dynamic equilibrium process [39]. The reaction between the cementitious material generated by hydration and water continues during the dry–wet cycle process [40], especially during the wet stage. At this point, the strength loss caused by the dry–wet cycle is not sufficient to offset the strength growth from the hydration reaction, resulting in an increase in the compactness and strength of the solidified soil [41].

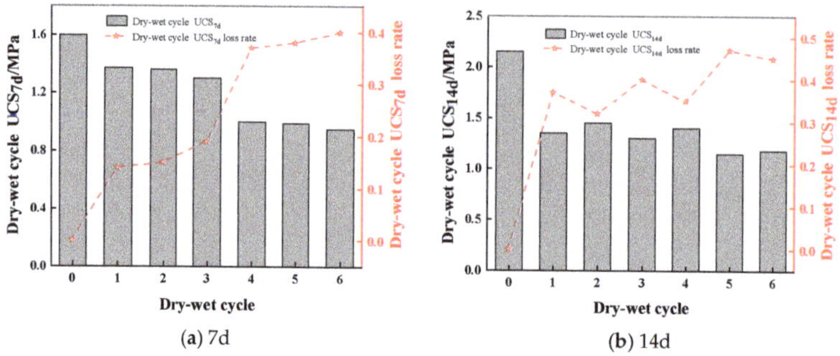

Figure 12. Influence of dry and wet cyclic effects on 7d and 14d unconfined compressive strength of MSWSS.

5.3. Mass and Volume Loss Rate Characteristics Under Dry–Wet Cycle

Under the condition of 0~6 dry–wet cycles, the curves of mass loss rate and volume loss rate for MSWSS at 7d and 14d are shown in Figures 13 and 14. Among them, the mass loss rate and volume loss rate are that with a certain number of cycles compared with the initial mass and volume without any cycle. The volume and mass loss rates at 7d and 14d are all below 1.5%. This indicates that the MSWSS has certain dry and wet resistance and good durability. The mass loss rate at 7d and 14d is negative before three dry–wet cycles, because the specimens store a small amount of free water during the dry–wet cycle. From the point of view of the volume variation, the volume loss rate at 7d shows a steady strong upward trend, while at 14d, it changes dramatically below three dry–wet cycles and becomes smooth at larger cycles. This is because the longer the curing time, the more thorough the hydration reaction and the denser the solidified soil. Therefore, the time point of the sharp volume loss is relatively backward, and the volume loss is severe when the dry–wet cycle is 3~4 times.

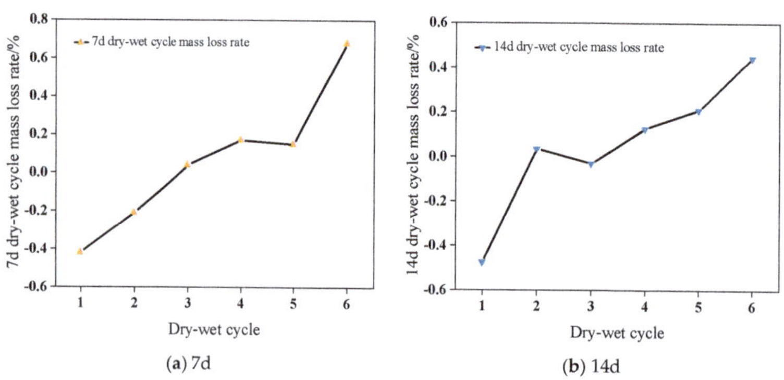

Figure 13. Variations of mass loss rate for MSWSS.

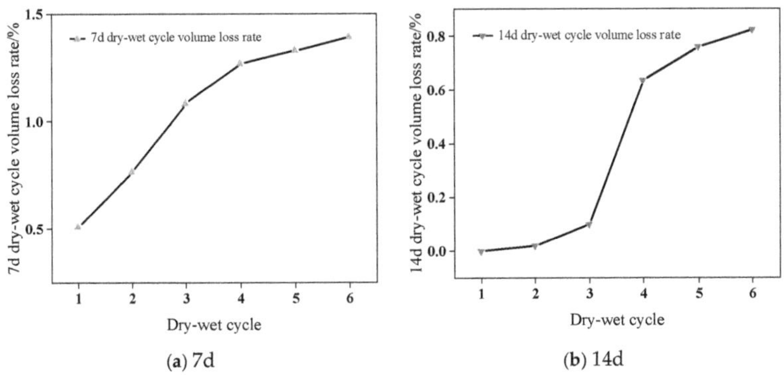

Figure 14. Variations of volume loss rate for MSWSS.

When the number of dry–wet cycles is large, the unconfined compressive strength of MSWSS is closely related to the dry–wet effect of the environment. With the increase in the number of dry–wet cycles, the appearance of the specimens at 14d shows almost no damage, indicating that the curing age has a great influence on the micropores of marine silt solidified soil. When the curing age is high, the dry–wet cycle mainly affects the micropore structure of the specimens.

6. Microscopic Mechanism Analyses

Figure 15 shows the microscopic morphology of pure marine-dredged mud, MWSS (waste steel slag content is 30%, particle size is 1.18~2.36mm) and MSWSS (optimal solidifying agent mix ratio) specimens magnified by 2000 times. Compared with (a) and (b), it can be found that the pure marine-dredged mud shows the phenomenon of internal looseness, large pores and widening cracks. While the structure of MWSS is obviously tighter, the waste steel slag particles provide a large number of 'anchoring surfaces' for marine-dredged mud with rough shape, and porous and multi-angular edges. This improves the pore structure of dredged mud and makes the compact. However, due to the lack of solidifying agent, there are still obvious cracks. Comparing (b) and (c), it can be clearly seen that the MSWSS has a denser structure. There are a large number of white flocculent cementation products adhered to the particles, which strengthens the bonding of the interface transition zone between waste steel slag and marine-dredged mud. So that the number and width of cracks are reduced, forming an integral structure.

In order to observe the micromorphological changes of MSWSS more carefully, the SEM analyses under 4000 and 8000 times is given in Figure 16. It can be seen clearly that a large number of flocculent, agglomerated and fibrous white gels are formed in the structure of MSWSS. The fibrous zeolite-like amorphous gel product named as hydrated calcium silicate (C-S-H) and plate-like $Ca(OH)_2$ crystals are observed. At this time, the waste steel slag particles, acting as 'anchoring surfaces' for cementation, are tightly wrapped by the solidifying gel product, which strengthens the interface transition zone of the two soil particles. Although there are still a small number of pores that have not been fully connected, the overall network gel has been formed. The fibrous C-S-H gel is interspersed in the microscopic pores, constraining the unhydrated waste steel slag and dredged mud particles in the form of a network. Thereby it can enhance the structural integrity and improve the macroscopic properties of the solidified soil.

(**a**) pure marine-dredged mud

(**b**) MWSS

(**c**) MSWSS

Figure 15. SEM analyses of pure mud, unsolidified MWSS and MSWSS at 2000×.

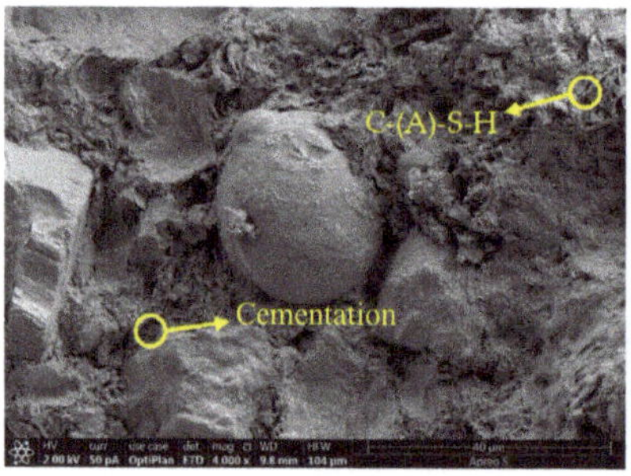

(**a**) 4000×

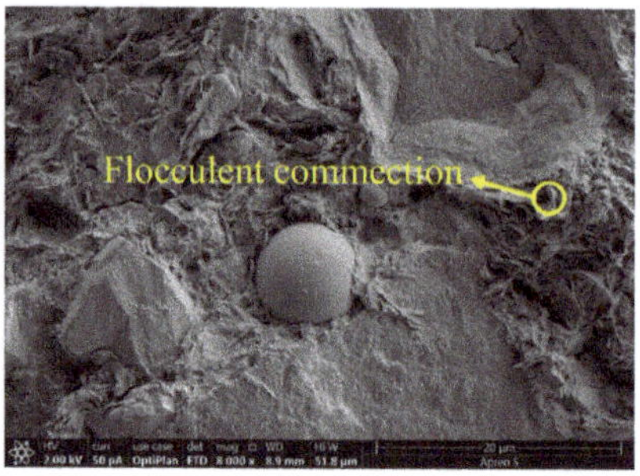

(**b**) 8000×

Figure 16. SEM analyses of MSWSS at 4000 and 8000 times.

The XRD analyses for pure marine-dredged mud and MSWSS specimens were carried out for main mineral compositions, shown in Figure 17. It can be seen that the marine-dredged mud mainly contains silicate minerals such as quartz (SiO_2) and calcium zeolite ($CaAl_2Si_3O_{10}{\bullet}3H_2O$). For the MSWSS, a large number of quartz diffraction peaks were found to disappear, indicating that reactions of the cement, lime and fly ash take place and generate hydration products including hydrated calcium silicate and hydrated calcium aluminate.

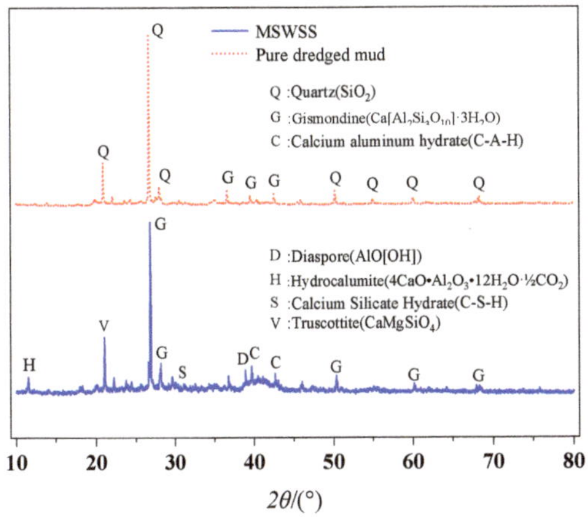

Figure 17. XRD pattern of MSWSS.

Specifically, according to the relevant studies [42,43], the XRD pattern of MSWSS shows new diffraction peaks of hydrated silicate (C-S-H) and hydrated aluminosilicate (C-(A)-S-H) at the position of $2\theta = 18°{\sim}35°$, and a new diffraction peak of hydrated aluminate (C-A-H) at the positions of $2\theta = 40.23°$ and 44.23°. These observations confirm the formation of hydration products such as C-A-H, C-S-H, and C-(A)-S-H during the solidifying process. This shows that in the composite cementitious system, the hydration reaction between cement, lime and fly ash is not just a simple superposition of their respective reactions. On the contrary, there is a synergistic effect between them. It can quickly form the skeleton structure of MSWSS, which is conducive to the further improvement of the strength. The existence of this synergistic effect also explains the significant interaction observed in the interaction analyses of cement, lime and fly ash in the response surface test. Therefore, the proportion of each component needs to be precisely controlled, and the mix ratio can be optimized by response surface analyses to obtain the best performance of MSWSS.

In order to verify the above analyses, SEM-EDS on the MSWSS was analyzed, as shown in Figure 18. According to Figure 18b,c, it is found that MSWSS contains a large amount of Al, Si and O elements at three points 1, 2 and 3 in Figure 18a. Especially at point 2, in addition to the above elements, a large amount of calcium (Ca) was detected. This finding shows that more reticular C-A-H and C-S-H structures are produced in MSWSS. With the filling and cementation effect of these hydration products, soil particles are connected to each other more closely, and a dense network structure is formed inside the soil, resulting in a whole network of gels.

Through the comprehensive analyses of SEM, SEM-EDS and XRD, we can fully confirm the existence of hydration products such as hydrated silicate, hydrated aluminate and hydrated calcium aluminosilicate, which play a role of adhesion and cementation in the main internal structural forms of MSWSS.

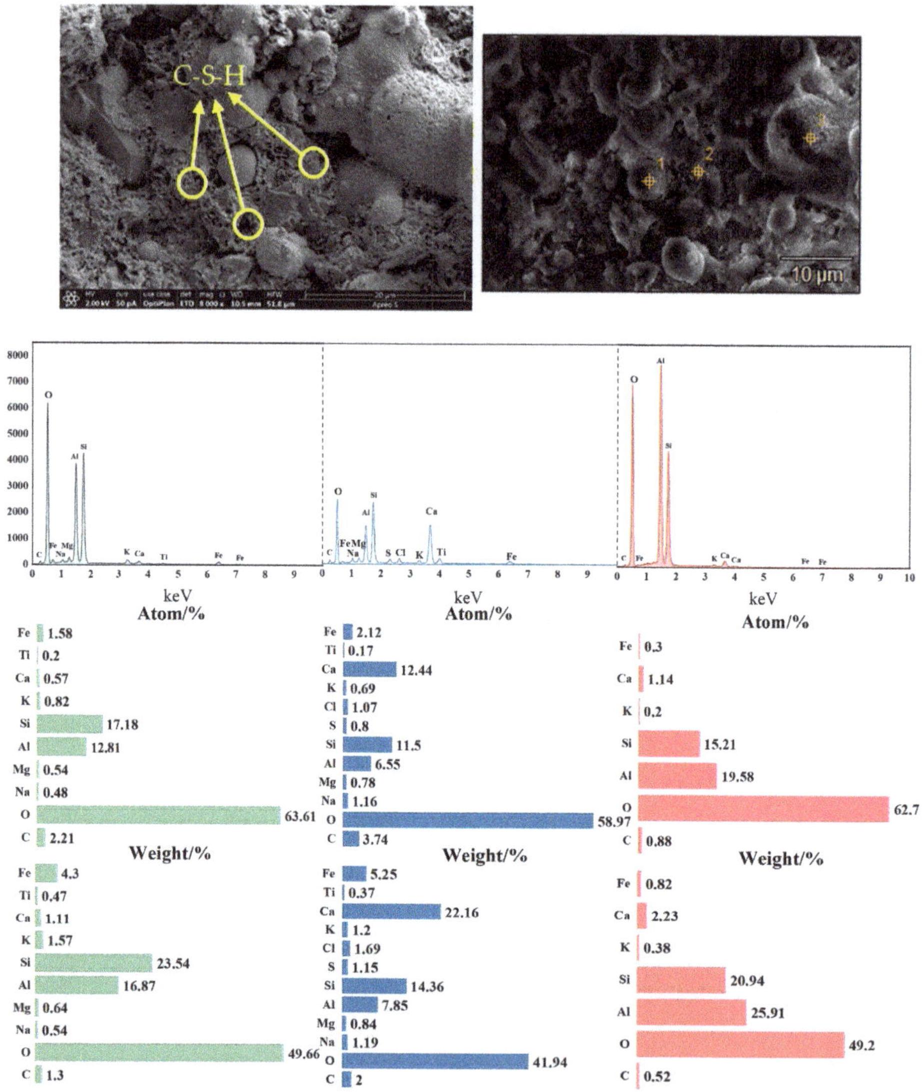

(**a**) Point 1 EDS analyses results (**b**) Point 2 EDS analyses results (**c**) Point 3 EDS analyses results

Figure 18. SEM-EDS analyses of MSWSS at 8000 times.

7. Conclusions

In order to improve the engineering performance of marine-dredged mud, waste steel slag and composite solidifying agents (cement, lime and fly ash) were used for physical and chemical improvement. Direct shear test under freeze–thaw cycles, unconfined compressive strength test based on response surface method, water stability test under dry–wet cycles and the microscopic mechanism tests were carried out to investigate the improvement effect. The main conclusions are as follows:

1. The 1.18~2.36 mm waste steel slag with 30% dosage has the best improvement effect on the shear strength of the marine-dredged mud.
2. The 7d and 14d unconfined compressive strength response surface regression models were established. And the interaction analyses show that cement and lime, lime and fly ash at the 7d age and cement and lime at the 14d age have significant influence.
3. The optimal mix ratio of the composite solidifying agent is 5.60%, 10.00%, and 7.96%, respectively, for cement, lime and fly ash. With this ratio, the unconfined compressive strength of MSWSS at 7d and 14d can reach 2.29 MPa and 2.71 MPa. The maximum error is 3.795%.
4. The water stability coefficients of 7d and 14d are 0.68 and 0.95, respectively, and the volume and mass loss rates under 0~6 dry–wet cycles are all below 1.5%, indicating that MSWSS has certain dry–wet resistance and good durability.
5. The microscopic analyses demonstrate that the improvement of waste steel slag lies in the provision of a large number of 'anchoring surfaces', which improves the pore structure of marine-dredged mud. The products such as hydrated calcium silicate and hydrated calcium aluminate by the solidifying agent play the role of filling and cementation, which makes the soil particles connect with each other to form a dense network structure.

Author Contributions: Conceptualization, Q.J.; methodology, Q.J. and Y.W.; software, Y.W.; validation, Q.J., Y.C., Y.W. and X.J.; formal analysis, Q.J. and Y.W.; investigation, Q.J. and Y.W.; resources, Y.C.; writing—original draft preparation, Q.J. and Y.W.; writing—review and editing, Q.J. and X.J.; visualization, Y.W. and X.J.; supervision, Q.J.; project administration, Y.C. All authors have read and agreed to the published version of the manuscript.

Funding: This research received no external funding.

Data Availability Statement: The data presented in this study are available on request from the corresponding author on reasonable request. The data are not publicly available due to privacy policies.

Conflicts of Interest: The authors declare no conflicts of interest.

References

1. Abidi, I.; Benamara, L.; Correia, A.A.S.; Pinto, M.I.M.; Cunha, P.P. Characterization of Dredged Sediments of Bouhanifia Dam: Potential Use as a Raw Material. *Arab. J. Geosci.* **2021**, *14*, 2631. [CrossRef]
2. Amar, M.; Benzerzour, M.; Kleib, J.; Abriak, N.-E. From Dredged Sediment to Supplementary Cementitious Material: Characterization, Treatment, and Reuse. *Int. J. Sediment Res.* **2021**, *36*, 92–109. [CrossRef]
3. Zhang, R.J.; Santoso, A.M.; Tan, T.S.; Phoon, K.K. Strength of High Water-Content Marine Clay Stabilized by Low Amount of Cement. *J. Geotech. Geoenviron. Eng.* **2013**, *139*, 2170–2181. [CrossRef]
4. Zhang, C.L.; Guan, F.F.; Li, L.; Huang, Y.H. The progress in The Reutilization Treatment and Disposal of Dredged Sediments in China. *Environ. Eng.* **2014**, *32*, 95–99.
5. Zhang, H.Q.; Xie, J.; Zhu, W.; Huang, Y.Z.; Shi, P. Present Situation of Dredged Materials Dumping and the Study of Transforming Dredged Mud into Regenerative Resources—Difficulties of Refuses Dumping in China Seas and Countermeasures to Deal with These Problems. *Mar. Sci. Bull.* **2004**, *6*, 54–60. (In Chinese)
6. Zhu, W.; Zhang, C.L.; Liu, H.L.; Gao, Y.F. The Status Quo of Dredged Spoils Utilization. *Environ. Sci. Ecotechnol.* **2002**, *4*, 39–41+50.
7. Phetchuay, C.; Horpibulsuk, S.; Arulrajah, A.; Suksiripattanapong, C.; Udomchai, A. Strength Development in Soft Marine Clay Stabilized by Fly Ash and Calcium Carbide Residue Based Geopolymer. *Appl. Clay Sci.* **2016**, *127–128*, 134–142. [CrossRef]
8. Li, L.; Zhang, H.; Xiao, H.; Pei, Y.; Wang, J. Mechanical and Microscopic Properties of Alkali-Activated Fly-Ash-Stabilised Construction and Demolition Waste. *Eur. J. Environ. Civ. Eng.* **2023**, *27*, 2661–2677. [CrossRef]
9. Yoobanpot, N.; Jamsawang, P.; Simarat, P.; Jongpradist, P.; Likitlersuang, S. Sustainable Reuse of Dredged Sediments as Pavement Materials by Cement and Fly Ash Stabilization. *J. Soils Sediments* **2020**, *20*, 3807–3823. [CrossRef]
10. Horpibulsuk, S.; Phetchuay, C.; Chinkulkijniwat, A. Soil Stabilization by Calcium Carbide Residue and Fly Ash. *J. Mater. Civ. Eng.* **2012**, *24*, 184–193. [CrossRef]
11. Cao, Z.; Shen, L.; Zhao, J.; Liu, L.; Zhong, S.; Yang, Y. Modeling the Dynamic Mechanism between Cement CO_2 Emissions and Clinker Quality to Realize Low-Carbon Cement. *Resour. Conserv. Recycl.* **2016**, *113*, 116–126. [CrossRef]
12. Wang, A.; Zhan, Q.; Dong, W.; Gu, W.; Zhou, J.; Pan, Z. Influence of Recycled Fine Aggregate Content on Properties of Soft Soil Solidified by Industrial Waste Residue. *Materials* **2022**, *15*, 7580. [CrossRef] [PubMed]

13. Mozejko, C.A.; Francisca, F.M. Enhanced Mechanical Behavior of Compacted Clayey Silts Stabilized by Reusing Steel Slag. *Constr. Build. Mater.* **2020**, *239*, 117901. [CrossRef]
14. Huang, W.; Zhou, M.H.; Yan, X.; Zhang, L.; Ye, Y.C. Properties of Steel Slag-Miscellaneous Fill Base Material. *Bull. Chin. Ceram. Soc.* **2023**, *42*, 196–204. (In Chinese)
15. Wang, B.R.; Zhang, L.Y.; Liu, X.F.; Duan, X.L.; Chen, X. Study on Road Performance and Engineering Application of Cement Steel Slag Stabilized Soil. *Bull. Chin. Ceram. Soc.* **2024**, *43*, 1472–1481. (In Chinese)
16. Kang, G.; Cikmit, A.A.; Tsuchida, T.; Honda, H.; Kim, Y. Strength Development and Microstructural Characteristics of Soft Dredged Clay Stabilized with Basic Oxygen Furnace Steel Slag. *Constr. Build. Mater.* **2019**, *203*, 501–513. [CrossRef]
17. Yao, J.; Qiu, H.; He, H.; Chen, X.; Hao, G. Application of a Soft Soil Stabilized by Composite Geopolymer. *J. Perform. Constr. Facil.* **2021**, *35*, 04021018. [CrossRef]
18. Wu, Y.; Qiao, X.; Yu, X.; Yu, J.; Deng, Y. Study on Properties of Expansive Soil Improved by Steel Slag Powder and Cement under Freeze-Thaw Cycles. *KSCE J. Civ. Eng.* **2021**, *25*, 417–428. [CrossRef]
19. Asadzadeh, S.; Khoshbayan, S. Multi-Objective Optimization of Influential Factors on Production Process of Foamed Concrete Using Box-Behnken Approach. *Constr. Build. Mater.* **2018**, *170*, 101–110. [CrossRef]
20. Shao, Y.; Sun, B.Z.; Zhou, D.Q.; Zhu, B.T. Experimental study of formula Optimization for Lacustrine Soft Soil Consolidation in Hefei City based on Box-Behnken method. *J. Exp. Mech.* **2017**, *32*, 286–294. (In Chinese)
21. Zhang, X.; Li, H.; Li, S.; Ding, Y.; Zhang, H.; Tong, Y.; Hua, S. Test and Microstructural Analysis of a Steel Slag Cement-Based Material Using the Response Surface Method. *Materials* **2022**, *15*, 3114. [CrossRef] [PubMed]
22. Li, S.; Zhang, H.R.; Wang, G.Y.; Deng, R.R. Experimental Study of Alkali-Activated Geopolymer Cured Silty Soil Based on Response Surface Method. *Bull. Chin. Ceram. Soc.* **2023**, *42*, 4438–4448. (In Chinese)
23. Wang, Y.H.; Zhu, L.Y.; Wang, C.; Wang, C.; Zhu, Y.Q. Experimental Study on Mineral Powder Composite Solidified Saline Soil Based on Response Surface Method. *Bull. Chin. Ceram. Soc.* **2024**, *43*, 1917–1927. (In Chinese) [CrossRef]
24. Wang, X.; Kim, S.; Wu, Y.; Liu, Y.; Liu, T.; Wang, Y. Study on the Optimization and Performance of GFC Soil Stabilizer Based on Response Surface Methodology in Soft Soil Stabilization. *Soils Found.* **2023**, *63*, 101278. [CrossRef]
25. Jin, J.; Qin, Z.; Yang, H.; Zuo, S.; Song, C. Designing a Composite Solidification Agent to Improve Mechanical Properties and Sulfate Resistance of Construction Clay-Residue. *J. Build. Eng.* **2022**, *56*, 104753. [CrossRef]
26. Wang, J.; Deng, Y.; Shao, Y.; Liu, X.; Feng, B.; Wu, L.; Zhou, J.; Yin, Y.; Xu, N.; Peng, H. Liquefaction Behavior of Dredged Silty-Fine Sands under Cyclic Loading for Land Reclamation: Laboratory Experiment and Numerical Simulation. *Environ. Earth Sci.* **2018**, *77*, 471. [CrossRef]
27. *JTG E40-2007*; Test Methods of Soils for Highway Engineering. Ministry of Transport: Beijing, China, 2007. (In Chinese)
28. Wang, A.H.; Duan, W.; Cai, G.J.; Huang, Z.W.; Zhang, Y.F.; Zhan, Q.W.; Rui, C.G. Experimental Study on Synergistic Strengthening of Muddy Clay Foundation with Industrial Waste and Recycled Fine Aggregate. *J. Eng. Geol.* **2023**, *409*, 133903. (In Chinese)
29. Gupta, A.; Biswas, S.; Arora, V.K. Ranking of Stabilizers to Stabilize/Solidify Dredged Soil as Highway Construction Material. *Mater. Today Proc.* **2021**, *43*, 1694–1699. [CrossRef]
30. *T 0843-2009*; Test Methods of Materials Stabilized with Inorganic Binders for Highway Engineering. Ministry of Transport: Beijing, China, 2009. (In Chinese)
31. Cai, Z.Y.; Zhu, R.; Huang, Y.H.; Zhang, C.; Guo, W.L. Centrifugal model tests on deterioration process of canal under cyclicaction of coupling wetting-drying and freeze-thaw. *J. Rock Mech. Geotech. Eng.* **2020**, *42*, 1773–1782. (In Chinese)
32. Duncan, J.M. State of the Art: Limit Equilibrium and Finite-Element Analysis of Slopes. *J. Geotech. Eng.* **1996**, *122*, 577–596. [CrossRef]
33. Kirkpatrick, S.; Gelatt, C.D.; Vecchi, M.P. Optimization by Simulated Annealing. *Science* **1983**, *220*, 671–680. [CrossRef] [PubMed]
34. Gao, Y.L.; Zhu, Z.H.; Meng, H.; Hu, X.L.; Li, Z.K. Synergistic Enhancement Mechanism of Calcium Carbide Residue-Desulfurization Gypsum-Steel Slag Modified Geopolymer. *J. Build. Mater.* **2023**, *26*, 870–878. (In Chinese)
35. Wang, Q.; Zhang, S.Q.; Zhang, J.Y. Research on Mechanical Properties and Microscopic Mechanism of Solidified Sludge with Industrial Waste Residue. *J. Anhui Univ. Sci. Technol. (Nat. Sci.)* **2024**, *44*, 38–45. (In Chinese)
36. *JTG D50-2017*; Industry Standards of the People's Republic of China. Ministry of Commerce: Beijing, China, 2017; ISBN 978-7-114-13760-0. (In Chinese)
37. Wang, X.Q.; Liu, N.; Li, Z.Q.; Han, Z.; Zou, W.L. Hydro-mechanical properties of sludge stabilized with magnesium oxychloride cement-based multi-cementitious materials under influences of drying-wetting cycles. *J. Rock Mech. Geotech. Eng.* **2023**, *45*, 2004–2013. (In Chinese)
38. Feng, Y.S. Solidification/Stabilization of Clay Soil Contaminated by Nickel and Zinc: Sustainable Binder Development and Performance Evaluation. Ph.D. Thesis, Southeast University, Nanjing, China, 2021. (In Chinese).
39. Liu, L. Study on Combined Remediation and Mechanical Properties of Chromium Contaminated Soil. Ph.D. Thesis, Liaoning University of Engineering and Technology, Fuxin, China, 2016. (In Chinese).
40. Cha, F.S.; Liu, J.J.; Xu, L.; Cui, K.R. Cyclic wetting and drying tests on heavy metal contaminated soils solidified/stabilized by cement. *J. Rock Mech. Geotech. Eng.* **2013**, *35*, 1246–1252. (In Chinese)
41. Wan, J.L.; Feng, Y.S.; Li, S.J.; Zhou, S.J.; Wang, S.; Du, Y.J. Leaching, physical and mechanical characteristics of nickel-zinc-contaminated clay solidified/stabilized by a novel steel slag-based binder subjected to wetting-drying cycles. *J. Rock Mech. Geotech. Eng.* **2021**, *43*, 213–216. (In Chinese)

42. Kunther, W.; Ferreiro, S.; Skibsted, J. Influence of the Ca/Si Ratio on the Compressive Strength of Cementitious Calcium–Silicate–Hydrate Binders. *J. Mater. Chem. A* **2017**, *5*, 17401–17412. [CrossRef]
43. Wang, Y.; Liu, X.; Zhang, W.; Li, Z.; Zhang, Y.; Li, Y.; Ren, Y. Effects of Si/Al Ratio on the Efflorescence and Properties of Fly Ash Based Geopolymer. *J. Clean. Prod.* **2020**, *244*, 118852. [CrossRef]

buildings

MDPI

Article

Study on Mechanical Properties of Road Cement-Stabilized Macadam Base Material Prepared with Construction Waste Recycled Aggregate

Yingjie Yuan [1,2], Xianhu Hu [3], Kai Wang [2], Zhi Liu [3], Mingchen Zhong [1] and Kun Meng [1,*]

[1] College of Transportation, Shandong University of Science and Technology, Qingdao 266590, China; 202001130308@sdust.edu.cn (Y.Y.); 202001130327@sdust.edu.cn (M.Z.)
[2] Qingdao Greensail Recycled Building Materials Co., Ltd., Qingdao 266043, China; qdlvfanwk@163.com
[3] Shandong Academy of Building Sciences Co., Ltd., Qingdao 266000, China; hxh87@163.com (X.H.); 201901130517@sdust.edu.cn (Z.L.)
* Correspondence: skd996565@sdust.edu.cn

Abstract: At present, construction waste recycled aggregates only partially replace natural aggregates to prepare road-based materials. This study addressed this limitation and experimentally investigated the mechanical properties of cement-stabilized macadam base materials utilizing a construction waste recycled aggregate. The feasibility of using these raw materials to prepare cement-stabilized macadam bases was established via experimental validation. Subsequently, compaction tests were conducted to ascertain the maximum dry density and optimum moisture content in the mixture. The mechanical characteristics were further examined using unconfined compressive strength tests, analyzing and discussing the influences of varying cement dosages and curing periods on the material strength. The results indicate that the properties of the recycled aggregates satisfied specification requirements, demonstrating satisfactory mechanical properties. The unconfined compressive strength with a 7-day curing period and a 5% cement content fulfilled the technical standards for expressway-grade heavy and extremely heavy traffic, while that with a 6% cement content (with an added curing agent) met these requirements after just 1 day. Additionally, the curing agent enhanced the early strength of the recycled aggregate base material. This study has broken through the technical bottleneck of low content of recycled aggregate, achieved 100% replacement of natural aggregate, and promoted the sustainable development of the industry.

Keywords: construction waste; road engineering; cement-stabilized macadam base; mechanical properties; comparative analysis

1. Introduction

Urbanization has surged with the rapid expansion of China's economy, leading to the rapid proliferation of demolition and reconstruction projects targeting numerous dilapidated structures, aged buildings, and obsolete infrastructure. This surge has, in turn, substantially increased the generation of construction waste. As of 2020, construction waste emissions have surpassed 5 billion tons, creating significant environmental challenges [1]. Concurrently, there has been exponential growth in the scale of infrastructural development. Gravel and other materials are commonly employed in cement-stabilized macadam bases in highway engineering construction. At the same time, due to the large amount of CO_2 produced in the production of cement, the use of a large amount of cement in the project has also caused irreversible damage to the ecological environment. However, the current scenario is marked by the soaring prices of natural resources, exacerbating the conflict between environmental conservation and the sustainable advancement of engineering projects. Consequently, it is necessary to explore alternatives to mitigate this issue.

Currently, there is an emphasis on utilizing construction waste in the production of cement-stabilized macadam base materials [2,3]. Recycled aggregates are prepared to re-place natural aggregates in the formulation of road cement-stabilized macadam bases via crushing and screening processes [4–7]. This transformation of construction waste into a valuable resource significantly enhances its utilization efficiency, curtails engineering costs, alleviates the strain on natural building material extraction, and mitigates environmental degradation [8].

A large number of scholars have fully studied the composition and physical properties of recycled aggregate from construction waste and reached a series of similar conclusions. Shah et al. [9] investigated the physical properties of recycled aggregates of concrete waste and revealed their excellent particle size distribution, durability, and shape. Kolay and Akentuna [10] and Bestgen et al. [11] conducted comparative analyses on the properties of recycled aggregates versus natural aggregates, establishing the similarity in their physical and mechanical characteristics. Hansen et al. [12] assessed the performance of recycled aggregates using primary concrete crushing tests with varying water–cement ratios. Their findings indicated a relationship between the water absorption rate and particle size, while the water–cement ratio showed no significant correlation with the water absorption rate. Hasaba et al. [13] observed that the water absorption of recycled aggregates of waste concrete ranging from 4.75 mm to 26.5 mm in particle size was approximately 7%. However, for recycled aggregates with particle sizes below 4.75 mm, the water absorption was approximately 11%, consistent with Hansen's results.

On the basis of previous studies, many researchers began to explore the preparation of stabilized materials from recycled aggregate and compared and analyzed some possible influencing factors. Jeselay H. C. Reis et al. [14], Ding et al. [15], and Esfahani [16] examined the influence of the recycled aggregate content on the strength characteristics of road base materials, noting a significant impact on their mechanical properties, but does not give an appropriate particle content index. Luo et al. [17] conducted unconfined compressive strength tests on construction waste materials, analyzing the effects of cement dosage and age on compressive strength. Yang et al. [18] studied the influence of the recycled aggregate content and different admixtures instead of natural aggregates on the crack resistance of recycled aggregate base materials. The researchers paved a test road to verify the material's feasibility. Nhieu et al. [19] studied the influences of the cement content, natural latex content, and other factors on the performance of a cement-stabilized recycled concrete aggregate as a pavement base material, analyzing the material's basic performance from a microscopic perspective. Due to the complex composition and other characteristics of recycled aggregate, the research results have limitations, and the main research object is not 100% content. Increasing the content of recycled aggregate plays a decisive role in the large-scale consumption of construction waste.

There is a lack of research on the 100% replacement of natural aggregates with construction waste recycled aggregates. Meng [20] investigated the feasibility of utilizing a 100% construction waste recycled aggregate in cement-stabilized materials. However, the analysis did not consider the various factors influencing the materials' mechanical properties. Rey et al. [21] used a construction waste recycled fine aggregate (mainly comprising a 0~8 mm recycled concrete aggregate) to prepare cement-stabilized base materials. The mechanical properties and durability were studied, and the effects of the particle size and aggregate source were analyzed. Therefore, focusing on construction waste with multiple components, this study developed a technology for the 100% replacement of natural aggregates with construction waste recycled aggregates to prepare road base materials and analyzed the influences of various factors, such as the curing age and cement content, on the materials' strength via relevant experiments. This study first evaluated the feasibility of preparing a cement-stabilized macadam base with selected raw materials via experiments. Subsequently, compaction tests were conducted to ascertain the maximum dry density and optimum moisture content with different mix proportions. Finally, unconfined compressive strength tests were employed to investigate the mechanical characteristics of recycled aggregate road base materials. This study analyzed and discussed the influences of differing dosages of cementitious materials (5% and 6%) and various ages (1 d, 3 d, 7 d, 14 d, and 28 d) on the strength of the recycled aggregate base materials. Furthermore, the influence of the curing agent content on the early strength of recycled aggregate base materials was analyzed by adding a curing agent to develop early-strength materials. At the same time, 144 standard specimens were prepared for the frost resistance test, and the frost resistance of the material was analyzed. This study provides a new and reliable method for the large-scale consumption of construction waste and has positive significance for promoting the sustainable development of infrastructural construction. After the large-scale promotion and application of this research, it can absorb construction waste on a large scale, protect the ecological environment, and 100% replace natural materials can relieve the mining pressure of natural materials and promote the sustainable development of the industry.

This study solved the problem of a 100% content of construction waste recycled aggregates in municipal road base materials. However, the cement content is high. Despite solving the environmental problem of construction waste, the excessive use of cement also causes environmental problems. The problem of reducing the cement content will be further studied to promote sustainable infrastructural development.

2. Research Scheme and Test Method

2.1. Testing Material

The recycled aggregate from construction waste employed in this study originated from Qingdao construction solid waste. The solid waste mainly comprised waste concrete, bricks, and a small number of ceramic tiles. It was categorized into three specifications based on particle size: a recycled fine aggregate with a particle size range of 0~5 mm, and two specifications of recycled coarse aggregate with particle size ranges of 5~10 mm and 10~25 mm, respectively, as depicted in Figure 1. The recycled aggregate of construction waste was equipped with a perfect aggregate production line, ensuring its storage stability, local availability, and material applicability. Shanshui brand 42.5 R ordinary Portland cement was utilized. A powdered inorganic curing agent produced by a manufacturer in Shandong was used. The initial setting time was 215 min, the final setting time was 295 min, the compressive strength was 43.5 MPa, and the flexural strength was 9.6 MPa. The curing agent was based on Portland cement and other active materials as the main components and mixed with various activators, water retention agents, and polymer materials.

(a) (b) (c)

Figure 1. Construction waste recycled aggregates: (**a**) 0~5 mm recycled fine aggregate; (**b**) 5~10 mm recycled coarse aggregate; and (**c**) 10~25 mm recycled coarse aggregate.

2.2. Research Program

In this study, a recycled road base material was designed utilizing construction waste mixed recycled aggregate with cement as the stabilizing agent. The aggregate consisted entirely of recycled construction waste aggregate. According to the requirements of mechanical properties and durability specifications, the design idea of gradually reducing the number of cementitious materials, and the local construction conditions, 5% and 6% cement contents were selected. Based on the local construction experience in Qingdao and the characteristics of the curing agent, a curing agent with a dosage of 2‰ is selected. This study's experimental group design was based on the characteristics of recycled aggregates and practical engineering experience. The performance indices of the raw materials were analyzed via various tests. Subsequently, compaction tests were conducted to determine the maximum dry density and optimum moisture content across the different mix ratios. Finally, the specimens were prepared according to the compaction test data, and the mechanical properties of the recycled aggregate road base material were studied with the unconfined compressive strength test. The influences of different dosages of cementitious materials (5% and 6%) and various ages (1 d, 3 d, 7 d, 14 d, and 28 d) on the early strength of the recycled aggregate base were analyzed and discussed, alongside the impact of increasing the curing agent dosage on the early strength of the recycled aggregate base material. At the same time, 144 standard specimens were prepared for the frost resistance test, and the frost resistance of the material was analyzed.

Four groups of samples, namely, A1, A2, B1, and B2, were designed. The cement–recycled aggregate mix ratios of the cement-stabilized recycled aggregate were set as 5:100 and 6:100. Additionally, a test group incorporating a curing agent was included, with the dosage set at 2‰ of the total mass of the cementitious material and recycled aggregate. The specific sample mixes are detailed in Table 1. In the compaction test, two groups were set up in the A1 and B1 test groups, five tests were set up for each group of dry density measurements, and ten test boxes were set up for each water content group. In the unconfined compressive strength test, 13 specimens were prepared for each of the 1 d, 3 d, 7 d, 14 d, and 28 d test conditions. The A2 and B2 groups with the curing agent were prepared with the same amount. A total of 260 specimens were set up in the unconfined compressive strength test. The study also conducted frost resistance tests, preparing 18 standard specimens for each group and standard curing for 28 days, of which 9 were freeze-thaw specimens and 9 were non-freeze–thaw comparison specimens. A total of 144 specimens were tested for frost resistance.

Table 1. Sample mix ratios.

Numbering		42.5 R Ordinary Portland Cement (%)	Curing Agent (‰)	Recycled Aggregate (%)		
				0~5 mm	5~10 mm	10~25 mm
A	A1	5	0	24	26	50
	A2	5	2	24	26	50
B	B1	6	0	24	26	50
	B2	6	2	24	26	50

2.3. Test Method

The performance of the raw materials was assessed according to the test methods outlined in the 'Highway Engineering Aggregate Test Procedure' (JTG E42-2005) [22], 'Highway Pavement Base Construction Technical Rules' (JTG/TF20) [23], and 'General Portland Cement' (GB175-2007) [24] to evaluate their suitability for road base applications. The compaction tests for the A1 and B1 test groups were conducted using the heavy compaction method outlined in the test procedure for inorganic binder-stabilized materials for high-way engineering (JTG E51-2009) [25]. The specimens for the unconfined compressive strength test were prepared using a pressure testing machine, as shown in Figure 2a. Building on this, the unconfined compressive strength test method for inorganic binder-stabilized materials outlined in the 'Highway Engineering Inorganic Binder-Stabilized Material Test Procedure'

(JTG_E51-2009) [25] was employed to ascertain the unconfined compressive strength of the inorganic mixtures with varying mix ratios (cement–recycled aggregate = 5:100 and 6:100) and different curing ages (1 d, 3 d, 7 d, 14 d, and 28 d), as displayed in Figure 2b. Based on actual engineering experience and construction methods, compare the construction costs of recycled aggregate and natural aggregate, and analyze the economic benefits. At the same time, the research carried out the frost resistance test according to the test specification to understand the material performance more comprehensively.

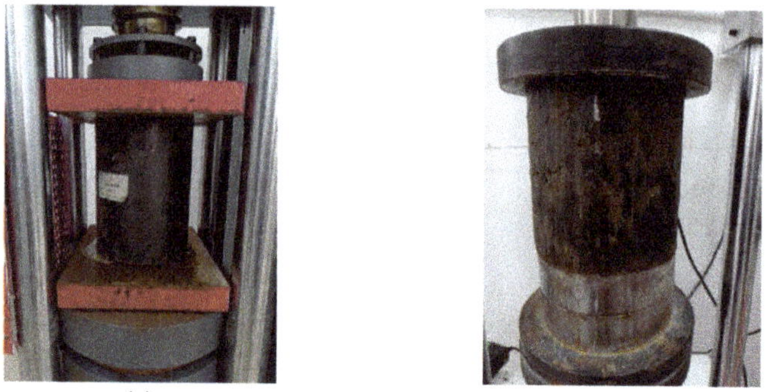

(a) (b)

Figure 2. Representative photos of test: (**a**) specimen compaction; (**b**) unconfined compressive strength test.

3. Raw Material Analysis

3.1. Recycled Aggregate

To detect and evaluate the physical properties of the recycled aggregates from construction waste, samples were extracted from several representative recycled aggregates following construction waste disposal. Upon observation of the aggregates' appearance characteristics, the recycled coarse aggregate, unlike the natural coarse aggregate, exhibited a substantial accumulation of aged cement mortar on its surface. The particles possessed more edges and corners, presenting a rough surface, and were interspersed with a higher proportion of brick particles. This discrepancy is fundamental to distinguishing recycled aggregates from natural aggregates.

As per the testing methodology specified in the 'Highway Engineering Aggregate Test Procedures' (JTG E42-2005), three specifications of recycled aggregate were sieved, and two groups of recycled aggregate for each specification were analyzed, to investigate the gradation of directly crushed recycled coarse aggregates. The average passing percentage across the three specifications is depicted in Figure 3. This figure demonstrates that the gradation composition of the recycled aggregate was suboptimal, with relatively concentrated particle sizes and approximately 90% of the 5~10 mm aggregate concentrated at approximately 5 mm.

There is currently no standardized basis for the grading criteria of recycled aggregate base subbase particles. A comprehensive approach was adopted, drawing on the grading requirements for cement-stabilized graded aggregates outlined in the 'Technical Rules for Construction of Highway Pavement Base' (JTG/TF20) and the principle of multiscale sol-id particle close packing. This approach fully integrated the characteristics of the recycled aggregate, including a complex composition, numerous microcracks, a high crushing index, and elevated water absorption.

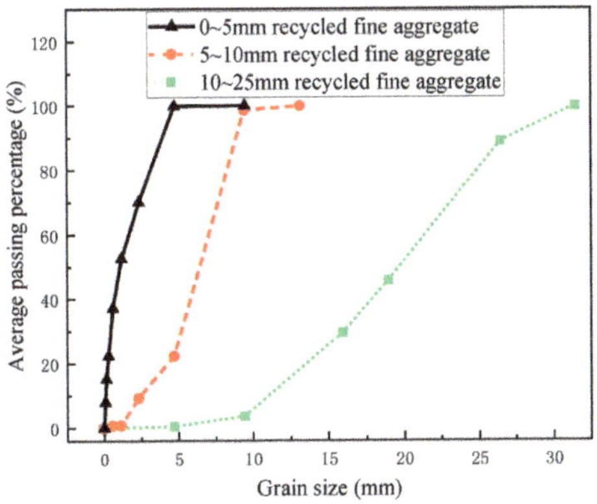

Figure 3. Screening results for three specifications of recycled fine aggregate.

The three specifications of recycled aggregate were intermixed to adjust the gradation, aiming to achieve a synthetic gradation closely aligned with the median gradation stipulated by the specifications. Based on the screening outcomes and specification mandates, the proportion of the synthetic graded crushed stone aggregate (0~5 mm:5~10 mm:10~25 mm = 24:26:50) was determined. The refined distribution of the recycled aggregate percentages is illustrated in Figure 4.

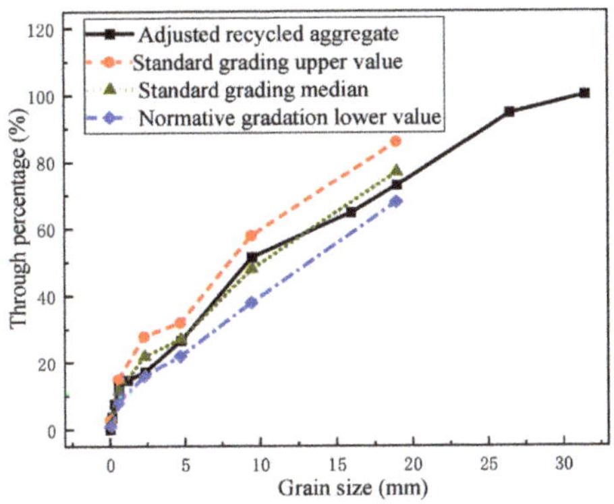

Figure 4. Adjusted recycled aggregate passing percentages.

According to the test method outlined in the 'Highway Engineering Aggregate Test Regulations' (JTG E42-2005), various properties including the crushing index, water absorption, water content, needlelike particle content, apparent density, bulk density, and mud content of the recycled aggregates were examined and compared against the technical criteria specified in the 'Highway Pavement Base Construction Technical Rules' (JTG/TF20-2015). The results are presented in Tables 2 and 3. The tables demonstrate that the physical properties of the recycled aggregate without cementitious binding materials meet the technical requirements of the 'Technical Specification for Construction of High-way Pavement Base' (JTG/TF20-2015) for base aggregates. This indicates its potential suitability as a substitute for natural aggregates in road base applications.

Table 2. Test results for physical properties of recycled coarse aggregates.

Index	Technical Requirement	5~10 mm Recycled Coarse Aggregate	10~25 mm Recycled Coarse Aggregate	Test Method
		Measurement Results		
Apparent density (kg/m^3)	—	2519	2505	T0304
Bulk density (kg/m^3)	—	1230	1220	T0309
Crush value (%)	≤22	—	20.9	T0316
Water absorption (%)	—	8.8	4.0	T0307
Water content (%)	—	1.9	0.79	T0305
Sil content (%)	≤1.2	0.2	0.2	T0310
Fat elongated particle content (%)	≤18	2.6	16	T0312

Table 3. Test results of physical properties of recycled fine aggregate.

Index	Technical Requirement	0~5 mm Recycled Fine Aggregate Measurement Results	Test Method
Apparent density (kg/m^3)	—	2431	T0328
Bulk density (kg/m^3)	—	1424	T0331
Crush value (%)	≤30	23.0	T0350
Water absorption (%)	—	8.9	T0330
Water content (%)	—	4.3	T0332
Sil content (%)	≤3.0	2.8	T0333

3.2. Cement

The cement samples' principal chemical compositions and mechanical and physical properties were evaluated following the procedures outlined in 'General Portland Cement' (GB175-2007). The cement's setting time and the cement mortar's strength were determined per GB/T1346 and GB/T17671, respectively. The outcomes are detailed in Tables 4 and 5. The tables show that the measured parameters align with the specifications outlined in 'General Portland Cement' (GB175-2007).

Table 4. Chemical composition of cement.

SiO_2	Al_2O_3	Fe_2O_3	CaO	MgO	SO_3	R_2O	Ignition Loss
22.13	5.80	3.93	61.22	1.86	2.62	0.67	1.52

Table 5. Physical and mechanical properties of cement.

Specific Surface Area (m^2/kg)	Density (g/cm^3)	Setting Time (min)		Break-Off Strength (MPa)		Compressive Strength (MPa)	
		Initial Set	Final Set	3 d	28 d	3 d	28 d
348	3.11	269	366	6.2	8.1	34.4	53.5

Compared with the technical specifications, the crushing value, water absorption, and other indicators of recycled aggregate meet the specifications. The properties of raw materials meet the specification requirements, and it is feasible to prepare road base stabilization materials.

4. Test Result Analysis

4.1. Moisture–Density Test

Two tests are set for each mix proportion, and the results are taken as the average of the two tests.

(1) Cement–synthetic recycled graded aggregate = 5:100

Based on the proportion of aggregate synthesis, the aggregate mass of a single specimen was determined to be 5500 g (0~5 mm: 1320 g; 5~10 mm: 1430 g; and 10~25 mm: 2750 g), while the cement mass was 275.0 g. Compaction tests were conducted with water contents ranging from 8% to 12%. The average values of the optimum water content and maximum dry density from the two tests were computed. The optimum water content was 10.55%, with a corresponding maximum dry density of 1.981 g/cm^3. The compaction curve depicted in Figure 5 exhibits an approximately parabolic trend.

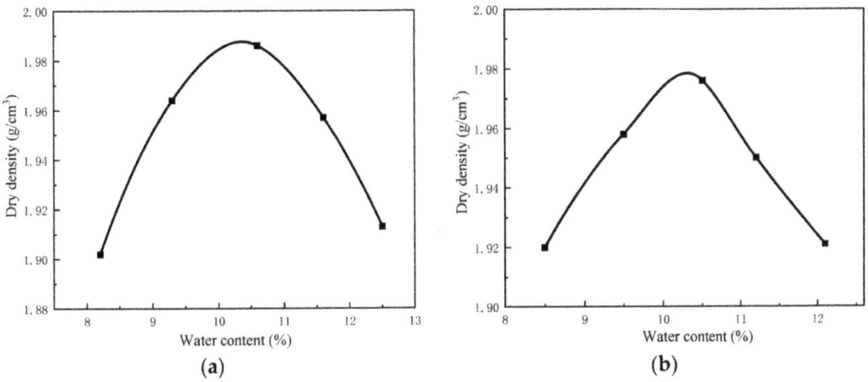

Figure 5. Compaction curve (cement content 5%): (**a**) the first group; (**b**) the second group.

(2) Cement–synthetic recycled graded aggregate = 6:100

According to the aggregate synthesis ratio, the aggregate mass of a single specimen was 5500 g (0~5 mm: 1320 g; 5~10 mm: 1430 g; and 10~25 mm: 2750 g), while the cement mass was 330.0 g. Compaction tests were conducted with water content ranging from 9.5% to 12.5%. The average values of the optimum water content and maximum dry density from the two tests were computed, resulting in an optimum water content of 10.82%. The maximum dry density was 2.006 g/cm^3. The compaction curve depicted in Figure 6 exhibits an approximately parabolic trend.

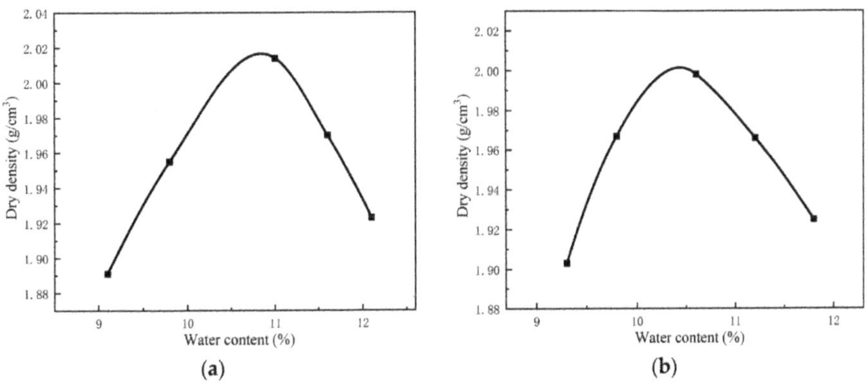

Figure 6. Compaction curve (cement content 6%): (**a**) the first group; (**b**) the second group.

4.2. Unconfined Compressive Strength Test

The test groups A1, A2, B1, and B2 were evaluated following the test method for the unconfined compressive strength of inorganic binder-stabilized materials outlined in the 'Test Specification for Inorganic Binder Stabilized Materials for Highway Engineering' (JTG_E51-2009 T0805). The failure modes of the 260 test specimens were approximately the same. Figure 7 depicts the two specimens with obvious failure modes. The surfaces of the specimens significantly fell off, after which the specimens had an 'upper and lower symmetrical triangle' shape. The middle parts also significantly fell off. The specimens contained oblique longitudinal cracks. A main crack ran throughout the whole of each specimen at approximately 70–80 degrees, which was key to the specimens' failure. The knocked-out specimens contained many microcracks, which caused the specimens to fall off again, and they could not bear the due pressure. The average values of all the results measured in each test condition of each group are shown in Table 6.

Figure 7. Crushing of representative test pieces (on the left is the A1 group test piece, and on the right is the B1 group test piece).

Table 6. Unconfined compressive strengths of cement-stabilized recycled graded aggregates (MPa).

Test Group Number / Instar	A1	A2	B1	B2
1 d	2.69	3.07	4.60	6.55
3 d	3.55	3.86	6.84	7.67
7 d	5.13	5.22	8.70	8.76
14 d	6.03	6.05	9.22	9.35
28 d	6.31	6.44	9.87	10.02

Figure 8 illustrates the strength comparison between the recycled aggregate bases at various ages with consistent curing agent quantities and varying cement amounts. The cement-stabilized recycled aggregate base demonstrated commendable mechanical properties. With a cement content of 5%, the 7-day unconfined compressive strength met the technical specifications for heavy and extremely heavy traffic on highways, as outlined in the 'Highway Pavement Base Construction Technical Rules' (JTG/TF20-2015). Upon increasing the cement content to 6%, the 1-day strength of group B1 without a curing agent approached 5 MPa. The 1-day strength of group B2 with a curing agent reached as high as 6.55 MPa. Increasing the cement content from 5% to 6% significantly enhanced the unconfined compressive strength of the cement-stabilized recycled aggregate base.

Using the 7-day age as an example, the strength of Group B1 surpassed that of Group A1 by 69.5%, while the strength of Group B2 exceeded that of Group A2 by 67.8%. Comparing Figures 8a and 8b, the increase in the unconfined compressive strength with age was more pronounced without a curing agent, whereas it remained relatively consistent over time with a curing agent. On the premise of meeting the technical performance and maintaining the ecological environment, 5% cement content is the best choice for the preparation of the material.

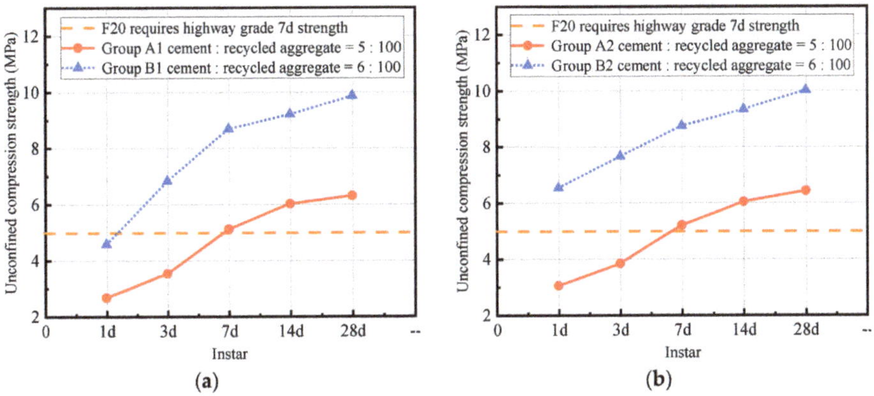

Figure 8. The influence of the cement content on the strength of the recycled aggregate base: (**a**) A1—B1 test group; (**b**) A2—B2 test group.

Figure 9 presents the strength comparison between the recycled aggregate bases at different ages with varying quantities of the curing agent and consistent cement amounts. Adding the curing agent effectively enhanced the early strength of the recycled aggregate base, with significant increases observed at 1 day and 3 days and marginal increases after 7 days. This phenomenon arises from the curing agent's ability to expedite cement hydration and react with cement to generate C-S-H and promote ettringite formation, thereby enhancing the base structure's compactness. Consequently, the later strength primarily relies on the cement, resulting in a minor increase in strength with the same cement quantity.

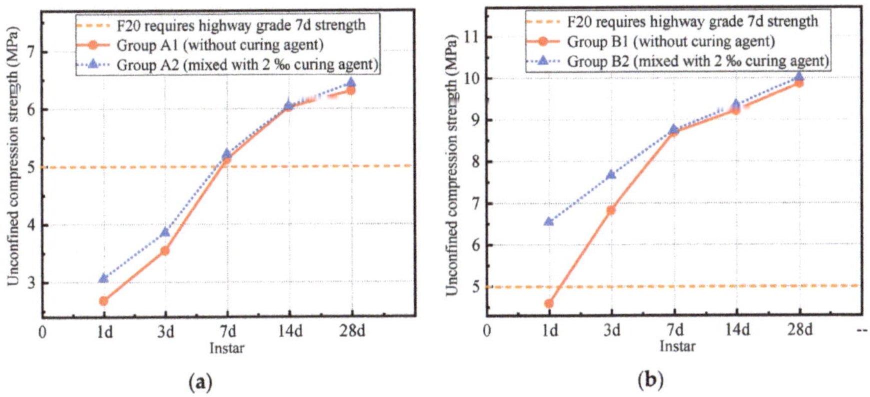

Figure 9. The influence of curing agent on the strength of recycled aggregate base: (**a**) A1–A2 test group; (**b**) B1–B2 test group.

Comparing Figure 9a with Figure 9b, adding the curing agent substantially boosted the early strength of the cement-stabilized recycled aggregate base with a cement content of 6%, with 42.3% and 12.1% strength increases at 1 day and 3 days, respectively. Conversely, the increase was relatively minor with a cement content of 5%. This discrepancy primarily stemmed from the curing agent's ability to promote a more comprehensive reaction of the cementitious system with a sufficient cement quantity. When 5% cement content is selected, the curing agent can not be added because the early strength improvement is relatively small. When 6% cement content is selected, whether to add a curing agent can be considered according to the actual demand. Even if no curing agent is added, the material can also be used as the base stabilization material of municipal roads. When the curing agent is added, the early performance of the material is far from meeting the technical requirements. In municipal road construction, 100% recycled aggregate can alleviate the shortage of natural materials and the ecological deterioration caused by quarrying.

Figure 10 shows the test results of frost resistance. It can be seen from Figure 10a that the mass loss rate of the base course specimens with two kinds of cement content after freezing and thawing is less than 5%. The original strength of the specimen with 6% cement content is higher, so the mass loss rate is lower than 5% cement content. The freeze-thaw mass loss rate of the test group with the curing agent was slightly lower than that of the test group without the curing agent, indicating that the curing agent can improve the compactness and cohesion of the recycled aggregate base. It can be seen from Figure 10b that the residual compressive strength ratio of recycled aggregate base specimens after freeze-thaw is greater than 70%. When the cement content is 6%, the residual compressive strength of the specimen after five freeze-thaw cycles at the age of 28 days is 81%, which is higher than that of the 5% cement content. There is a direct linear relationship between the residual compressive strength and the original strength. The test group with high initial strength has high strength after freeze-thaw. The addition of the curing agent improves the frost resistance of recycled aggregate base. The materials with two kinds of cement content showed good frost resistance.

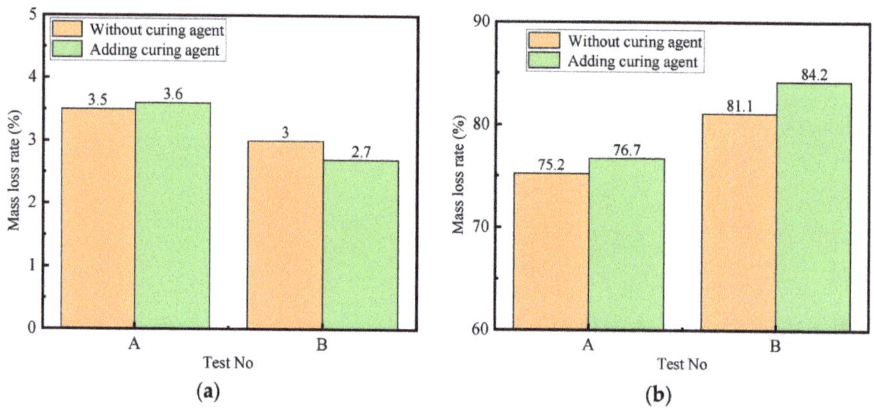

Figure 10. Frost resistance test results: (**a**) mass loss rate of the specimen after freezing and thawing; (**b**) Residual compressive strength ratio of the specimen after freezing and thawing.

Qingdao municipal road engineering mainly uses a semirigid base, and the thickness of the structural layer is large, which has a large demand for aggregate. The use of construction waste recycling materials to prepare various materials for municipal road engineering has the characteristics of a large amount of construction waste, strong operability, and stable product performance. Taking the construction of a two-way 8-lane urban expressway as an example, the road project is estimated based on the design thickness of the Cement-Stabilized Graded Crushed Stone Subbase of 20 cm and the design thickness of the base course of 20 cm. It can absorb about 30,000 tons of recycled aggregate per kilometer, saving 27,000 tons of natural coarse aggregate and 3000 tons of natural fine aggregate of the same quality. According to the price data of natural sand and stone in China in June 2023, the average price of natural coarse aggregate (granite 10–25 mm) is 95 yuan/ton, the average price of natural coarse aggregate (granite 5–10 mm) is 95 yuan/ton, and the average price of natural fine aggregate (0–5 mm) is 115 yuan/ton. The total amount of natural sand and stone used in the project is about 2.91 million yuan. However, at present, the comprehensive estimated unit price of such recycled aggregate is 60 yuan/ton, and the total amount of recycled sand and stone used in the project is about 1.8 million yuan, reducing the cost of raw materials by 38.1% year-on-year, with significant economic benefits.

5. Conclusions

A construction-waste-recycled aggregate can replace only part of a natural aggregate. To address this problem, the mechanical properties of cement-stabilized macadam base materials prepared via the 100% replacement of a natural aggregate with a construction-waste-recycled aggregate were examined in this study. Physical and chemical tests on the raw materials were carried out to evaluate the feasibility of various raw materials for preparing a cement-stabilized macadam base. A compaction test was carried out to determine the maximum dry density and optimum moisture content in different mix ratios. Finally, an unconfined compressive strength test was carried out by preparing a test piece to study the mechanical characteristics of the recycled aggregate road base material. The influences of different dosages of cementitious materials and different ages on the strength of the recycled aggregate base were analyzed and discussed, alongside the improvement in the early strength of the base material with the curing agent. At the same time, the standard specimens were prepared for the frost resistance test. The conclusions are as follows:

1. The physical and chemical properties of the raw materials meet the requirements of relevant specifications and are feasible for preparing road cement-stabilized macadam base materials.
2. The material exhibits good mechanical properties. With a cement content of 5%, the 7-day unconfined compressive strength meets the technical requirements for heavy and extremely heavy traffic on highways. With a cement content of 6%, the 1-day strength approaches the technical requirements. The higher the cement content, the better the mechanical properties.
3. Incorporating a curing agent effectively enhanced the early strength of the recycled aggregate base. The unconfined compressive strength significantly increased with the increase in the cement content. This phenomenon is attributed to the more comprehensive reaction of the cementitious system facilitated by a sufficient cement content.
4. The material has good frost resistance, with a mass loss rate of less than 5% after freeze-thaw, and a residual compressive strength ratio of over 70%.

This study achieved the 100% replacement of a natural aggregate with a recycled aggregate of construction waste to prepare road base materials, meeting the requirements of relevant specifications. Nevertheless, the cement content is high, which increases the environmental problems caused by cement despite solving the problem of environmental pollution from construction waste. Future research should focus on reducing cement content, reducing dependence on cement materials, and reducing the large amount of CO_2 generated from cement production. In this study, it can be seen that curing agents can improve the mechanical properties and frost resistance of materials. Therefore, new types of curing agents can be explored to achieve the preparation of base stabilization materials with low cement content. Under the premise of lower cement content, using a large amount of recycled aggregate to prepare stable materials, large-scale consumption of construction waste, breaking the technological bottleneck of industry development, and promoting green and sustainable development of the industry. While consuming construction waste on a large scale, attention should also be paid to the necessity of low cement content.

Author Contributions: Conceptualization, Y.Y.; methodology, X.H.; software, K.W.; validation, Z.L., M.Z. and Y.Y.; formal analysis, X.H.; investigation, Y.Y.; resources, X.H.; data curation, M.Z.; writing—original draft preparation, K.M.; writing—review and editing, M.Z.; visualization, K.W.; supervision, K.M.; project administration, K.M.; funding acquisition, K.M. All authors have read and agreed to the published version of the manuscript.

Funding: This research was funded by the National Science Foundation for Young Scientists of China (grant number 52108326) and the China Postdoctoral Science Foundation, under grant number 2023M742144.

Data Availability Statement: The original contributions presented in the study are included in the article, further inquiries can be directed to the corresponding author.

Conflicts of Interest: Authors Yingjie Yuan and Kai Wang were employed by the company Qingdao Greensail Recycled Building Materials Co., Ltd. Authors Xianhu Hu and Zhi Liu were employed by the company Shandong Academy of Building Sciences Co., Ltd. The remaining authors declare that the research was conducted in the absence of any commercial or financial relationships that could be construed as a potential conflict of interest.

References

1. Yue, P. Environmental Pollution Hazards of Urban Construction Wastes and Related Resource Recycling Measures: A Case Study in Central China. *Nat. Environ. Polution Technol.* **2018**, *17*, 777–782.
2. Xuan, D.; Molenaar, A.; Houben, L. Evaluation of Cement Treatment of Reclaimed Construction and Demolition Waste as Road Bases. *J. Clean. Prod.* **2015**, *100*, 10077–10083. [CrossRef]
3. Agrela, F.; González-Gallardo, F.; Rosales, J.; Tavira, J.; Ayuso, J.; Cabrera, M. Complete Real-Scale Application of Recycled Aggregates in a Port Loading Platform in Huelva, Spain. *Materials* **2020**, *13*, 2651. [CrossRef] [PubMed]
4. Tavira, J.; Jiménez, J.R.; Ayuso, J.; Sierra, M.J.; Ledesma, E.F. Functional and Structural Parameters of a Paved Road Section Constructed with Mixed Recycled Aggregates from Non-selected Construction and Demolition Waste with Excavation Soil. *Constr. Build. Mater.* **2018**, *164*, 57–69. [CrossRef]
5. Giwangkara, G.G.; Mohamed, A.; Khalid, N.H.A.; Nor, H.M.; Hainin, M.R.; Jaya, R.P.; Sani, W.N.H.M.; Ismail, C.R.; Aziz, M.M.A. Recycled Concrete Aggregate as a Road Base Material. *IOP Conf. Ser. Mater. Sci. Eng.* **2019**, *527*, 012061. [CrossRef]
6. Lopez-Uceda, A.; Ayuso, J.; Jiménez, J.R.; Galvín, A.P.; Del Rey, I. Feasibility Study of Roller Compacted Concrete with Recycled Aggregates as Base Layer for Light-traffic Roads. *Road Mater. Pavement Des.* **2020**, *21*, 276–288. [CrossRef]
7. Javad MH, K.; Zeinab, J.; Rassoul, A.; Gelsefidi, S.A.M. Feasibility of Using Recycled Combined Construction and Demolition Waste for Road Base and Subbase in Fooladshahr, Isfahan Province, Iran: A Case Study. *J. Transp. Eng. Part B Pavements* **2021**, *147*. [CrossRef]
8. Zhao, Y.; Goulias, D.; Tefa, L.; Bassani, M. Life Cycle Economic and Environmental Impacts of CDW Recycled Aggregates in Roadway Construction and Rehabilitation. *Sustainability* **2021**, *13*, 8611. [CrossRef]
9. Shah, S.S.; Azman, M.; Salama, A.E. Mechanical and Physical Properties of Recycled Concrete Aggregates for Road Base Materials. *J. Phys. Conf. Ser.* **2021**, *1973*, 012236.
10. Kolay, P.K.; Akentuna, M. Characterization and Utilization of Recycled Concrete Aggregate from Illinois as a Construction Material. In *Geo-Congress 2014: Geo-characterization and Modeling for Sustainability*; ASCE: Reston, VA, USA, 2024; pp. 3561–3570. [CrossRef]
11. Bestgen, J.O.; Hatipoglu, M.; Cetin, B.; Aydilek, A.H. Mechanical and Environmental Suitability of Recycled Concrete Aggregate as a Highway Base Material. *J. Mater. Civ. Eng.* **2016**, *28*, 04016067. [CrossRef]
12. Hansen, T.C.; Narud, H. Strength of Recycled Concrete Made from Crushed Concrete Coarse Aggregate. *Concr. Int. Des. Constr.* **1983**, *5*, 79–83.
13. Hasaba, S.; Kawanmura, M.; Toriik, K.; Takemoto, K. Drying Shrinkage and Durability of The Concrete Made of Recycled Concrete Aggregate, Trans of The Janpan Concrete Institute. *Trans. Jpn. Concr. Inst.* **1981**, *3*, 55–60.
14. Reis, J.H.; Soares Silva, S.; Ildefonso, J.S.; Yshiba, J.K. Evaluation of Soil, Cement and Construction and Demolition Waste (CDW) Mixtures for Use in Road Pavement Base and Sub-Base Applications. *Key Eng. Mater.* **2014**, *634*, 247–255. [CrossRef]
15. Ding, L.; Zhang, J.; Feng, B.; Li, C. Performance Evaluation of Recycled Asphalt Mixtures Containing Construction and Demolition Waste Applicated as Pavement Base. *Adv. Civ. Eng.* **2020**, *11*, 8875402. [CrossRef]
16. Esfahani, A.M. Evaluating the Feasibility, Usability, and Strength of Recycled Construction and Demolition Waste in Base and Subbase Courses. *Road Mater. Pavement Des.* **2020**, *21*, 156–178. [CrossRef]
17. Luo, X.; Liu, G.; Zhang, Y.; Meng, T.; Zhan, L. Estimation of Resilient Modulus of Cement-treated Construction and Demolition Waste with Performance-related Properties. *Constr. Build. Mater.* **2021**, *283*, 122107. [CrossRef]
18. Yang, S.; Yang, Z.; Chen, K.; Zhang, L.; Chu, J. Using case study on the performance of rural cement stabilizing construction waste recycling road base. *Constr. Build. Mater.* **2023**, *405*, 133351. [CrossRef]
19. Nhieu, D.V.; Hoy, M.; Horpibulsuk, S.; Karntatam, K.; Arulrajah, A.; Horpibulsuk, J. Cement—Natural rubber latex stabilised recycled concrete aggregate as a pavement base material. *Road Mater. Pavement Des.* **2023**, *24*, 1636–1650. [CrossRef]
20. Meng, T.; Lian, S.; Ying, K.; Yu, H. Feasibility Study of Cement-stabilized Materials Using 100% Mixed Recycled Aggregates from Perspectives of Mechanical Properties and Microstructure. *Rev. Adv. Mater. Sci.* **2021**, *60*, 490–502. [CrossRef]
21. Del Rey, I.; Ayuso, J.; Barbudo, A.; Galvín, A.P.; Agrela, F.; De Brito, J. Feasibility Study of Cement-treated 0–8 mm Recycled Aggregates from Construction and Demolition Waste as Road Base Layer. *Road Mater. Pavement Des.* **2016**, *17*, 678–692. [CrossRef]

22. *JTG E42-2005;* Highway Engineering Aggregate Test Procedure. Ministry of Transport of the People's Republic of China: Beijing, China, 2005.
23. *JTG/TF20-2015;* High-Way Pavement Base Construction Technical Rules. Ministry of Transport of the People's Republic of China: Beijing, China, 2015.
24. *GB 175-2007/XG1-2009;* Common Portland Cement. Ministry of Industry and Information Technology of the People's Republic of China: Beijing, China, 2007.
25. *JTG E51-2009;* Highway Engineering Inorganic Binder-Stabilized Material Test Procedure. Ministry of Transport of the People's Republic of China: Beijing, China, 2009.

buildings

MDPI

Article

Study on the Repair Effect of Self-Healing Cementitious Material with Urea-Formaldehyde Resin/Epoxy Resin Microcapsule

Hanqing Mao [1], Xuemei Cao [2,3], Minru Guo [2,3], Chaozhe Jiang [4] and De Chen [2,3,5,*]

1 Qinghai Provincial Department of Transportation, Xining 810001, China
2 School of Civil Engineering, Southwest Jiaotong University, 1426 Civil Building, West Park of Testing High-Tech Zone, Chengdu 610031, China
3 Key Laboratory of High-Speed Railway Engineering, Ministry of Education, Southwest Jiaotong University, 1426 Civil Building, No. 111, North 1st Section of Second Ring Road, Chengdu 610031, China
4 School of Transportation, Southwest Jiaotong University, No. 111, North 1st Section of Second Ring Road, Chengdu 610031, China; jiangchaozhe@swjtu.cn
5 School of Mechanical Engineering, Institute of Tribology, Southwest Jiaotong University, No. 111, North 1st Section of Second Ring Road, Chengdu 610031, China
* Correspondence: chendeswjtu@outlook.com

Abstract: Recent studies on microencapsulated self-healing cementitious materials have primarily focused on the particle size and preparation methods of the microcapsules. However, there has been limited attention paid to the microscopic aspects, such as the selection of curing agents and the curing duration of these materials. In this study, urea-formaldehyde resin/epoxy resin E-51 microcapsules were synthesized through in situ polymerization. This research investigates the feasibility of self-healing from a molecular mechanism perspective and evaluates the repair performance of microencapsulated self-healing cement mortar with varying microcapsule concentrations, curing agent types, and curing ages. The findings demonstrate that the microcapsule shells bond effectively with the cementitious matrix, with radial distribution function peaks all located within 3.5 Å. The incorporation of microcapsules enhanced the tensile strength of the modified cement mortar by 116.83% and increased the failure strain by 110%, indicating improved adhesion and mechanical properties. The restorative agent released from the microcapsule core provided greater strength after curing compared to the uncured state. Although the overall strength of the microencapsulated self-healing cement mortar decreased with higher microcapsule concentrations, the repair efficiency improved. The strength recovery rate of 28-day aged modified cement mortar had a significant improvement with the addition of X and Y curing agents, respectively.

Keywords: self-repair; microcapsules; molecular dynamics; cement mortar; mechanical properties

1. Introduction

Since its inception in 1824, Portland cement has become the most extensively used construction material globally, favored for its cost-effectiveness and durability [1]. With China's rapid economic and technological advancements, the demand for cement concrete in infrastructure projects such as highways, high-speed railways, and airports has surged. Despite its high compressive strength, cement concrete is susceptible to cracking under external loads and environmental factors. These cracks not only propagate and form significant structural weaknesses but also facilitate the ingress of aggressive ions like Cl^-, SO_4^{2-}, and CO_3^{2-}, which accelerate the degradation of both the cement matrix and the embedded steel reinforcement [2]. The corrosion products can reach 4–6 times the volume of the rebar and the volume increase causes internal tension to the concrete on the surface of the corrosion material; when the tensile stress exceeds the tensile strength of the concrete, the concrete will further produce cracking and spalling damage [3].

It can be seen that the presence of cracks is a key factor affecting the service life of cement concrete [4]. In order to extend the service life of cement concrete and to ensure its safe service during its economic life, the periodic inspection and maintenance of cement concrete is usually required. Figure 1 illustrates the relationship between maintenance and structural performance during the service life of cement concrete. At present, there are more mature repair techniques for the developed macro cracks in cement concrete, such as the grouting method, secondary reinforcement, and the application of surface functional coatings mentioned in EN1504:2004/2013, ACI562-16, and ACI546R-14 [5–7].

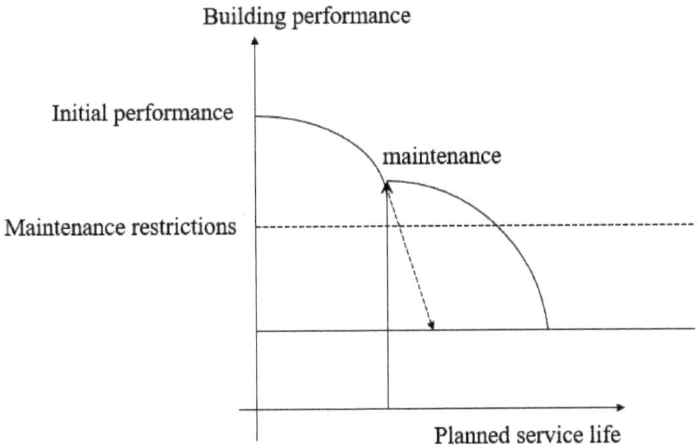

Figure 1. Relationship between building performance and its service life [8].

However, the existing cement concrete crack repair techniques have certain defects. First, the grouting method can only repair larger macro cracks and easy to run slurry problems affecting the structural aesthetics at the same time; therefore, the repair effect is difficult to guarantee [9]. Second, the secondary reinforcement method is only for the repair of larger cracks and the deterioration of the shedding part of the repair process in order to enhance the bonding effect between the old and new concrete. It is used to expand the cracked surface, increase the roughness, aid with reinforcement planting, improve the strength of the concrete and other measures, labor intensity, high construction requirements, pollution and other problems [10]. Coating surface functional coatings can somewhat enhance the performance of cement concrete resistance to infiltration, corrosion, and other properties. High construction requirements, pollution and other problems [10] must also be considered. Coating surface functional coatings can improve the permeability of cement concrete, corrosion resistance, and other properties to a certain extent, but coating surface functional coatings can not achieve the repair of internal cracks in cement concrete and there are problems such as rapid aging and the possibility of this falling off [11]. For example, in the European Union, 20% of cement concrete repairs fail within 5 years, 55% fail within 10 years, and 25% fail within 25 years of service, and the major cause leading to the failure of cement concrete repairs is cracking damage [12]. This, in turn, leads to a continuous cycle of repetitive repairs and life cycle cost accumulation. It is estimated that cement concrete production cost is 65–80 USD/m^3, but its maintenance and repair cost during its service life is about 147 USD/m^3 [13]. In the United States [14], the maintenance of cement concrete structures for transportation infrastructure such as bridges and pavements costs nearly $4 billion per year; in Australia [15], the maintenance and repair cost of cement concrete for transportation infrastructure is close to $550 million per year; and in Europe, nearly 50% of the annual construction budget is spent on the maintenance and repair of existing cement concrete infrastructure.

The traditional crack repair technology of cement concrete is difficult to meet the development needs of low-carbon and long-life cement concrete. Especially for cement concrete in harsh service environments and under the strong action of traffic dynamic load, it is more necessary to control the emergence and expansion of microcracks in cement concrete from the source. Inspired by the self-healing phenomenon of natural biological materials, people began to explore a variety of biomimetic self-healing intelligent materials, using self-healing technology, which can automatically detect and eliminate the hidden danger of micro-cracks in the material in a timely manner to maintain the performance of the material, ensure its safety, and prolong its service life. Since self-healing technology was introduced into cementitious materials, the cementitious self-healing materials have made great developments. The existing bionic intelligent self-repair cementitious materials are mainly divided into hollow fiber self-repair cementitious materials, mineral self-repair cementitious materials, microbial self-repair cementitious materials, shape memory alloy self-repair cementitious materials, electrochemical deposition self-repair cementitious materials, microcapsule self-repair cementitious materials and so on, according to the principle of repairing technology, among which microcapsule self-repair method is the most promising method for realizing engineering large-scale repair in the bionic intelligent self-repair methods of many cementitious materials. Among them, the microcapsule self-repair method is a bionic intelligent self-repair method which is most promising for realizing the large-scale popularization and application of engineering.

Therefore, this paper carries out research on the performance of self-healing cement mortar with microcapsules. Firstly, using the in-situ polymerization method, the microcapsules for self-repairing cement mortar were prepared; then, using molecular dynamics theory, molecular dynamics characterization between the capsule core/capsule wall-cement matrix was carried out; finally, the influence of the design parameters on the repairing effect of microcapsule self-repairing cement mortar was studied. The findings aim to provide a foundation for the optimal design of microcapsule-enhanced self-healing cement concrete and offer insights into the development of low-carbon, long-life concrete solutions.

2. Materials and Methods

2.1. *Raw Material Characteristics*

2.1.1. Microcapsule

The core material of the microcapsule, serving as the primary repair agent, should exhibit excellent stability, fluidity, alkali resistance, and rapid reaction capabilities. The microcapsule's wall not only stores the core material but also needs to bond effectively with the cementitious material, requiring properties such as robust sealing and adhesion. Commonly used in civil engineering for crack repair, Bisphenol A type epoxy resin E-51 offers good adhesion, corrosion resistance, high strength, and minimal shrinkage upon curing. Urea-formaldehyde resin, chosen for the capsule wall, provides superior mechanical properties, heat resistance, and compatibility with cementitious materials, synthesized via solution polymerization in an aqueous dispersion medium. The core material was a diluted mixture of epoxy resin E-51 and n-butyl glycidyl ether in a 100:17.5 ratio. Triethanolamine solution was used as the catalyst, sodium dodecylbenzenesulphonate solution as the active agent (0.5% concentration), and custom-developed X-type and Y-type curing agents were employed to enhance the toughness of the bonding materials.

2.1.2. Cement Mortar

The experiment utilized ordinary silicate cement (P.O42.5 grade) detailed in Tables 1 and 2, which outline its primary components and properties. River sand with a fineness modulus of 2.6 (Zone II) served as the fine aggregate, with key technical specifications in Table 3. The study employed homemade A-type and B-type curing agents, along with urea-formaldehyde resin/epoxy resin E-51 microcapsules.

Table 1. Basic properties of P.O42.5 grade ordinary silicate cement.

Water Consumption for Standard Consistency (%)	Flexural Strength (MPa)		Compressive Strength (MPa)	
	3 d	28 d	3 d	28 d
28	≥3.5	≥6.5	≥17	≥42.5

Table 2. Basic indexes of P.O42.5 grade ordinary silicate cement.

Performance Indicators	Density (g/cm^3)	80 μm Sieve Residue (%)	Content of CaO (%)	Content of SiO_2 (%)	Content of Al_2O_3 (%)	MgO in Clinker (%)	Fineness (%)	Stability	Condensation Time (min)	
									Condensation	Congeal
standardised requirements	-	-	-	-	-	-	≤10.0	eligible (voter etc.)	≥180	≤600
experimental	3.15	0.3	62.4	21.7	6.6	2.24	3.5	eligible (voter etc.)	210	250

Table 3. Technical specifications of fine aggregates.

Characterisation	Apparent Density (kg/m^3)	Bulk Density (kg/m^3)	Voids (%)	Moisture Content (%)	Mud Content (%)
request	>2500	>1350	<47	4%~6%	≤2.0
measured value	2614	1761	35.5	4.35%	0.4

2.2. Preparation Method

2.2.1. Microcapsule Preparation Process

The preparation of urea-formaldehyde resin-epoxy resin E-51 (UF-E) microcapsules is shown in Figure 2. The preparation involves several stages: (1) Prepolymerization: urea and a 37 wt% formaldehyde solution are mixed in a 2:3 molar ratio. After dissolving the urea, triethanolamine is gradually added to adjust the pH to 8–9. The mixture is then stirred at 70 °C for 40 min and cooled to room temperature. (2) Emulsification: a 0.5% sodium dodecylbenzene sulfonate solution is prepared with distilled water and the pH is adjusted to 7. This solution is mixed with the diluted epoxy resin E-51, stirred at 50 °C for 1 h to form a homogeneous emulsion. (3) Acidification: the pH is adjusted to 2–4 using dilute sulfuric acid over 2 h, maintaining a temperature of 50 °C. (4) Vesicle Wall Formation: as copolymerization progresses, the urea-formaldehyde resin precipitates and deposits on the emulsion droplets, forming and thickening the capsule wall. (5) Curing and Precipitation: the mixture is maintained at 50 °C and stirred for 1.5 h. Deionized water is added gradually to manage the viscosity. The resin solidifies into a non-soluble mesh, which is then filtered, dried, and sieved to obtain the microcapsules.

Figure 3 shows the surface morphology of microcapsules at different pH values and core/wall mass ratios. As can be seen from Figure 3, the lower the pH value, the denser and smoother is the capsule wall of the UF-E microcapsules. The particle sense gradually weakened, the three-dimensional network structure formed by the copolymerization reaction was more solid and close, and the capsule wall was more solid. When the mass ratio of the capsule core/capsule wall was 1:1, the polymerization reaction rate of the urea-formaldehyde resin was moderate, the deposition on the surface of the capsule core droplets was more uniform, the surface of the formed capsule wall was dense and smooth, the degree of agglomeration of the urea-formaldehyde resin was low, and the encapsulation effect of the capsule core material was good, with the encapsulation rate reaching 55.59%. There was also less adhesion between the microcapsules. Therefore, for the microcapsules with an average particle size of 495.56 μm prepared when the reaction system was pH = 2 the reaction temperature was kept at 50 °C, and those with a mass ratio of capsule core/capsule wall of 1:1 were more suitable for the self-repair of cement-based materials.

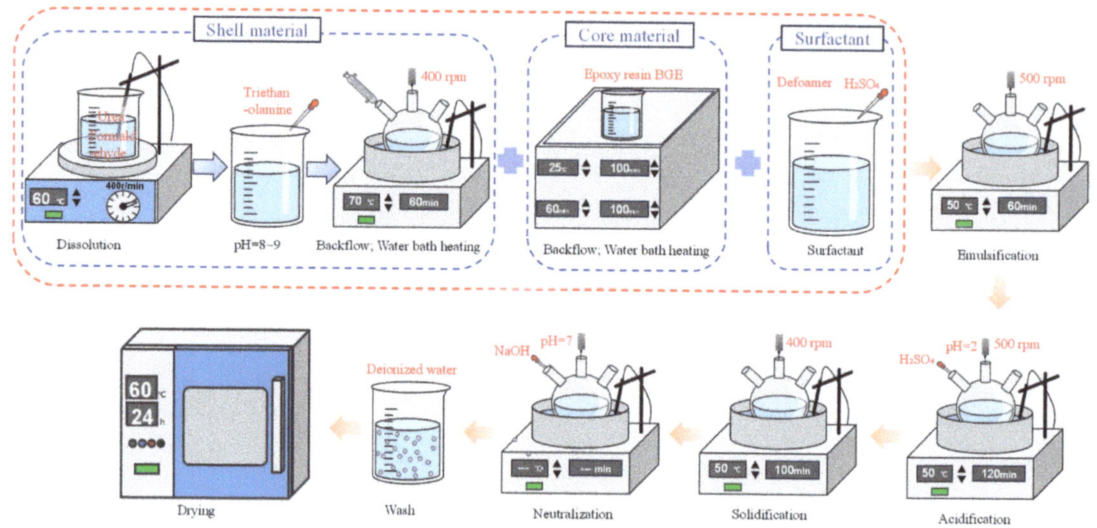

Figure 2. Schematic diagram of the process of preparing UF-E microcapsules.

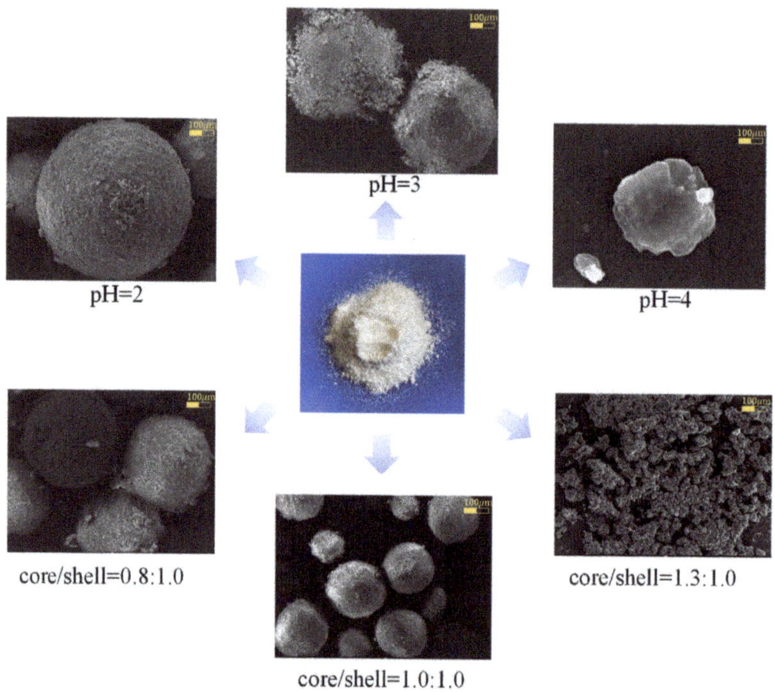

Figure 3. Surface morphology of microcapsules at different pH values and core/wall mass ratios.

2.2.2. Cement Mortar Sample Preparation

(1) Mixing ratios

To isolate the self-healing effects of the microcapsule bonding material, the experiment excluded coarse aggregates and other admixtures. The cement mortar's fixed mixing ratio was water:cement:sand = 1:2:6. Microcapsules were added based on a percentage of the cement's mass, with each curing agent (X and Y) constituting 50% of the microcapsules' mass in the same batch. The mixing ratio of the self-healing cement mortar was as shown in Table 4.

Table 4. Mortar mixing ratios.

Groups	Water/g	Cement/g	Sand/g	Microencapsulation/g	X/g	Y/g
W1	225	450	1350	0	0	0
W2-1	225	450	1350	9	4.5	0
W2-2	225	450	1350	9	0	4.5
W3-1	225	450	1350	18	9	0
W3-2	225	450	1350	18	0	9
W4-1	225	450	1350	27	13.5	0
W4-2	225	450	1350	27	0	13.5

(2) Cement mortar test piece preparation

The mixing process began with blending the cementitious material, curing agents, and water in a mortar mixer. The microcapsules were introduced once the cement slurry was uniform, followed by sand after 2 min, and then mixed for an additional 3 min. The homogeneous mixture was then cast into 40 mm × 40 mm × 160 mm molds in two stages. After 24 ± 2 h, the specimens were demolded and maintained at standard conditions (20 °C ± 2 °C and 90% RH) until the specified test age, followed by natural drying in a ventilated, shade-free environment (as shown in Figure 4).

Figure 4. Self-healing mortar repair performance and effect of specimens during indoor film curing.

2.3. *Test Methods*

2.3.1. Molecular Dynamics Simulation Method

The interfacial properties between the microcapsule shell and the cement matrix, as well as between the encapsulated restorative and the cement matrix, are crucial for the self-healing efficacy of microcapsule-based systems. Gaining insights into these microencapsulated interfaces, particularly at the microscopic level, is challenging through experiments alone. Molecular dynamics (MD) simulations, which employ classical mechanics to model atomic motions and statistical mechanics to assess system properties, are instrumental in this research. This study utilizes MD to analyze the interfacial properties between the capsule components and the cement matrix, focusing on key hydration products like calcium silicate hydrate (C-S-H).

(1) Cyst wall-cement matrix modelling

The modeling process for the vesicle wall-cement matrix is depicted in Figure 5. Tobermorite 14Å, with its interlayer structure resembling C-S-H, is used for modeling. The chemical formula for tobermorite 14Å is Ca_5 Si O_{616} $(OH)_2$-$7H_2$ O. A urea-formaldehyde resin chain with seven repeating units (n = 7) is modeled to represent the organic component. The composite structure of C-S-H and urea-formaldehyde resin is then constructed as a super unit, optimizing the molecular arrangement for stability and interaction analysis.

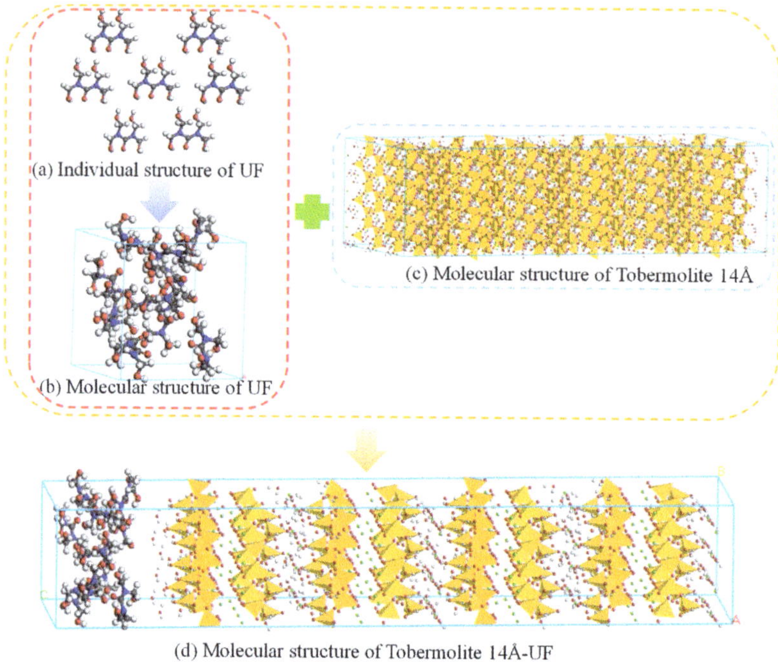

Figure 5. Flow of cyst wall-cement matrix model construction.

(2) Sac-core-cement matrix modelling

The modeling process for the sac-core-cement matrix, shown in Figure 6, involves reducing the viscosity of E-51 epoxy resin using a diluent (BGE) to facilitate its penetration into cracks. The molecular structures of both the uncured and cured states of the epoxy resin are modeled in layers with tobermorite 14Å, reflecting the structural complexity of C-S-H. The entire composite structure is optimized for stability and analyzed under uniaxial tensile conditions to evaluate the interfacial properties and the effects of different curing agents.

(3) Indicators for the evaluation of calculation results.

The evaluation of MD simulation results utilizes several indicators:

- Bulk and Young's Moduli: these measure the material's deformation properties.
- Radial Distribution Function (RDF): this function indicates the probability density of atomic proximity, providing insights into the potential for bonding at the interface.
- Relative Concentration: this metric assesses the concentration of each molecular component along an axis, revealing interactions within the layered structure.
- Stress-Strain Curves: these curves demonstrate how the microcapsule repair materials influence the mechanical strength of the cement matrix.

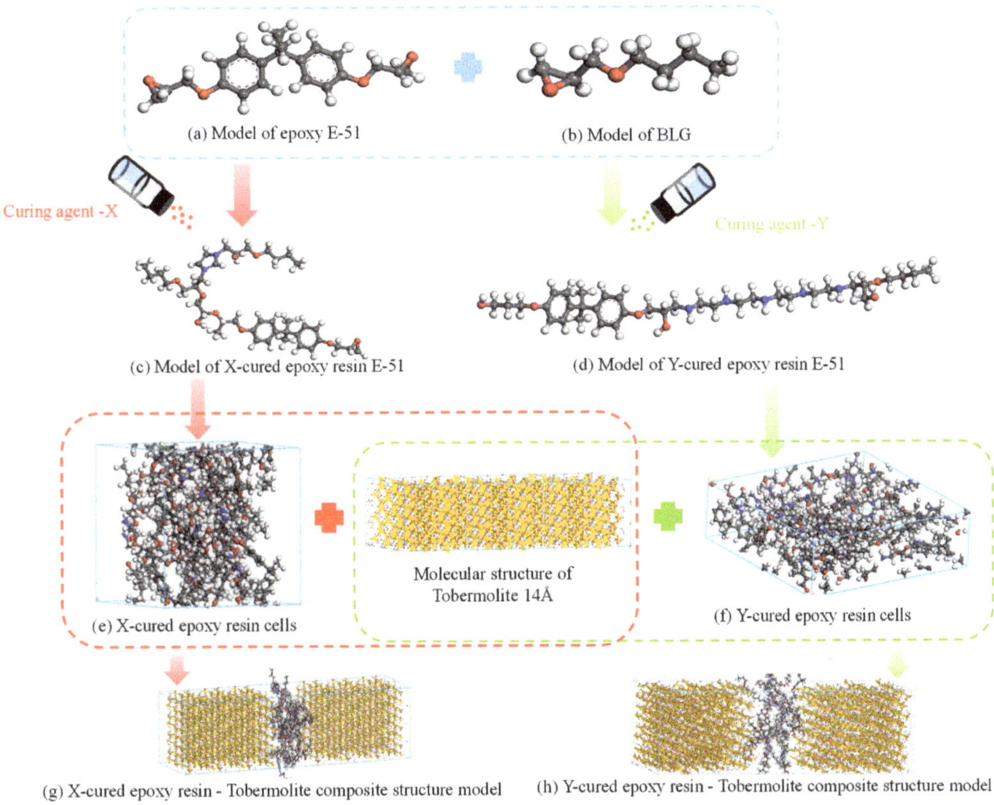

Figure 6. Flow of the sac-core-cement matrix modelling process.

2.3.2. Self-Healing Function Detection

To evaluate the self-repair capabilities of the microcapsules in the cement mortar, it was necessary to pre-damage the specimens by introducing microcracks. This simulates the natural damage that triggers the microcapsules' self-healing response, differentiating it from the natural healing mechanisms of standard cementitious materials.

The three-point bending method is a primary technique for inducing controlled structural cracks into cementitious materials due to its simplicity and the ability to control crack size. In this study, we performed a three-point bending test on prismatic cementitious specimens to introduce pre-damage. The deformation of each specimen was measured using two linear variable differential transformers (LVDTs) positioned at the center of the specimen's bottom on both the front and back sides. The crack size was regulated by monitoring the bending strain at the specimen's base.

Previous research indicates that pre-compression with at least 60% of the maximum compressive stress (σ_{max}) can generate micro-cracks within the structure. Additionally, the stress at the crack tip can rupture the microcapsules, releasing a repair agent to fill these micro-cracks. To ensure adequate tip stress for microcapsule rupture, the specimens were pre-compressed to 70% of their damage load. The pre-damage procedure involved loading specimens, aged to their maintenance period, on an electro-hydraulic servo universal testing machine. The loading rate was maintained at 2400 N/s $\pm$ 200 N/s. After reaching 70% of the original specimen's destructive load, the loading was halted, maintained for 180 s, and then released to complete the pre-damage process. Post-damage, the specimens were stored in a standard curing box for 7 and 28 days.

Ultrasonic testing, which is more effective in hardened concrete (4000~5000 m/s) compared to water (1480 m/s) or air (350 m/s), was employed to assess the repair efficacy of microcapsules (as shown in Figures 7 and 8). The study designed three experimental conditions: original, damaged, and repaired, as depicted in Figure 9. The original condition involved testing specimens immediately after 28 days of standard curing post-molding. The damaged condition involved testing immediately post-pre-damage, and the repaired condition involved testing immediately after 28 days of standard post-repair curing.

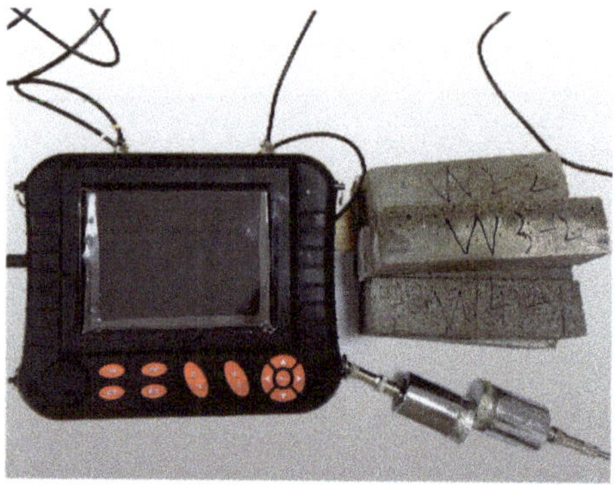

Figure 7. Ultrasonic tester for self-repairing cement mortar specimens.

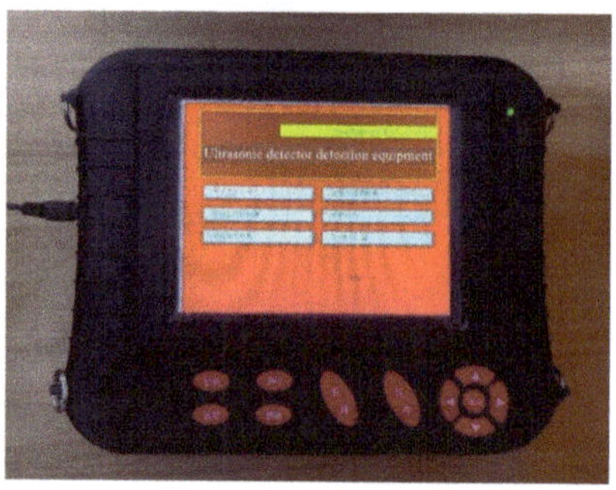

Figure 8. Ultrasonic detector detection equipment.

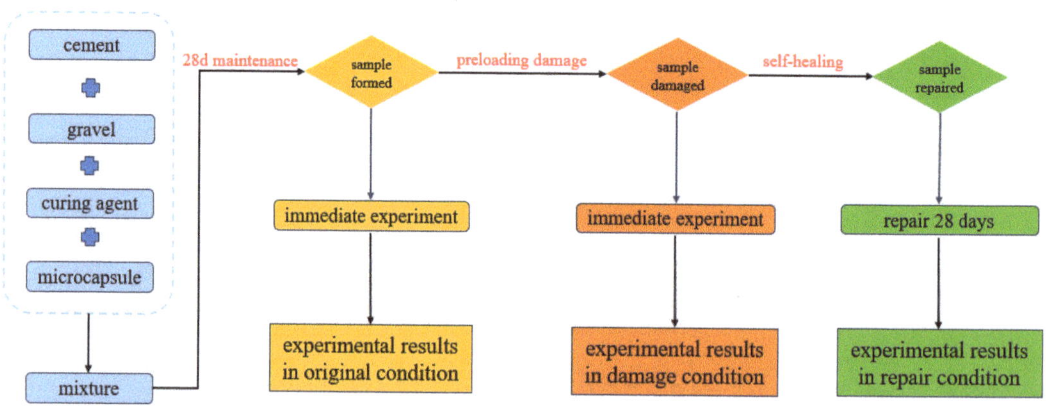

Figure 9. Self-healing mortar specimen ultrasonic testing flowchart.

2.3.3. Test Programme for the Effectiveness of the Repair

To evaluate the impact of microcapsules on the internal damage repair and mechanical property recovery of cement mortar specimens, flexural and compressive tests were conducted. Variables included the age of curing (7, 14, 28 days), specimen conditions (original, damaged, repaired), and repair age (7, 28 days), following the ISO 679:2009 [16]. The original compressive strength (F_0), immediate post-damage strength (F_s), and post-repair strengths at 7 days (F_7) and 28 days (F_{28}) were recorded in MPa. The effectiveness of the repair was quantified using strength recovery and repair rates, calculated as per Equations (1)–(4).

$$\text{7 d strength recovery rate}: \eta_{i7,x} = \frac{F_7}{F_0} \times 100\% \quad (1)$$

$$\text{28 d strength recovery rate}: \eta_{i28,x} = \frac{F_{28}}{F_0} \times 100\% \quad (2)$$

$$\text{7 d strength repair rate}: \eta_{S7,x} = \frac{F_7 - F_S}{F_s} \times 100\% \quad (3)$$

$$\text{28 d strength repair rate}: \eta_{S28,x} = \frac{F_{28} - F_S}{F_s} \times 100\% \quad (4)$$

$\eta_{i7,x}$ and $\eta_{i28,x}$ are the strength recovery rates in % of after 7 d or 28 d of repair of self repairing mortar specimens after pre-damage at x% microencapsulation dosage respectively. $\eta_{S7,x}$ and $\eta_{S28,x}$ are the strength recovery rates in % after 7 d or 28 d repair at of pre-damaged self-healing mortar with x% of microencapsulation, respectively.

2.3.4. Pore Size Distribution Test Method

A MicroMR23-025V NMR tester (Suzhou Newmax Analytical Instruments Co., Ltd., Suzhou, China) was used to determine the pore size distribution of the comparison samples and the self-repairing microencapsulated cement mortar after standard maintenance for 28 d. The size of the mortar specimen was 30 mm × 15 mm × 10 mm. The size of the mortar specimen for NMR test was 30 mm × 15 mm × 10 mm. Before the test, the surface of the mortar specimen was wiped clean and then placed in a vacuum drying oven for 24 h. Then, the sample was taken out and placed in a vacuum pressurized device to be saturated with water. The vacuum time was 60 min, the pressurized strength was 20 MPa, and the saturation time was 24 h. Then, the specimen was taken out and put into the instrument to carry out the test of the pore size distribution, and the resonance frequency was 23.40%. The resonance frequency was 23.40 MHz, the temperature was 32.00 ± 0.02 °C, the diameter of the probe was 25 mm, and the pore size distribution was calculated by applying Equation (5).

$$\frac{1}{T_2} = \rho\left(\frac{S}{V}\right)_{pone} \quad (5)$$

where: T_2—relaxation time;

ρ—surface relaxation rate (70 μm/ms);

s/v—pore surface area volume ratio.

2.3.5. Rapid Chloride Diffusion Method

In accordance with the rapid chloride migration coefficient method (RCM method), the chloride diffusion coefficient of the comparison sample and the self-repairing microcapsule cement mortar after 28 d of standard maintenance were tested by using the RCM-6T chloride diffusion coefficient tester (Beijing Numerical Intelligence Yilong Instrument Co., Ltd., Beijing, China). Before testing, the mortar specimen block was cut into ϕ 100 mm × 50 mm samples. The chloride diffusion coefficient of the mortar was calculated according to Equation (6).

$$D_{RCM} = \frac{0.0239 \times (273+T)L}{(U-2)t}\left(X_d - 0.0238\sqrt{\frac{(273+T)LX_d}{U-2}}\right) \quad (6)$$

where: D_{RCM}—diffusion coefficient of chloride ions in mortar (10^{-12} m^2/s);
U—loading voltage (V);
T—mean value of the initial and ending temperatures of the anode (°C);
L—thickness of the mortar specimen (mm);
X_d—average depth of penetration of chloride ions (mm);
t—duration of the whole experimental test.

3. Results and Discussion

3.1. Interfacial Contact Properties and Microscopic Remediation

3.1.1. Cyst Wall-Cement Matrix

(1) Moduli of the composite structure

The bulk, shear, and elastic moduli of the bladder-cement matrix composite are detailed in Table 5. Notably, the Young's modulus is highest in the y-direction, likely due to the stability of silicate chains and the disruption of the hydrogen bonding network in other directions. The moduli for the T14Å-urea-formaldehyde ester composite are lower than those of pure tobermorite 14Å, indicating that microcapsule addition increases porosity and slightly diminishes the material's mechanical properties. This reduction in elastic modulus at the microscopic level is attributed to increased interatomic distances, leading to decreased binding energy. However, this also suggests enhanced ductility due to greater elastic deformation.

Table 5. Bulk modulus, shear modulus, and elastic modulus of T14Å-urea-formaldehyde ester models.

Modelling	Bulk Modulus (GPa)	Shear Modulus (GPa)	Modulus of Elasticity (GPa)		
			x	y	z
T14Å-urea formaldehyde ester	26.8332	9.0122	24.6834	37.5336	20.878

(2) Radial Distribution Function (RDF)

As shown in Figure 10, the RDF of tobermorite 14Å-urea-formaldehyde resin exhibits sharp peaks below 3.5 Å, indicating strong chemical bonding with the cementitious material, predominantly through electrostatic interactions between Ca^{2+}, O^{2-}, and N^- ions. This bonding is crucial for the material's structural integrity, contrasting with the weaker Coulombic and Van der Waals' forces.

(3) Stress-strain behavior

The stress-strain curve for the z-direction (Figure 11) reveals three distinct phases: elastic, yield, and failure. The tensile strength of the tobermorite 14Å-urea-formaldehyde resin composite reaches 2.19 GPa, surpassing the 1.01 GPa of pure tobermorite 14Å. This enhancement is attributed to the urea-formaldehyde resin's ability to mitigate crack propagation, thereby increasing the material's strength. Additionally, the failure strain of the composite is 0.44, compared to 0.20 for pure tobermorite 14Å, indicating improved ultimate strain due to the inclusion of microcapsules.

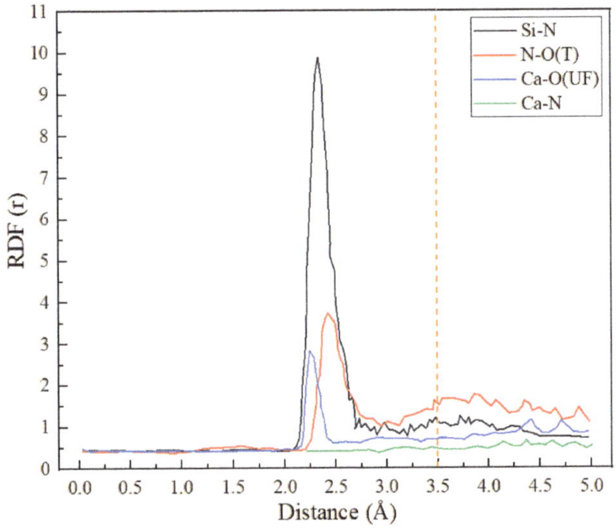

Figure 10. Radial distribution function of 7 Tobey mullite 14Å-urea-formaldehyde resin.

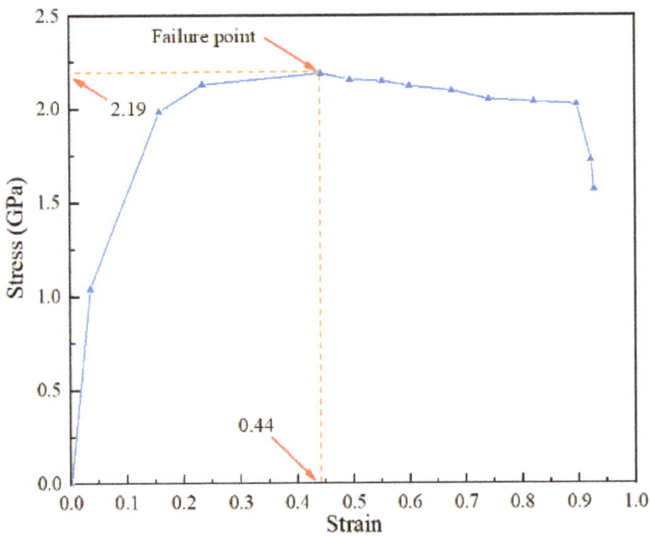

Figure 11. Stress-strain behavior for 8 Tobey mullite 14Å-urea-formaldehyde resin model.

3.1.2. Core-Cement Matrix

(1) Relative concentration

Figure 12 illustrates the relative concentrations in the composite model, highlighting a reduction in the model's length along the z-direction. Specifically, Figure 12a shows atom distribution in the second layer ranging from 50.74 to 61.92 Å, with no distribution in the first and third layers. Conversely, Figure 12b reveals atom distribution in the second layer from 26.81 to 34.33 Å, and in the first and third layers from 0 to 34.98 Å and from 25.35 to 59.23 Å, respectively. These distributions indicate that the atoms in the three-layer structure have interpenetrated post-simulation, suggesting enhanced interactions between the core material and the cement matrix.

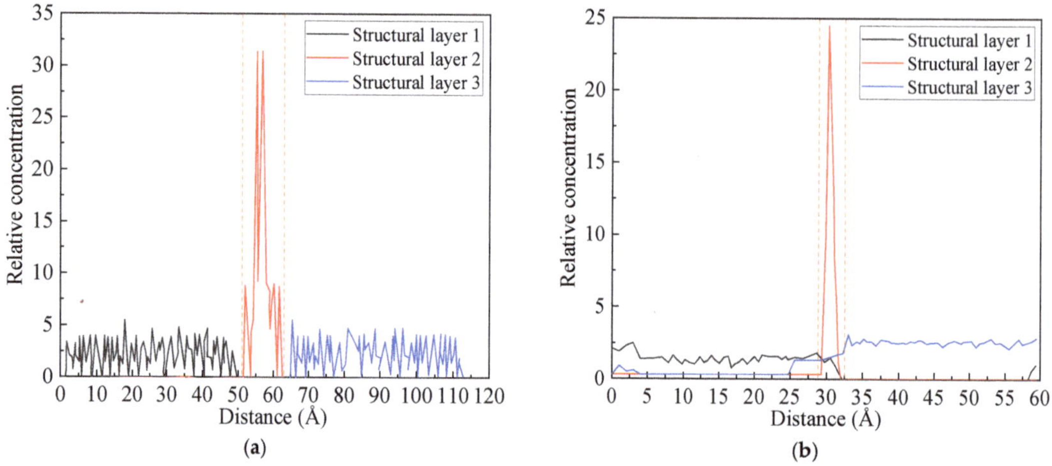

Figure 12. Relative concentration distributions before and after molecular dynamics calculations (structural layers 1 and 3 are C-S-H layers and structural layer 2 is an epoxy resin layer): (**a**) Relative concentration distribution of the model before simulation, (**b**) Relative concentration distribution of the model after simulation.

(2) Stress-strain curve

Figure 13 presents the stress-strain curves for three composite models derived from simulations. The maximum strains observed are 0.19, 0.34, and 0.27 for models with n = 0, X, and Y, respectively. Notably, the strain in the cured epoxy-cementitious material (with X or Y) is higher compared to the pure Tobermorillonite 14Å model (0.20), while the uncured model shows a slight reduction in strain. This underscores the significant impact of epoxy resin curing on the mechanical properties of microencapsulated self-healing cementitious materials.

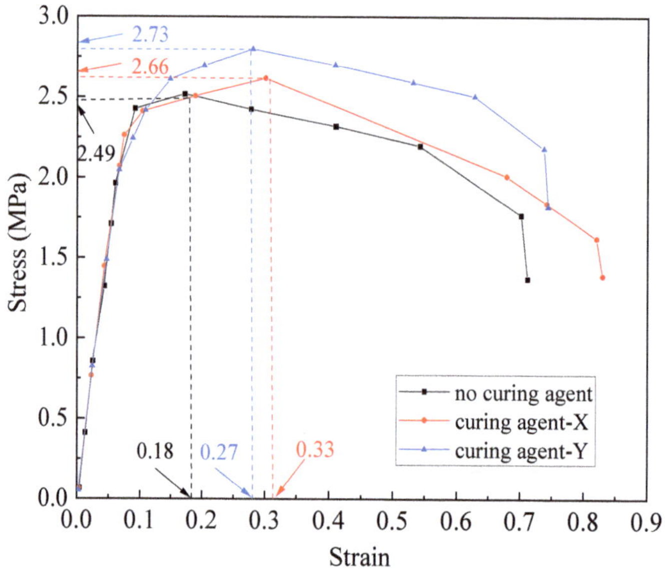

Figure 13. Stress-strain curves for three composite models.

(3) Radial Distribution Function

Figure 14 displays the radial distribution functions (RDF) for the three models. The interaction between Tobermorite 14Å and epoxy resin primarily occurs through Ca ions in Tobermorite 14Å and N and O atoms in the epoxy resin. In the uncured n = 0 model, where N atoms are absent, the interaction is limited to Ca-O, resulting in weaker interatomic forces compared to the other models. The first RDF peak positions for Ca(T)-O(ex) atoms in the n = 0, X, and Y models are at 2.33 Å, 2.17 Å, and 2.16 Å, respectively, indicating stronger interactions in the cured models. Additionally, the interaction between Ca ions and N atoms is enhanced in the cured models, particularly when Y is used as the curing agent, leading to more pronounced directional interactions.

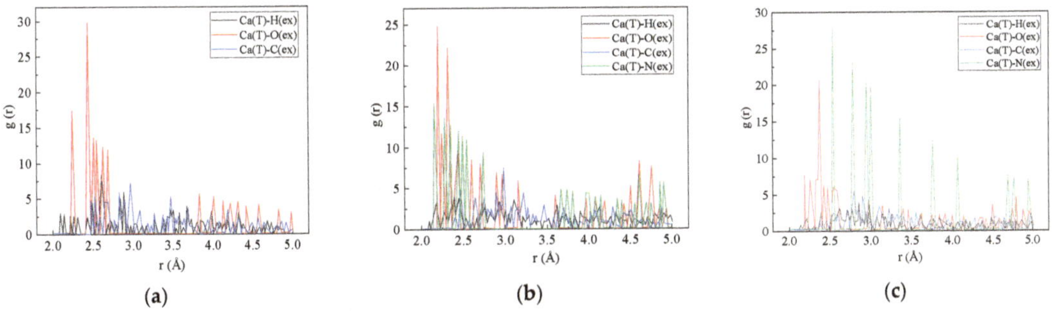

Figure 14. Radial distribution functions for three models: (**a**) n = 0, (**b**) X (**c**) Y.

3.2. Structural Damage Repair Effect

The impact of microcapsule doping on ultrasonic propagation speed is depicted in Figure 15. It is observed that the speed decreases as the microcapsule concentration increases, with the lowest speed at 6% doping. This reduction is attributed to the microcapsules causing phenomena such as the reflection and bypassing of the ultrasonic waves within the cement mortar, which increases the path of propagation and attenuates the energy, thereby reducing the wave speed.

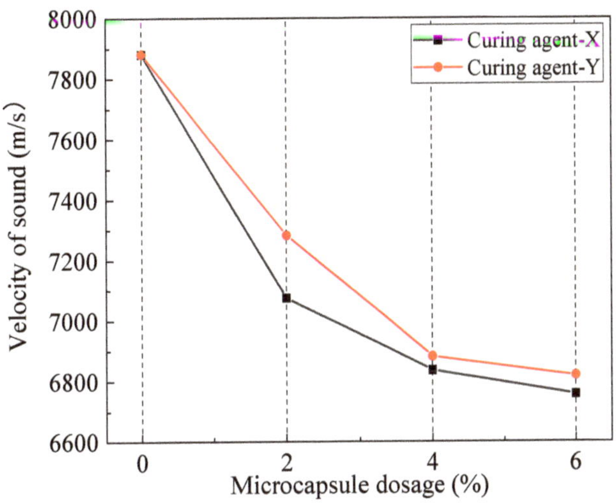

Figure 15. Ultrasonic wave velocity of microencapsulated self-healing cement mortar under original working conditions.

Figures 16 and 17 show the ultrasonic waveforms for undamaged cement mortar specimens under normal conditions, displaying regular and complete sinusoidal patterns, indicating no internal structural damage.

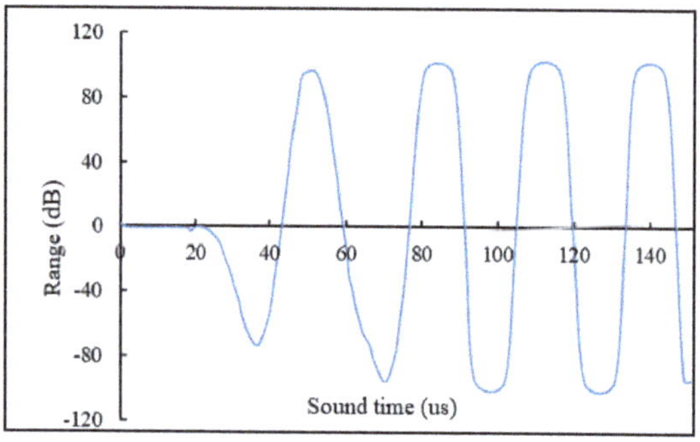

(a) 0%

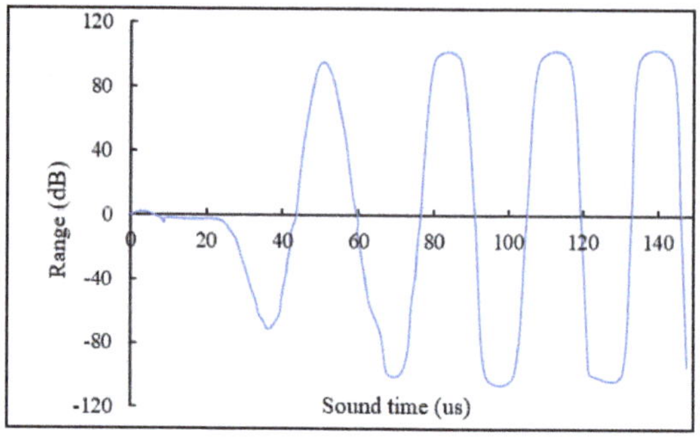

(b) 2%

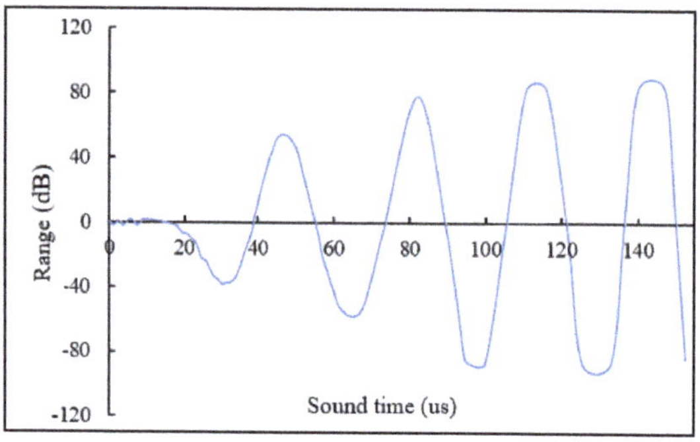

(c) 4%

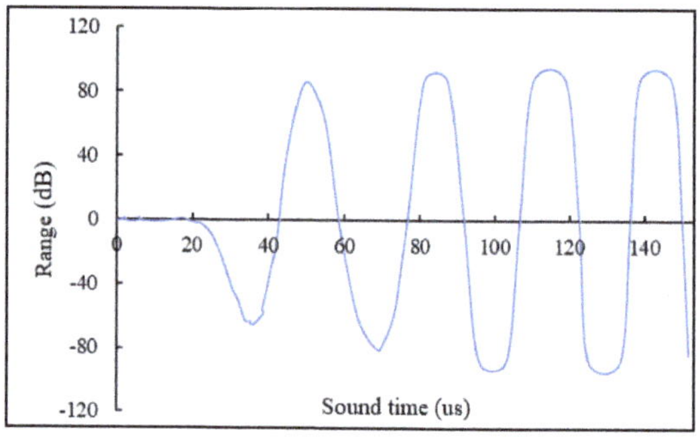

(d) 6%

Figure 16. Waveforms of microencapsulated self-repairing cement mortar under original working conditions (X).

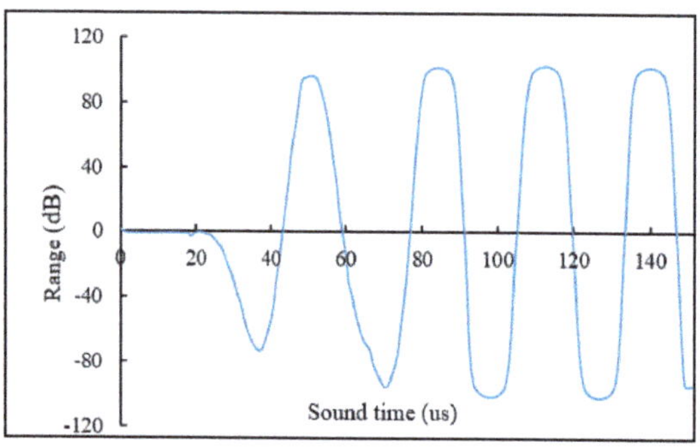

(a) 0%

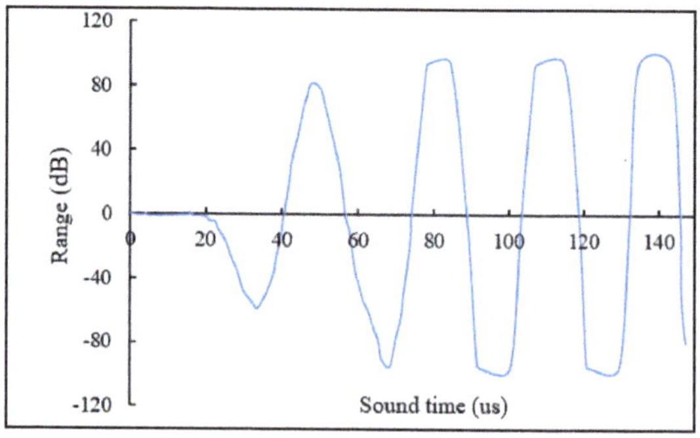

(b) 2%

Figure 17. *Cont.*

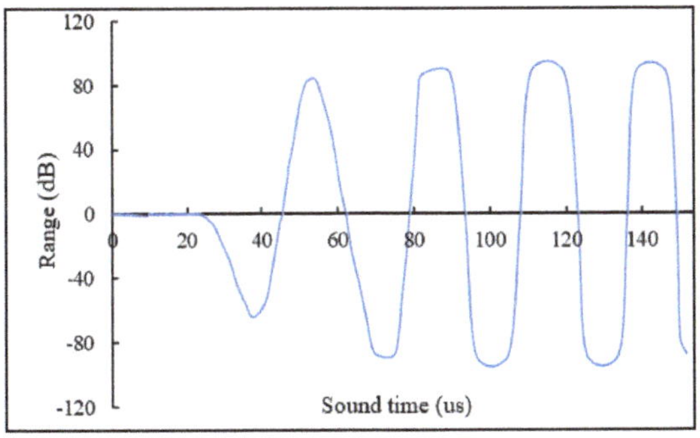

(c) 4%

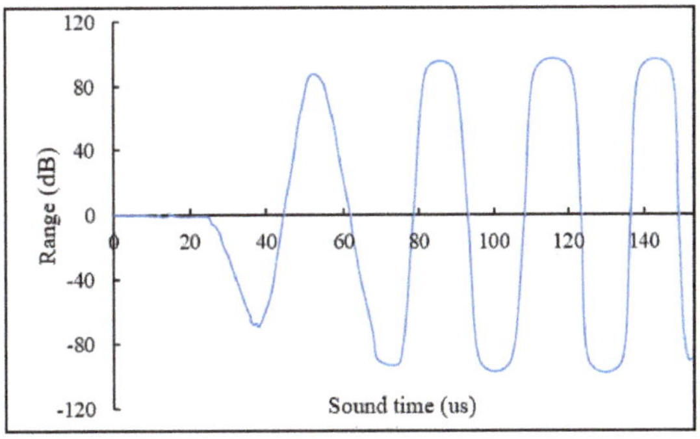

(d) 6%

Figure 17. Waveforms of microencapsulated self-repairing cement mortar under original working conditions (Y).

From Figures 18–20, it can be seen that the ultrasonic sound velocity of the damaged cement mortar specimen decreased significantly compared with that of the original working condition, that the waveforms were also distorted to different degrees, and that the waveforms were incomplete. As shown in Figures 18 and 19, the most significant difference is the change of the direction of the first wave of the waveform. The first wave is changed from downward to upward, which fully proves that defective damage has been produced inside the specimen structure.

Under 28 days of repair conditions, the acoustic properties of microencapsulated self-healing cement mortar specimens are shown in Figures 21–23. Although the ultrasonic speed has not fully returned to the original condition, the waveforms are more uniform and complete, and the speed is improved. This suggests that the microcapsules continue to function under varying temperatures, filling microcracks and maintaining the integrity of the repair material, which exhibits good low-temperature toughness and resistance to temperature-induced stress.

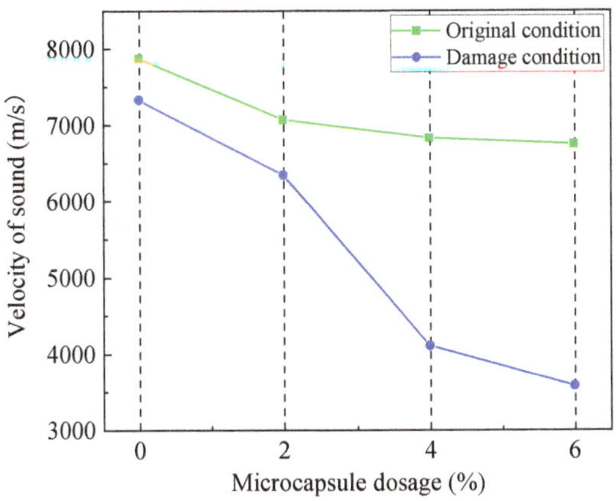

(a) Curing Agent-X

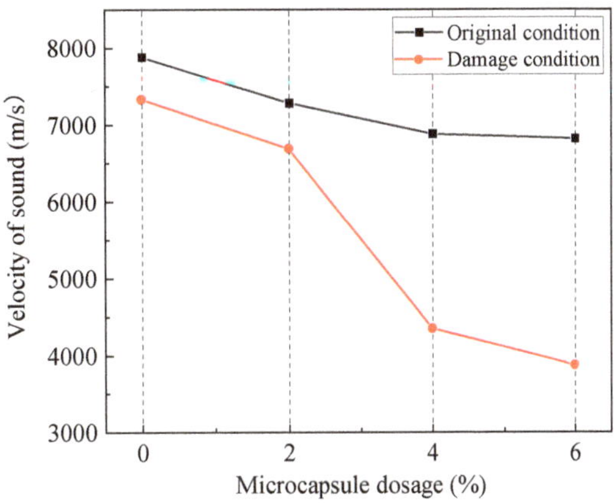

(b) Curing Agent-Y

Figure 18. Comparison of ultrasonic velocities at 5 raw/damaged working conditions.

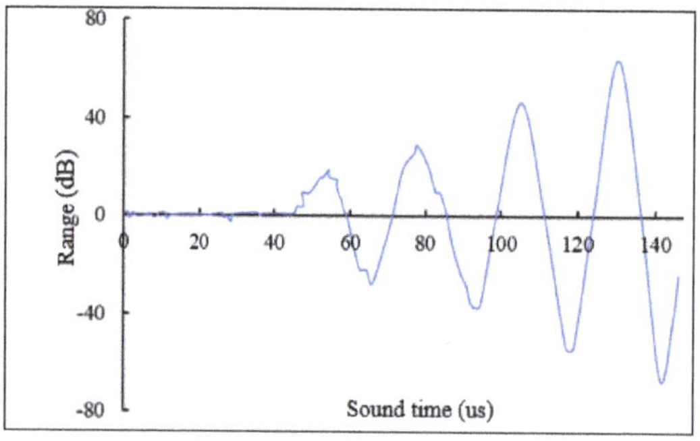

(a) 0%

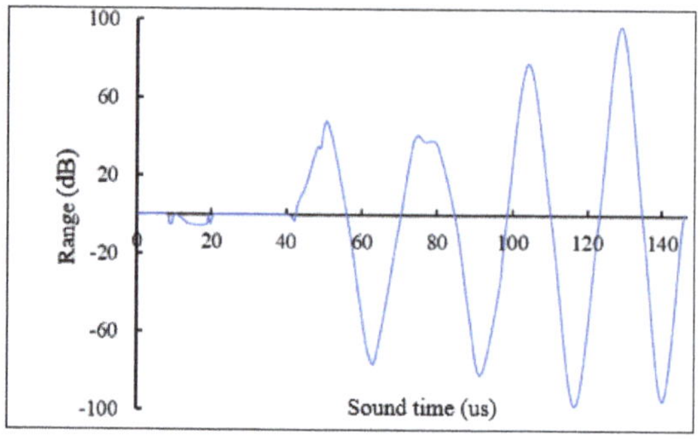

(b) 2%

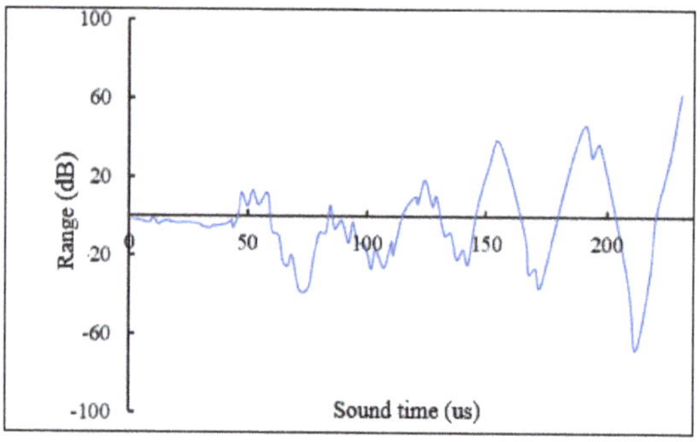

(c) 4%

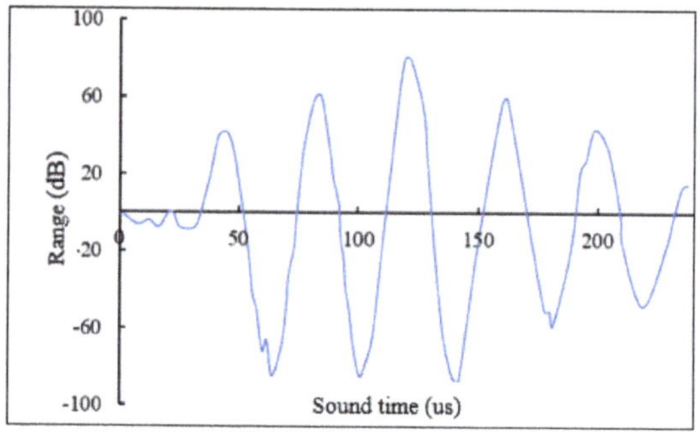

(d) 6%

Figure 19. Waveforms of microencapsulated self-healing cement mortar under damage conditions (curing agent-X).

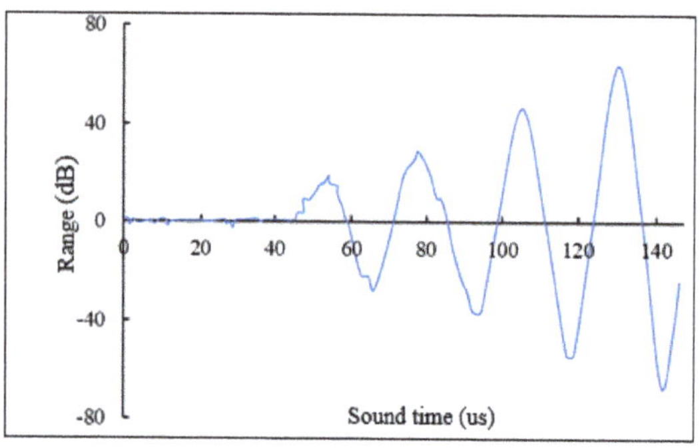

(a) 0%

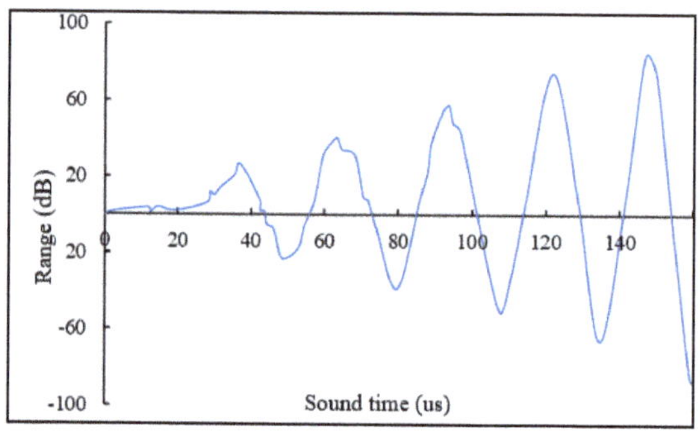

(b) 2%

Figure 20. *Cont.*

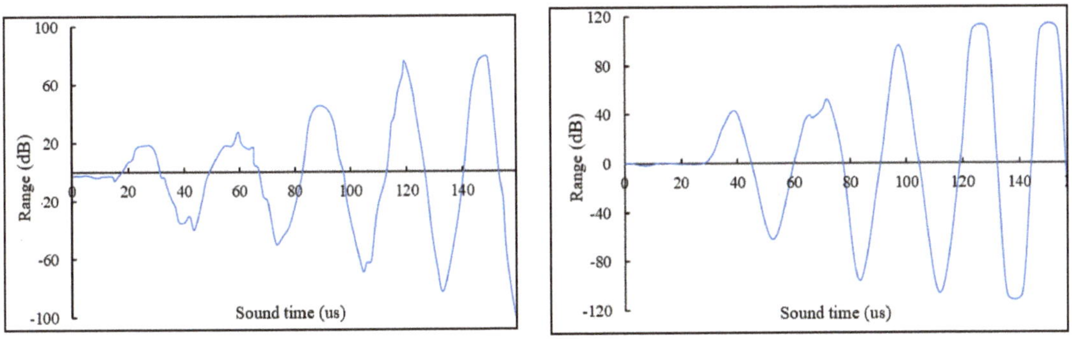

(c) 4% (d) 6%

Figure 20. Waveforms of microencapsulated self-healing cement mortar under damage conditions (curing agent-Y).

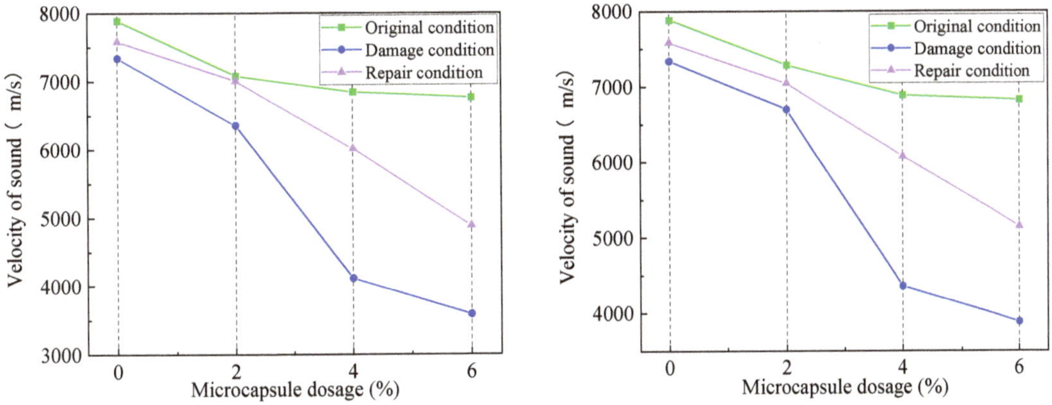

(a) (b)

Figure 21. (**a**,**b**) Comparison of ultrasonic velocities under original/damaged/repaired working conditions.

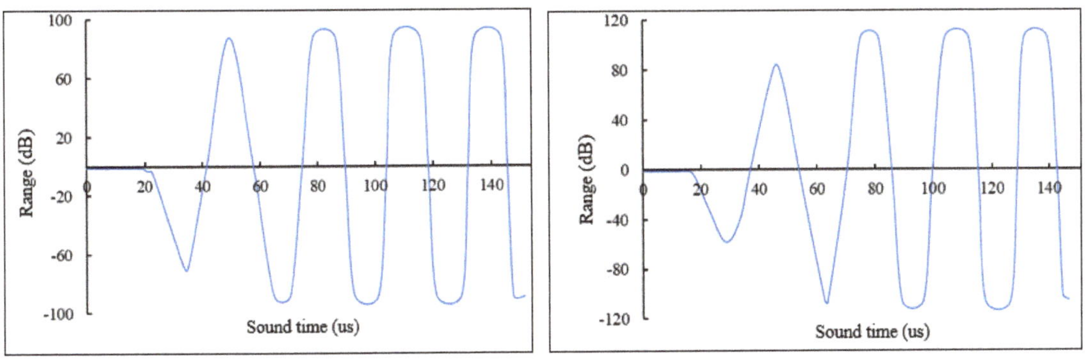

(a) 0% (b) 2%

Figure 22. *Cont.*

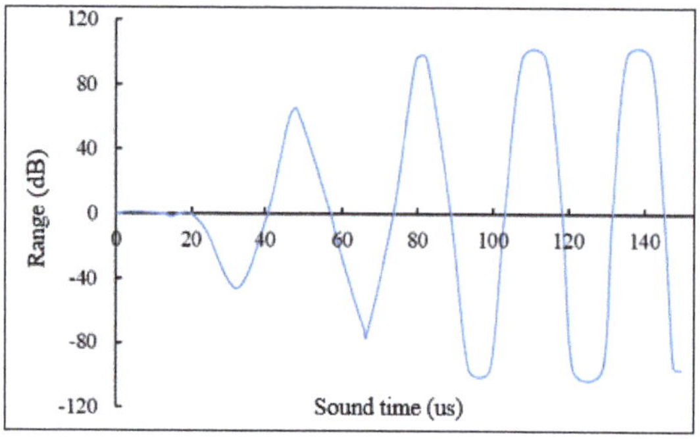

(**c**) 4%

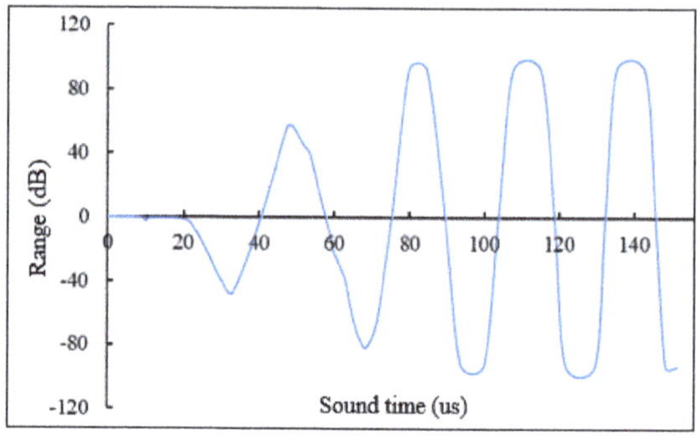

(**d**) 6%

Figure 22. Waveforms of microencapsulated self-healing cement mortar under repair conditions (curing agent-X).

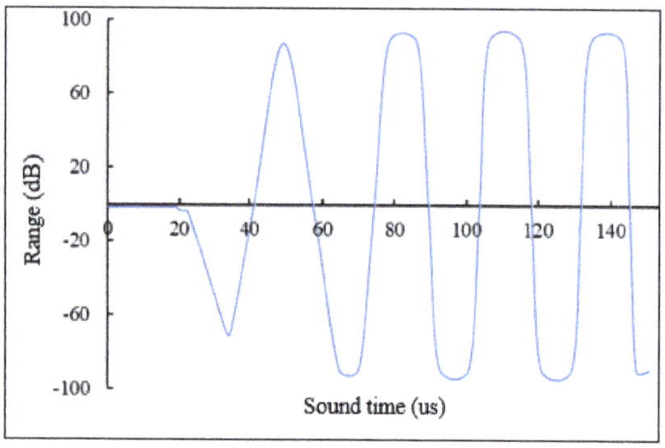

(**a**) 0%

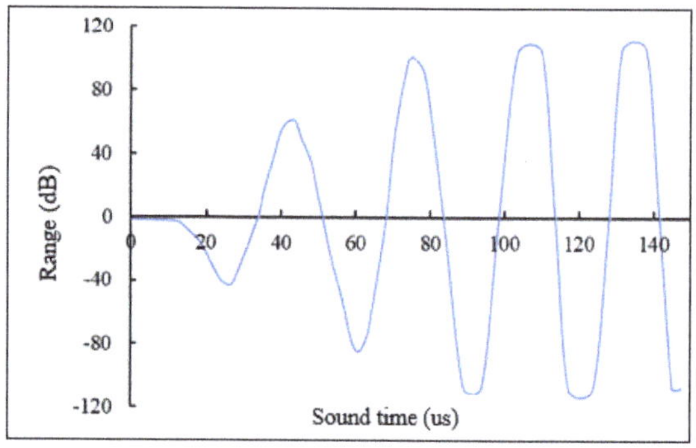

(**b**) 2%

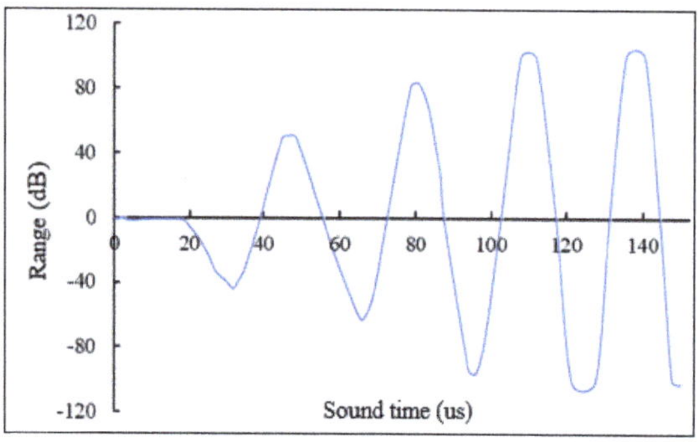

(**c**) 4%

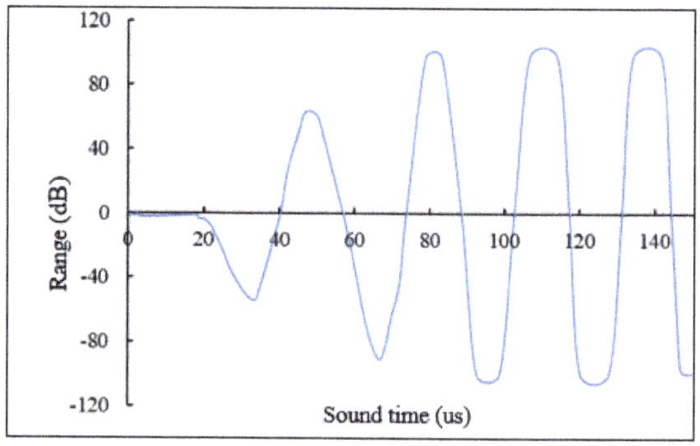

(**d**) 6%

Figure 23. Waveforms of microencapsulated self-healing cement mortar under repair conditions (curing agent-Y).

3.3. *Effect of Initial Strength of Cement Mortar*

The stress-time curves for flexural and compressive tests of specimens at various ages are shown in Figures 24 and 25.

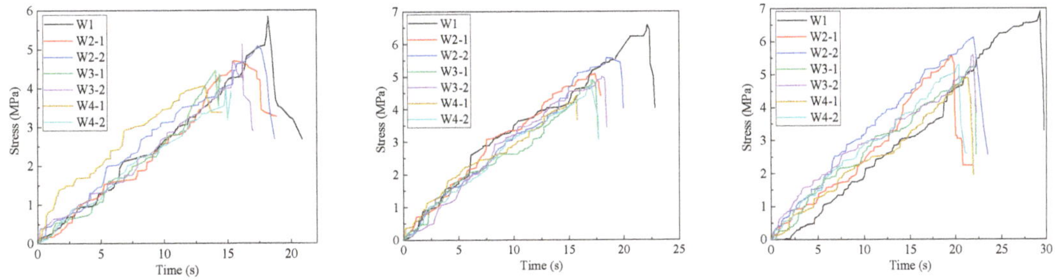

(**a**) 7 d (**b**) 14 d (**c**) 28 d

Figure 24. Initial flexural strength of each test group at different ages.

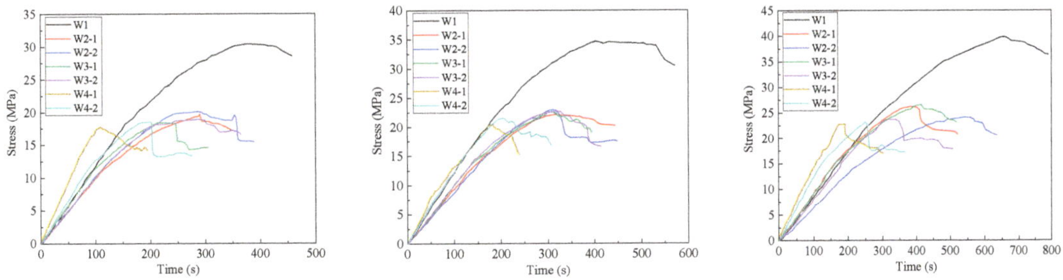

(**a**) 7 d (**b**) 14 d (**c**) 28 d

Figure 25. Initial compressive strength of each test group at different ages.

These curves reveal that the initial strength of both the compressive and flexural strength decreases with the incorporation of microcapsules, and the decrease in the initial strength of the specimen increases gradually with the increase in the amount of microcapsule incorporation, as shown in Figure 23. The effect is more pronounced on compressive strength than on flexural strength.

The decrease in the initial strengths is due to the mechanical properties of the microcapsules being inferior to those of the aggregate, reducing the overall component ratio in the cement matrix. The microcapsules act as fillers, creating numerous pores within the matrix. Microscopic analysis shows that the modulus values of the cementitious-urea-formaldehyde ester composite are lower than those of pure cement, indicating a reduction in the bonding energy and overall mechanical properties due to increased porosity. However, this also results in increased elastic deformation, enhancing the material's ductility.

Figures 26 and 27 also indicate that the initial strength of the cement mortar increases with specimen age, attributed to the enhanced hydration and densification of the cement matrix over time, which is further aided by the filling effect of the microcapsules.

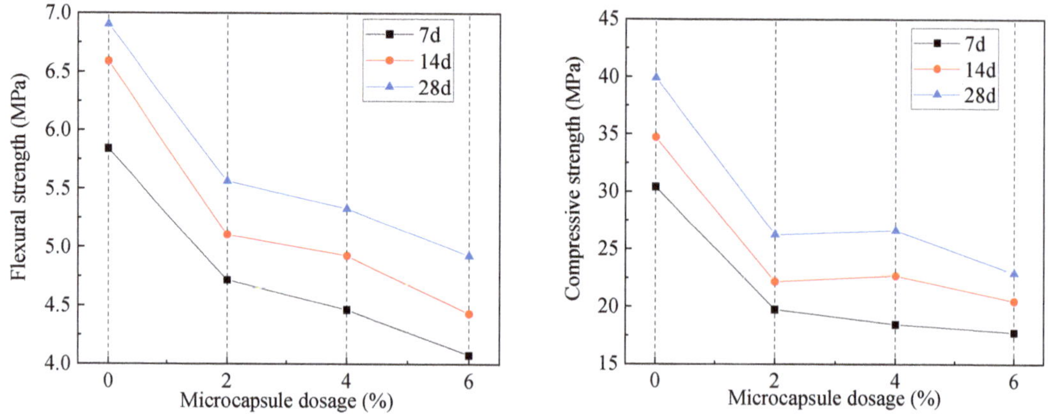

(**a**) Flexural strength (**b**) Compressive strength

Figure 26. X initial strength at various ages with different microcapsule dosage.

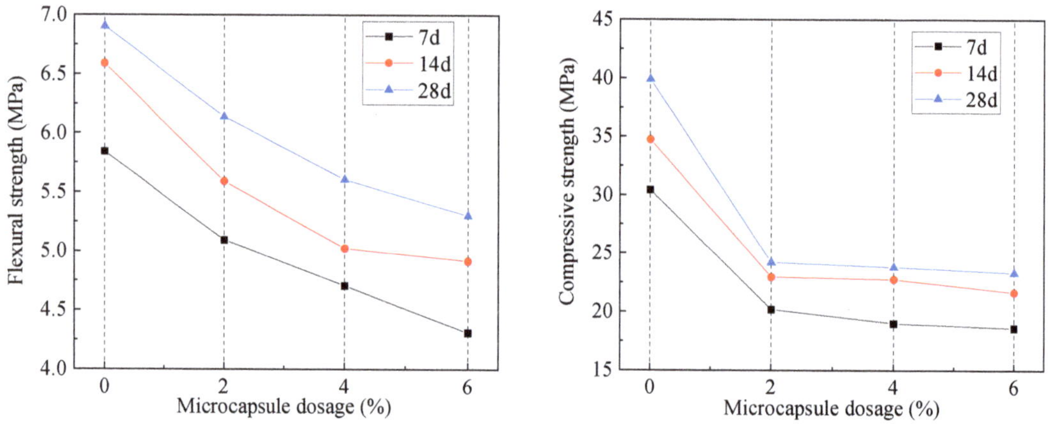

(**a**) Flexural strength (**b**) Compressive strength

Figure 27. Y initial strength at various ages with different microcapsule dosages.

3.4. *Influence of Cement Mortar Strength on the Effectiveness of Repair*

3.4.1. Experimental Group without Microcapsule Incorporation

The strength recovery and repair rates of the W1 group specimens after 7 and 28 days of maintenance are depicted in Figure 28.

These rates decline as the maintenance period increases. At 7 and 14 days, the strength values showed some recovery after maintenance, with recovery rates of 107.81% and 115.03% at 7 days, and 102.65% and 106.46% at 14 days, respectively. The repair rates were 8.97% and 16.27% at 7 days, and 4.13% and 8.0% at 14 days. However, at 28 days, the strength decreased rather than increased post-maintenance, indicating a negative repair rate.

Cementitious materials exhibit inherent self-repair capabilities, though these are limited. Early in the maintenance period, unhydrated cement particles can rapidly hydrate upon damage, aiding strength recovery. Beyond 28 days, the hydration level is high, reducing the availability of unhydrated particles to contribute to repair, and thus failing to restore strength after damage.

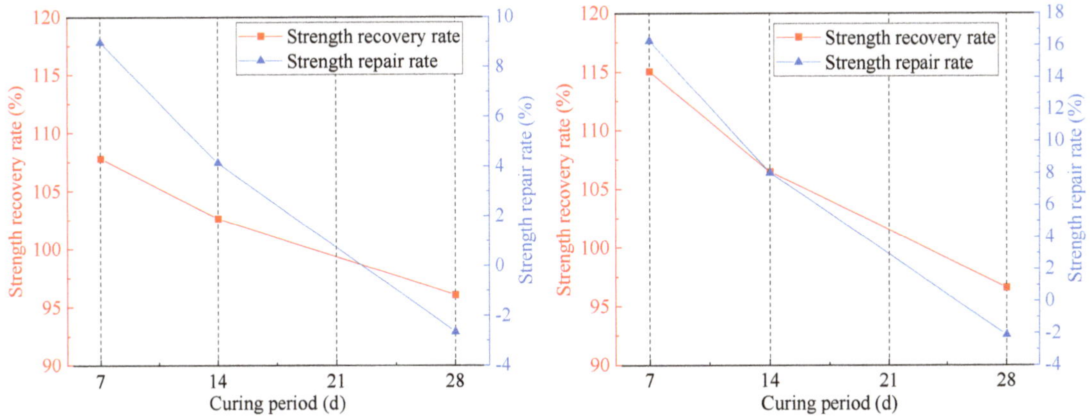

(**a**) 7 d (**b**) 28 d

Figure 28. Strength recovery rate and strength repair rate of group W1 specimens after conservation repair.

3.4.2. Test Group with Microencapsulation

Figures 29–32 illustrate the strength repair effects of self-repair mortar specimens with microcapsules across various maintenance periods.

The microcapsules release a core repair agent upon damage, which reacts with the curing agents in the cementitious material to form a bond and restore strength. This process is confirmed by molecular mechanisms showing independent and concurrent repair actions from unhydrated cement hydration and microcapsule-based repair.

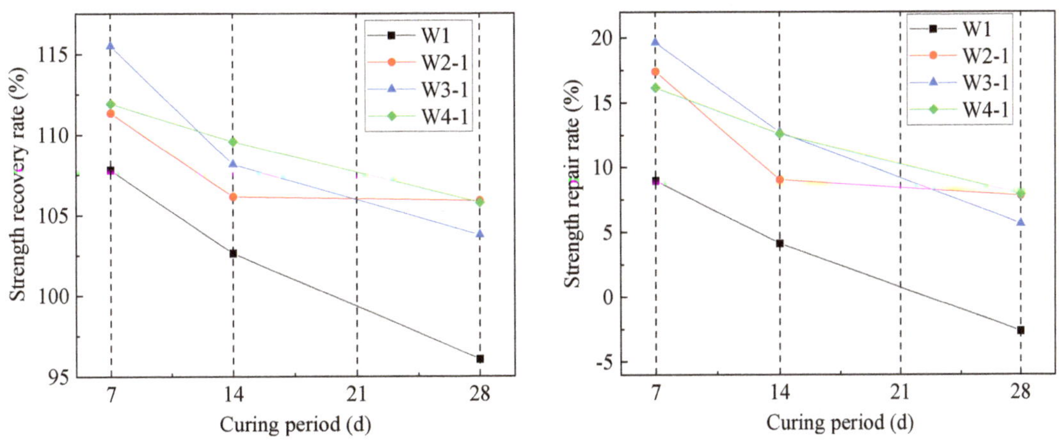

(a) Strength recovery rate (**b**) Strength repair rate

Figure 29. Strength recovery and repair rate of specimens of different ages after 7 d self−repair (X).

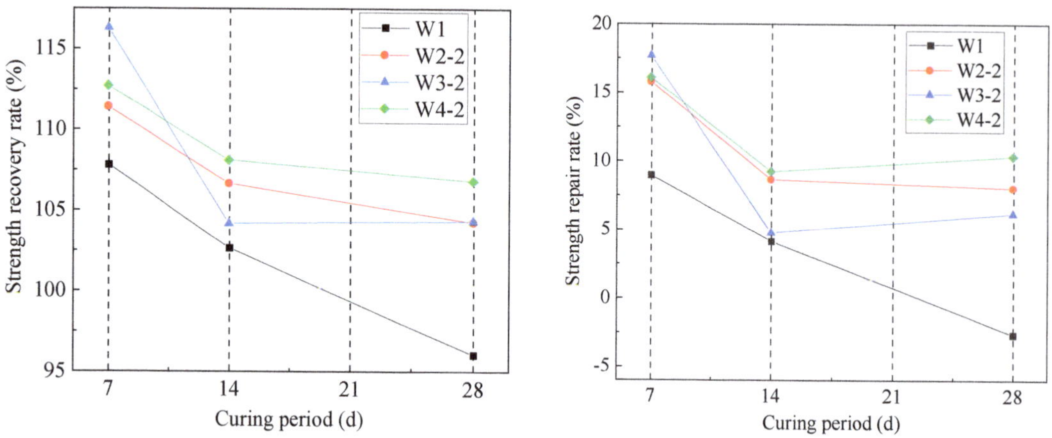

(**a**) Strength recovery rate (**b**) Strength repair rate

Figure 30. Strength recovery and repair rate of specimens of different ages after 7 d self−repair (Y).

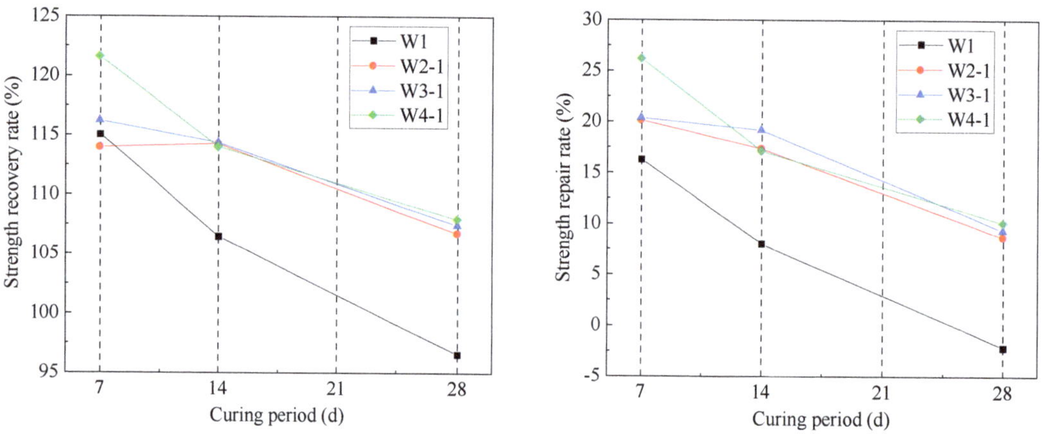

(**a**) Strength recovery rate (**b**) Strength repair rate

Figure 31. Strength recovery and repair rate of specimens of different ages after 28 d self−repair (X).

In groups W2-1, W3-1, and W4-1, using curing agent-X, the strength recovery rates after 7 and 28 days were 111.34% and 113.98%, 115.49% and 116.22%, and 111.91% and 121.63%, respectively. The rates were more stable over time compared to the non-microcapsule groups, likely due to the microcapsules compensating for the reduced number of unhydrated cement particles in the older specimens.

Groups W2-2, W3-2, and W4-2, using curing agent-Y, showed higher strength recovery rates than those using X, attributed to Y's liquid form ensuring more uniform distribution and integration within the mortar.

The optimal microcapsule dosages varied with age and curing agent type. For agent X, the best results were at 4.0% dosage at 7 days, 6% at 14 days, and 2% at 28 days. For agent Y, 4% was optimal at 7 days, with 6% being most effective at both 14 and 28 days.

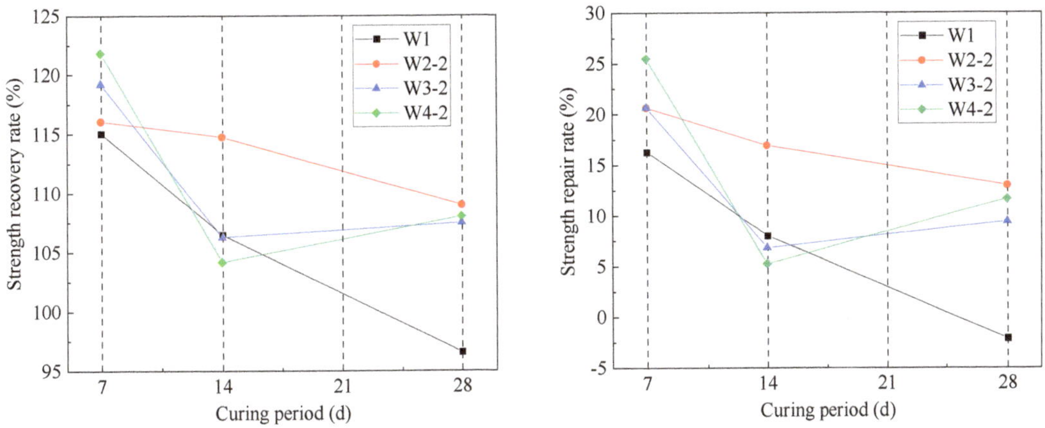

(**a**) Strength recovery rate (**b**) Strength repair rate

Figure 32. Strength recovery and repair rate (Y) of specimens of different ages after 28 d selfhealing.

3.5. Microscopic Morphology of Cement Mortar Containing Microcapsules

The scanning electron microscopy results of cement mortar containing self-healing microcapsules are shown in Figure 33.

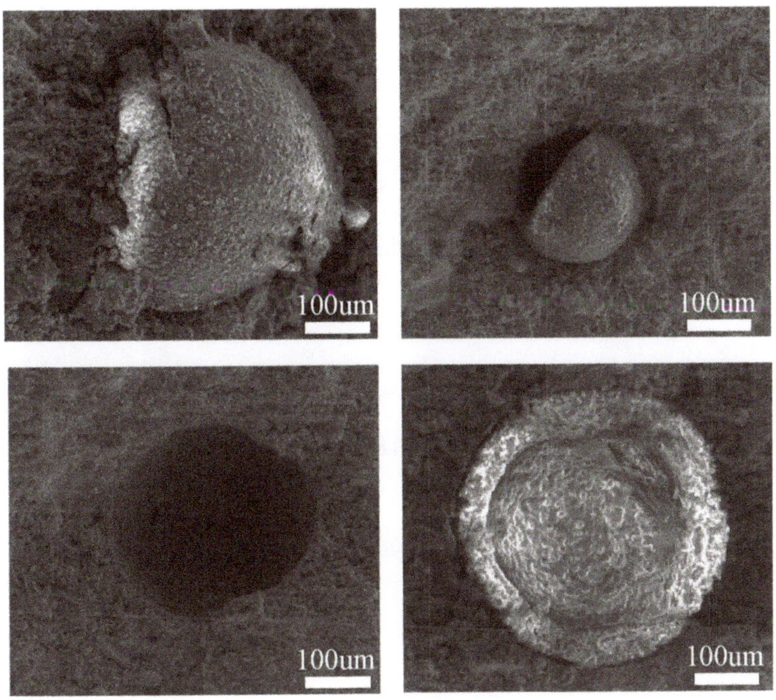

Figure 33. Scanning electron microscopy results of cement mortar containing microcapsules.

It can be seen in the figure that the self-repairing microcapsules are well embedded inside the structure of the cement matrix and well preserved without rupture during the mixing and molding process of the cement mortar. At the same time, the microcapsules can rupture at the right time when the cement structure is damaged, releasing the self-repairing agent stored inside for rapid repair of the damaged microcracks.

3.6. Effect of Microcapsules on the Pore Structure of Cement Mortar

The pores in concrete are categorized as harmless pores (below 20 nm), less harmful pores (20 nm~100 nm), harmful pores (100 nm~200 nm), and polyhazardous pores (above 200 nm).

Figure 34 shows the pore size distribution of cement mortar doped with different dosages of self-repairing microcapsules after 28 days of standard curing. As shown in Figure 34, the proportion of harmless or less harmful pores (pore size distribution of 0–0.1 μm) in the comparison sample (W1) after 28 days of standard maintenance was 65.17%, while the proportion of pores larger than 0.1 μm was 34.83%.

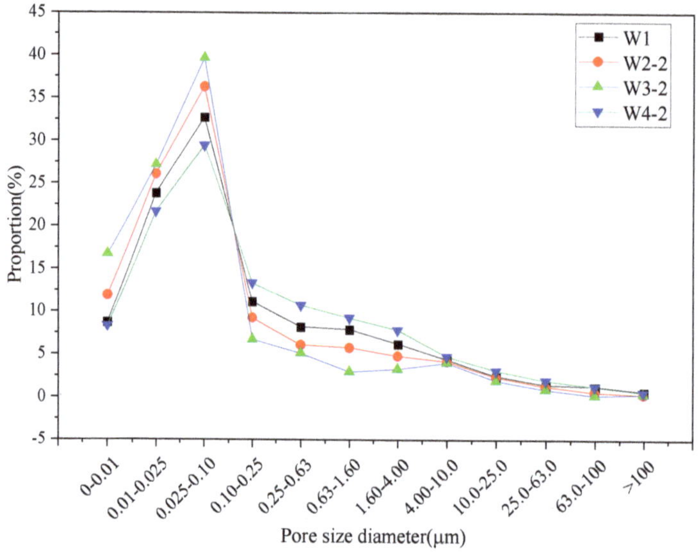

Figure 34. Pore size distribution of cement mortars with different dosages of microencapsulated cement mortars.

The proportion of harmless pores or less harmful pores in cement mortars (W2-2, W3-2 and W4-2) mixed with microcapsules according to the standard curing of 2%, 4% and 6% of the cement mass for 28 days was 74.11%, 83.38% and 59.3%, respectively, and the proportion of pores larger than 0.1 μm was 25.89%, 16.62% and 40.7%, respectively. As can be seen from Figure 34, compared with the comparison sample (W0), the proportion of pores larger than 0.1 μm in cement mortar (W2-2) with 2% of cement mass dosed with microcapsules decreased by 25.67%, and the proportion of pores larger than 0.1 μm in the cement mortar (W3-2) with 4% of cement mass dosed with microcapsules decreased by 52.28%. This indicates that the proportion of pores larger than 0.1 μm in the cement mortar decreases with the increase of microcapsule dosage from 0 to 4%, which implies that the densification of the cement mortar is improved. This is because the cement mortar is a kind of mixture prepared by a variety of different particle size materials; after molding, there will be a certain number of pores if the appropriate amount of microcapsules is added in the preparation process. Some of the pores inside the cement mortar can be filled so that the internal structure of the cement mortar is more dense, thus reducing the proportion of holes with a diameter greater than 0.1 μm in the cement mortar. From Figure 15, it can also be found that when the dosage of microcapsules in the cement mortar is increased from 4% to 6% (W4-2), the proportion of holes larger than 0.1 μm in the cement mortar rises by 16.85%

compared to the comparison sample (W1). This indicates that when more microcapsules were added, the number of internal pores in the cement mortar was increased to a certain extent, which decreased the compactness of the cement mortar.

3.7. Effect of Microcapsules on the Impermeability of Cement Mortar

Figure 35 shows the chloride diffusion coefficients of cement mortars with different dosages of self-repairing microcapsules after 28 days of standard curing. As can be seen from Figure 35, the chloride ion diffusion coefficient of the comparison sample (W1) after 28 days of standard curing was 20.32×10^{-12} m^2/s, whereas the chloride ion diffusion coefficients of the cement mortars doped with 2%, 4%, and 6% of microcapsules by weight of cement (W2-2, W3-2 and W4-2) were 18.58×10^{-12} m^2/s, 15.95×10^{-12} m^2/s, and 22.18×10^{-12} m^2/s after 28 days of standard curing, respectively. The chloride diffusion coefficients of cement mortar (W2-2) with 2% of microcapsule doping decreased by 8.56% and that of cement mortar (W3-2) with 4% of microcapsule doping decreased by 21.51% compared with the comparison sample (W1). This is because cement mortar is a mixture of materials with different particle sizes, and a certain number of pores will exist inside the cement mortar after molding and maintenance. Adding the appropriate amount of microcapsules in the preparation process can fill some of the pores inside the cement mortar and increase the compactness of the internal structure of the cement mortar, which improves the impermeability of the cement mortar and leads to a decrease in the chloride ion diffusion coefficient of the cement mortar. However, when the doped microcapsules rose to 6% (W4-2), the chloride ion diffusion coefficient of the cement mortar increased by 9.15% compared with the comparison sample. This indicates that when excessive microcapsules are added to cement mortar, the number of internal pores of cement mortar is increased to a certain extent which decreases the densification of cement mortar, thus leading to a decrease in the impermeability of cement mortar.

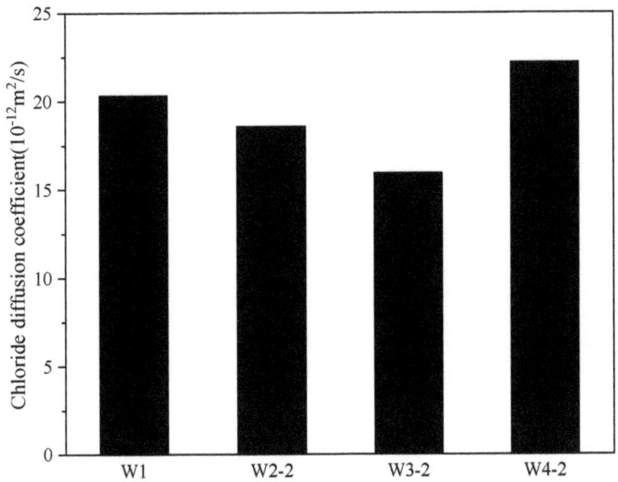

Figure 35. Chloride diffusion coefficients of cement mortars with different dosages of microencapsulated cements.

4. Conclusions

(1) Urea-formaldehyde resin/epoxy resin E-51 microcapsules were successfully synthesized via in-situ polymerization with optimal conditions identified for particle size, morphology, and encapsulation rate. The microcapsules made under the optimal conditions had a dense and smooth morphology, with an encapsulation rate of 55.59% at an average particle size of 495.56 μm.

(2) The microcapsule wall and the cured core material exhibit strong interfacial contact with the cementitious material, enhancing ductility and providing significant strength support, whereas the uncured core material diminishes mechanical properties.
(3) Microcapsule-enhanced cement mortar demonstrates a superior self-repair capability compared to ordinary cement mortar, although its initial strength is reduced with higher microcapsule dosages.
(4) The repair effectiveness of microencapsulated self-healing cement mortar diminishes over time due to the ongoing hydration of cement particles and the release of the repair material from the capsule core. However, the self-healing effect remains stable over a longer period compared to ordinary cement mortar, which shows a nearly linear decline in repair capability.
(5) The incorporation of microcapsules will have a significant effect on the flexural strength and compressive strength of the mortar. The effect on the compressive strength is greater the larger the dosage, the initial strength value of the cement mortar decreases more; the loss of strength of the cement mortar when using the Y curing agent is less than that of the use of the X curing agent, and the strength of the cement mortar with the use of the Y curing agent is more recovered.

Author Contributions: Conceptualization, D.C. and C.J.; Investigation, X.C. and M.G.; Writing—review & editing, H.M. All authors have read and agreed to the published version of the manuscript.

Funding: This research was funded by Qinghai Provincial Transportation Technology Project (Grant No. HYZD-KT-2023-01), Inner Mongolia Autonomous Region science and technology planning project (Grant No.2021GG0038), and the APC was funded by Qinghai Provincial Department of Transportation.

Data Availability Statement: The data presented in this study are available from the first and corresponding author upon request. The data are not publicly available due to the policy of the data provider.

Conflicts of Interest: The authors declare no conflicts of interest.

References

1. Jonkers, H.M. *Self Healing Concrete: A Biological Approach*; Springer Series in Materials Science; Springer: Dordrecht, The Netherlands, 2007; Volume 100, pp. 195–204.
2. Wedding, P.A.; Cady, P.D.; Weyers, R.E. Chloride Penetration and the Deterioration of Concrete Bridge Decks. *Concr. Aggreg.* **1983**, *5*, 81–87.
3. Marsavina, L.; Audenaert, K.; De Schutter, G.; Faur, N.; Marsavina, D. Experimental and Numerical Determination of the Chloride Penetration in Cracked Concrete. *Constr. Build. Mater.* **2009**, *23*, 264–274. [CrossRef]
4. Chahal, N.; Siddique, R.; Rajor, A. Influence of bacteria on the compressive strength, water absorption and rapid chloride permeability of fly ash concrete. *Constr. Build. Mater.* **2012**, *28*, 351–356. [CrossRef]
5. *BS EN 1504*; Products and Systems for the Protection and Repair of Concrete Structures-Definitions, Requirements, Quality Control and Evaluation of Conformity Concrete Injection. BSI Standards: Hong Kong, China, 2013.
6. *ACI 562-16*; Code Requirements for Assessment, Repair, and Rehabilitation of Existing Concrete Structures and Commentary (Metric). ACI: Farmington Hills, MI, USA, 2016.
7. *ACI 546R-14*; Guide to Concrete Repair. ACI: Farmington Hills, MI, USA, 2014.
8. Nishiwaki, T.; Mihashi, H.; Jang, B.-K.; Miuea, K. Development of self-healing system for concrete with selective heating around crack. *J. Adv. Concr. Technol.* **2006**, *4*, 267–275. [CrossRef]
9. Fengbin, Y.; Juanru, J.; Xuejun, C. Preferred design methods for reinforcement of concrete structures. *Constr. Technol.* **2016**, *45*, 22–24.
10. Lubelli, B.; Nijland, T.G.; Van Hees, R.P.J. Self-healing of lime based mortars: Microscopy observations on case studies. *Heron* **2011**, *56*, 81–97.
11. Yuhong, Q.; Guoliang, Z.; Jinfeng, C. Research progress of anticorrosive coatings for concrete. *Coat. Ind.* **2018**, *48*, 63–71.
12. Concrete Repairs. *Performance in Service and Current Practice (EP 79)*; Building Research Establishment: Garston, UK, 2007.
13. Taylor, H.F.W. Proposed Structure For Calcium Silicate Hydrate Gel. *J. Am. Ceram. Soc.* **1986**, *69*, 464–467. [CrossRef]
14. Pacheco-Torgal, F.; Labrincha, J.A.; Leonelli, C.; Palomo, A.; Chindaprasirt, P. *Handbook of Alkali-Activated Cements, Mortars and Concretes*; Woodhead Publishing: Delhi, India, 2015.

15. Bureau of Transport and Communications Economics. *The Cost of Maintaining the Australian National Highway System*; Australian Government Publishing Service: Canberra, Australia, 1992.
16. *ISO 679:2009-05*; Cement—Test Methods—Determination of Strength. ISO: Geneva, Switzerland, 2009.

buildings

MDPI

Article

Analysis and Comparison of Three Bending Tests on Phosphogypsum-Based Material According to Peridynamic Theory

Haoyu Ma [1,2], Kai Zhang [3], Sheng Liang [4], Jiatian Dong [1,2], Xiangyang Fan [3] and Xuemei Zhang [1,*]

1 School of Transportation and Logistics Engineering, Wuhan University of Technology, No. 1178 Heping Avenue, Wuhan 430063, China; mahaoyu@whut.edu.cn (H.M.); 284769@whut.edu.cn (J.D.)
2 Hubei Highway Engineering Research Center, Wuhan University of Technology, No. 1178 Heping Avenue, Wuhan 430063, China
3 Jiangxi Transportation Science Research Institute Co., Ltd., Nanchang 330025, China; kaizhang@whut.edu.cn (K.Z.); fanxiangyang@whut.edu.cn (X.F.)
4 Shenzhen Road & Bridge Group Co., Ltd., Shenzhen 518024, China; sheng_liang@whut.edu.cn
* Correspondence: xuemeiz@whut.edu.cn

Abstract: Phosphogypsum-based materials have gained much attention in the field of road infrastructure from the economic and sustainable perspectives. The Three-point bending test, the Four-point bending test and the Semi-circular bending test are three typical test methods applied for fracture energy measurement. However, the optimal test method for fracture energy evaluation has not been determined for phosphogypsum-based materials. To contribute to the gap, this study aims to analyze and compare the three test methods for fracture energy evaluation of phosphogypsum materials based on the peridynamic theory. For this purpose, the load–displacement, vertical displacement–Crack Mouth Opening Displacement (CMOD) and fracture energy of the phosphogypsum-based materials were measured and calculated from the three test methods. The simulated load–displacement and vertical displacement–CMOD by PD numerical models, with different fracture energy as inputs, were compared to the corresponding tested values according to simulation error results. The results showed that the Four-point bending test led to minimized errors lower than 0.189 and indicators lower than 0.124, demonstrating the most optimal test method for the fracture energy measurement of phosphogypsum-based material. The results of this study can provide new methodological references for the selection of material fracture energy measurement tests.

Keywords: phosphogypsum-based material; bending test; road infrastructure; fracture energy measurement; peridynamic theory

1. Introduction

Phosphogypsum, as a raw material of base layers, is one of the primary concerns for road technology from the economic and sustainable perspectives [1,2]. Most studies have focused on the mix design and mechanical characterization of phosphogypsum. Some scholars verified that phosphogypsum mixed with crushed rocks, clay, lime, fly ash and curing agents in certain proportions could be a substitute for traditional semi-rigid bases [3–5]. However, limited analysis has been carried out on the crack resistance of phosphogypsum-based base materials, while extensive investigations have been conducted on the performance of traditional semi-rigid base materials [6,7]. To promote the widespread application of phosphogypsum as a semi-rigid base for road construction, it is imperative to concisely characterize the crack resistance of phosphogypsum-based material.

The fracture energy is a critical parameter of materials in evaluating the ability to resist damage-induced cracking. Moreover, Mode I cracks are the most common form of base materials. Thus, Mode I fracture energy represents the focus of fracture research [8].

Currently, three experimental methods are conducted for obtaining Mode I fracture energy, including direct tension testing [9], indirect tension testing [10], and bending testing [11,12]. Among the three test methods, an appropriate cracking test method for accurately measuring the fracture energy of phosphogypsum has not yet been determined. In this study, identifying the most suitable method for acquiring fracture energy will not only provide a reliable experimental methodology, but also propose a guideline for further research on the fracture failure mechanisms of phosphogypsum and the rational design of structures.

Theoretically, fracture energy derived from crack initiation tests on standardized cross-sections should exhibit uniformity. However, the results obtained from different tests vary in practice [13]. Such discrepancies may stem from dissimilarities in energy dissipation, deviations in crack propagation from the principal tensile stress direction, and other factors [14,15]. To address these root causes, Zhang et al. [16] investigated kinetic energy dissipation, crack propagation trajectories, and deformation energy beyond the fracture surface during loading. These critical sources of error were thoroughly scrutinized, serving as the foundation for selecting experimental methods. Nonetheless, this research paradigm, grounded in extensive theoretical groundwork and experimental methodologies, presented formidable challenges and substantial cost.

To tackle these deficiencies, another approach to evaluate optimal fracture energy was proposed in this study, grounded in numerical simulation. Accurate fracture energy values, as inputs of damaged numerical models, can significantly enhance the accuracy of numerical simulation outcomes, rendering them more faithful to practical damage scenarios [17]. If fracture energy data measured from a specific test type can effectively accommodate all fracture test methods within numerical models, it means the test method is the most optimal method for fracture energy evaluation. This approach, underpinned by insights gleaned from numerical model results fueled by fracture energy, explores the veracity of fracture energy data with avoiding the test errors.

Regarding traditional damaged numerical models, numerical outcomes are influenced not only by fracture energy but also by additional factors, including tensile strength, criteria for crack propagation deflection, and handling of singular regions [18,19]. This fact indicates that fracture energy is insufficient as the sole input parameter for traditional numerical models. In light of this challenge, this study adopts peridynamic (PD) as the analytical framework, as the fracture energy is the only input characterizing the entire material damage process [20]. This theory has developed from its initial applicability to models for linear elastic–brittle materials to diverse theoretical models suitable for non-linear elastic–brittle, elastic–plastic, viscoelastic, and viscoplastic materials [21–23]. Moreover, it has demonstrated successful application in the fracture testing of engineering materials such as asphalt mixtures and cement concrete with excellent simulation outcomes [24,25].

Based on the above background, few studies focused on selecting a suitable fracture energy testing method for phosphogypsum-based materials among different test modes. Thus, this study addresses this gap, as well as delving into the difference in fracture energy between these tests, based on PD simulation. In this study, suitable test methodologies for phosphogypsum materials across various fracture energy testing scenarios are delineated, employing numerical simulations and aiming to provide an optimal fracture energy testing method. The fracture energy of phosphogypsum-based specimens will be measured by the Three-point bending test [26], the Four-point bending test [27], and the Semi-circular bending test [28] and will be the input for peridynamic numerical simulations. The optimal fracture energy testing method will be determined by analyzing the simulation error between tested and simulated results.

2. PD Theory

Peridynamic is a method based on the concept of non-local interactions, and describes the mechanical behavior of materials by solving spatial integral equations. To provide theoretical and technical support for determining the optimal test method in the following sections, this section briefly introduces the theory of PD and its numerical implementation process. Firstly, the form of the motion equations is introduced. Secondly, the specific form of the force density function in the motion equations is provided, which are the fundamental equations of the PD theory. Finally, the methods for solving the equations in the numerical computation process are explained.

2.1. Kinematic Equation

The two-parameter bond-based PD model was selected due to its capacity to address the limitations associated with fixed Poisson's ratios in traditional bond-based models. Additionally, it offers notable computational advantages compared to state-based models [29].

As depicted in Figure 1a, at any time instance t, each particle x_k within a spatial domain R encompasses a family domain with a radius of δ. Within this domain, other material points are interconnected to the central point in a bond-like fashion, facilitating the transmission of forces. The specific formulation of motion equations for the two-parameter model is derived from the principle of virtual work, shown in Equation (1) [30]:

$$\begin{aligned} \rho\frac{\partial^2 u(x_k,t)}{\partial t^2} &= \int_H f_c(u(x_j,t)-u(x_k,t),x_j-x_k,\theta_k,\theta_j)dV_{x_j}+b(x_k,t), \\ I\frac{\partial^2 \theta(x_k,t)}{\partial t^2} &= \int_H f_\kappa(u(x_j,t)-u(x_k,t),x_j-x_k,\theta_k,\theta_j)dV_{x_j}+M(x_k,t) \end{aligned} \quad (1)$$

where ρ represents material density, I denotes moment of inertia; u signifies the displacement of material points, and θ represents the rotation angle of material points; b stands for body force density, and M represents external torque; f_c and f_κ, respectively, represent normal bond force and tangential bond force.

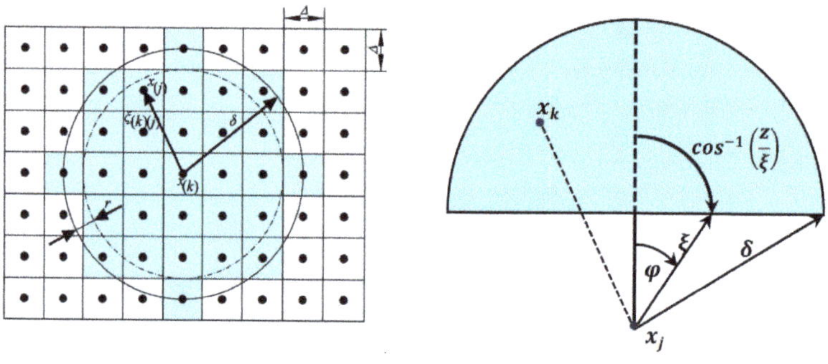

(a) (b)

Figure 1. PD model. (**a**) The schematic diagram of the PD constitutive model. (**b**) The integration domain of bond fracture.

2.2. Force Function

In this study, the phosphogypsum-based materials used for road base layer were assumed as ideal elastic–brittle materials [31]. Subsequently, the ensuing discussion introduces the theoretical framework of the elastic–brittle model within the context of the two-parameter bond-based PD.

In bond-based PD models, the constitutive force function solely delineates the radial displacement and radial force interactions among material points, thereby inadequately capturing the comprehensive motion state of these points. In contrast, the two-parameter model discerns between shape alterations and volume variations by concurrently incorporating axial forces resulting from axial deformation and bending moments arising from rotational deformation between material points. This approach effectively addresses the constraint of fixed Poisson's ratio [29]. The definitions of the normal bond force and tangential bond force are as follows:

$$\begin{aligned} \vec{f}_c\langle u' - u, X' - X\rangle &= c\vec{s}\langle u' - u, X' - X\rangle, \\ \vec{f}_\kappa\langle u' - u, X' - X\rangle &= \kappa\vec{\beta}\langle u' - u, X' - X\rangle \end{aligned} \tag{2}$$

where $\vec{f}_c$ and $\vec{f}_\kappa$ represent normal bond force and tangential bond force, respectively, and c and κ are the PD parameters in axial direction and tangential direction, respectively; $\vec{s}$ represents the bond stretch, and $\vec{\beta}$ represents the change in the bond rotation.

Following the principle of equal strain energy density, the relationship expression between PD parameters and Young's modulus E, as well as Poisson's ratio ν, can be derived. The specific expressions are detailed in Table 1, with a comprehensive derivation process provided in ref. [32].

Table 1. Parameters of the PD Model.

Model Parameters	Three-Dimensional	Plane Stress	Plane Strain
c	$\dfrac{6E}{\pi\delta^3(1-2v)(1+v)}$	$\dfrac{6E}{\pi\delta^3 t(1-v)}$	$\dfrac{6E}{\pi\delta^3 t(1+v)(1-2v)}$
k	$\dfrac{E(1-4v)}{4\pi\delta^2(1-2v)(1+v)}$	$\dfrac{E(1-3v)}{6\pi\delta t(1-v^2)}$	$\dfrac{E(1-4v)}{6\pi\delta t(1-2v)(1+v)}$

2.3. Failure Criterion

Combining Silling and Askari's derivation of fracture criteria [33], the fracture energy of the material is expressed as follows:

$$G^{PD} = 2h\int_0^\delta \int_z^\delta \int_0^{\arccos\frac{z}{\xi}} \left(\frac{1}{2}cs_0^2\xi + \frac{1}{2}\kappa\beta_0^2\xi\right)\xi d\varphi d\xi dz \tag{3}$$

where S_0 represents the critical bond stretch, and β_0 represents the critical bond rotation, as illustrated in Figure 1b.

For a Mode I crack, the material undergoes tensile failure, and the energy at this juncture is solely attributed to the energy derived from normal bond deformation. Consequently, the critical bond stretch for tensile failure is determined as follows:

$$s_0 = \sqrt{\frac{4G_I}{hc\delta^4}} \tag{4}$$

where G_I represents the experimentally measured type I fracture energy of the material, and h represents the material thickness.

In summary, three mandatory mechanical parameters, the Young's modulus, Poisson's ratio, and fracture energy, are required for PD numerical model simulations. The acquisition of these parameters will be elaborated upon in Section 3.

2.4. Numerical Method

To obtain the solution of the PD motion equations, it is necessary to discretize the spatial domain occupied by the material into a finite number of material points during the numerical computation process. Thus, the spatial integral form of the motion equations is transformed into a discrete sum form. For the two-parameter model, according to Equation (1), its motion equations can be discretized as [34]:

$$\begin{aligned} \rho\frac{\partial^2 u(x_i^n,t)}{\partial t^2} &= \sum_{j=1}^{p} f_c\left(u\left(x_j^n,t\right) - u(x_i^n,t), x_j^n - x_i^n, \theta_i^n, \theta_j^n\right)V_{x_j} + b(x_i^n,t), \\ I\frac{\partial^2 \theta(x_i^n,t)}{\partial t^2} &= \sum_{j=1}^{p} f_\kappa\left(u\left(x_j^n,t\right) - u(x_i^n,t), x_j^n - x_i^n, \theta_i^n, \theta_j^n\right)V_{x_j} + M(x_i^n,t) \end{aligned} \tag{5}$$

The time iteration in this study adopts an explicit finite difference method, where the differentiation of displacement with respect to time is converted into a finite difference formula. The velocity and acceleration at any time step are represented by the following equations:

$$\begin{aligned} \frac{\partial^2 u_i^n}{\partial t^2} &= \frac{u_i^{n+1} - 2u_i^n + u_i^{n-1}}{\Delta t^2}, \\ \frac{\partial u_i^n}{\partial t} &= \frac{u_i^{n+1} - u_i^{n-1}}{2\Delta t} \end{aligned} \tag{6}$$

Based on Equation (6), after obtaining the acceleration of a material point at the nth time step, the velocity and displacement at the next time step can be computed.

$$\begin{aligned} \frac{\partial u_{x_i}^{n+1}}{\partial t} &= \frac{\partial^2 u_{x_i}^n}{\partial t^2}\Delta t + \frac{\partial u_{x_i}^n}{\partial t}, \\ u_{x_i}^{n+1} &= \frac{\partial u_{x_i}^n}{\partial t}\Delta t + u_{x_i}^n \end{aligned} \tag{7}$$

where Δt represents the time step length during the numerical computation process.

The selection of time step length is crucial in explicit time differencing. To obtain convergent computational results, the time step must satisfy certain stability conditions. The choice in this study is based on the work of Silling and Askari [30].

2.5. Simulation Result Processing Method

In order to simulate the bending experiment of phosphogypsum-based material in PD theory, virtual material layers are set at the top and bottom of the sample. During the loading process, the virtual material layers will bear the reaction forces applied by the actual materials, and the relationship between the axial force P and the vertical component F_y^i of the PD force density is expressed as follows:

$$P = \left|\sum_{i=1}^{N} F_y^i V_i\right| \tag{8}$$

where N represents the total number of material points in the virtual layer.

3. Methodology

In order to ascertain the fracture energy of phosphogypsum and furnish the material parameters for the numerical model delineated in Section 4, this section introduces the preparation of phosphogypsum specimens, the mechanical test process, and establishment of numerical models.

3.1. Materials and Sample Preparation

In this study, the phosphogypsum-based material applied for the road base layer is the pre-treated material with optimized water content, as determined and studied within the framework of the project titled "Research on Environmentally Friendly and Durable Phosphogypsum-Based Assembled Road Pavement Material and Structure".

As presented in Figure 2, the preparation process of this material was divided into five steps: First, the phosphogypsum powder was completely mixed with water for about 1.5 min until the lumps disappear. Second, the mixed phosphogypsum material was poured into beam and cylindrical molds and pressed for 15 min into a mold. Third, using a demolding machine, beam specimens with dimensions of 50 mm × 50 mm × 200 mm and cylindrical specimens with dimensions of Φ150 mm × 100 mm were formed. Fourth, the demolded specimens were cut into beam-shaped and semi-circular specimens with dimensions of 50 mm × 50 mm × 100 mm and Φ150 mm × 40 mm, respectively. Then, pre-cut cracks of 10 mm and 15 mm, respectively, were made at the midspan of the beam-shaped and semi-circular specimens for better control of the crack propagation path. Lastly, all specimens were cured at 20 °C for seven days before testing.

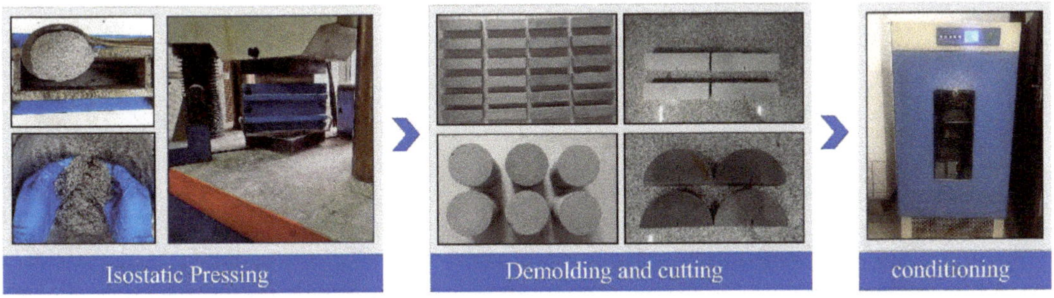

Figure 2. The flowchart of specimen molding process.

3.2. Tests

The bending test is a commonly utilized and effective method for evaluating the fracture performance of brittle materials [35]. In this study, Three-point bending, Four-point bending, and Semi-circular bending tests were performed for characterizing the fracture energy of phosphogypsum-based materials. Additionally, the uniaxial compression tests were conducted to measure Young's modulus and Poisson's ratio of the phosphogypsum-based materials. Each type of experiment is conducted with six parallel tests, and the three groups with the least variability are selected as the experimental results. The instruments, test parameters, and procedures employed in the compression and bending tests are outlined as follows.

3.2.1. Bending Test

The bending tests were performed by a servo-pneumatic Mechanical Testing System (MTS) with a capacity of 50 KN, produced by Cooper Technology (Borken, Germany). The MTS exerted a controlled linear axial load with a steady rate of 0.2 mm/min to ensure the quasi-static loading.

The illustration of the bending tests is shown in Figure 3. The beam specimen dimensions refer to the recommendations for fracture energy testing in the "Test Methods of Materials Stabilized with Inorganic Binders for Highway Engineering" (JTG 3441-2024) [36]. The semi-circular specimen dimensions follow the International Society for Rock Mechanics (ISRM) guidelines for Semi-Circular Bend (SCB) testing [37]. During the loading process, vertical displacement and vertical load were recorded in the MTS, CMOD was measured by a YYJ-12/10 clip gauge produced by the Central Iron & Steel Research Institute, Beijing, which was used to form the load–displacement curve and the displacement–CMOD curve. The fracture energy is the area below the load–displacement curve until the specimen is broken, and describes the energy released during cracking.

$$G_f = \frac{W}{A_{lig}} \quad (9)$$

where W is the fracture's work and Alig is the ligament area, defined as:

$$A_{lig} = (r - a)xh \quad (10)$$

where r is the specimen radius or height, in m, a is the notch length, in m, and h specimen thickness, in m.

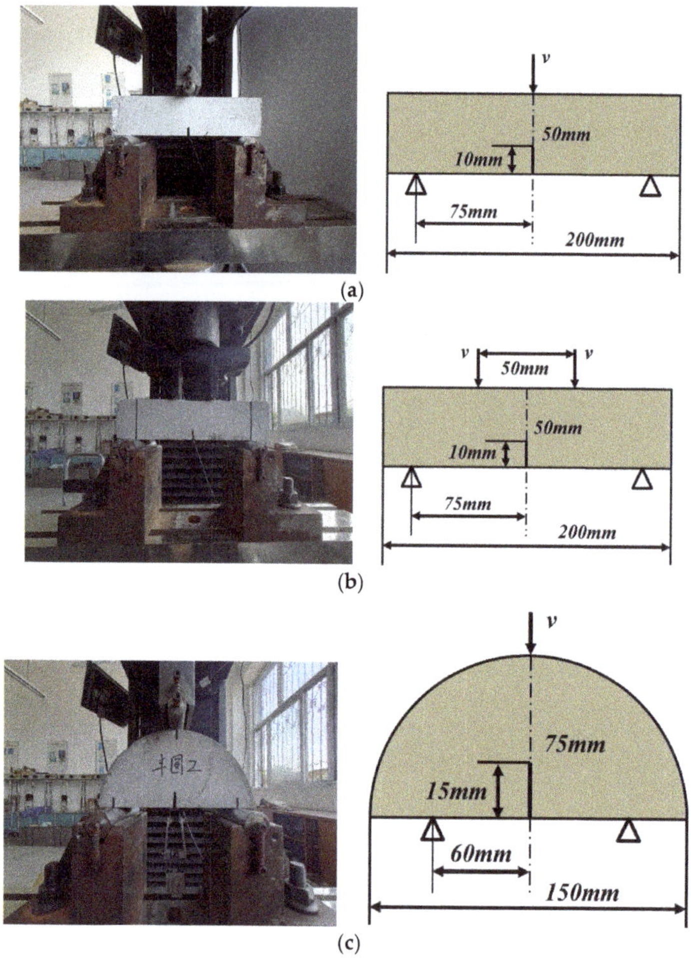

Figure 3. Various types of bending tests: (**a**) Three-point bending test; (**b**) Four-point bending test; (**c**) Semi-circular bending test.

3.2.2. Uniaxial Compression Test

The uniaxial compression test was performed using the Wdw-300 Series Computerized Electronic Universal Testing Machine manufactured by Kexin (Shenzhen, China). The machine is capable of exerting a linear axial load. Four strain gauges were affixed vertically and horizontally on opposite sides of the specimen. According to the "Test Methods of Materials Stabilized with Inorganic Binders for Highway Engineering" (JTG 3441-2024), the UC modulus test was conducted in a displacement-controlled loading mode at a constant rate of 1 mm/min. The decision to use prismatic specimens is due to challenges in measuring Poisson's ratio with cylindrical specimens.

The illustration of the UC test is shown in Figure 4. The UC test determines the Young's modulus and Poisson's ratio uniaxially, as it measures the stress $\sigma y(t)$ and the strain $\varepsilon x(t)$ and $\varepsilon y(t)$, which is expressed as follows:

$$E = \frac{\sigma_y(t)}{\varepsilon_y(t)}, \nu = -\frac{\varepsilon_x(t)}{\varepsilon_y(t)} \tag{11}$$

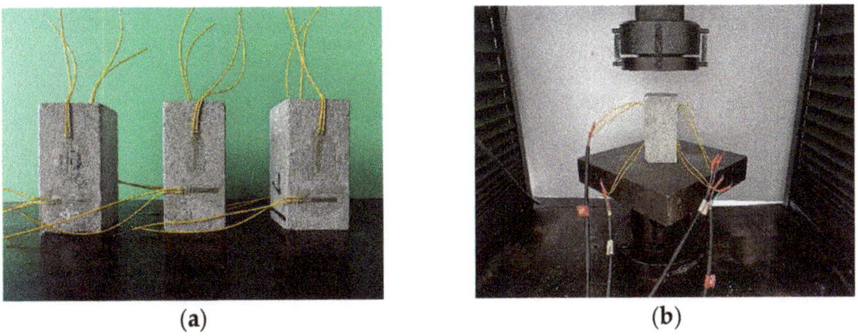

Figure 4. Process of material parameter testing: (**a**) Prismatic specimen; (**b**) Uniaxial compression test.

In the UC test, the value of axial stress, axial strain and transverse strain were recorded, which constitute the stress–strain curves and bi-directional strain curves. The value of stiffness from the linear segment of the stress–strain curve determined the elastic modulus of the UC test specimen at a certain test condition. The peak value of axial strain and transverse strain were chosen for calculating the Poisson's ratio.

3.3. Numerical Model Establishment

To fulfill the aim of identifying the optimal fracture energy testing approach in this study, this section established distinct PD numerical models for the Three-point bending test, the Four-point bending test, and the Semi-circular bending test. The fracture energy data garnered from varied bending tests served as the input parameters in their respective numerical models. Subsequently, the simulated outcomes were compared to the actual test results for analyses.

The virtual experiments conducted in this study were addressed as plane stress problems. Two rectangular plates measuring 50 mm × 200 mm and one semi-circular plate with a diameter of Φ150 mm were configured. They were discretized into particles, forming a simple cubic lattice with Δx = 0.5 mm and δ = 3.015Δx. Cracks of 10 mm and 15 mm were, respectively, set at the midspan of the rectangular plates and the semi-circular plate by truncating the PD bonds. The schematic diagram of the virtual model is shown in Figure 5.

To achieve a more precise simulation of the phosphogypsum-based material in this study, all material property parameters utilized in the model were derived from the experimental tests conducted in Section 3.2. The loading and support areas consisted of 6 × 6 particles. For boundary conditions, a constant loading rate of 0.2 mm/min was applied downward at the upper boundary. Through a combined optimization test ensuring integral stability and computational efficiency, the time step was established as $\Delta t = 1.33 \times 10^{-6}$ s in this study. Considering the fracture energy results from the three types of bending tests, the critical bond stretch was computed by Equation (4). Subsequently, three critical bond stretch values were sequentially input into every numerical model to explore the simulation outcomes of each type of virtual test under varied fracture energy inputs. In this study, the load–displacement curve and the displacement–CMOD curve were used for test scheme comparison.

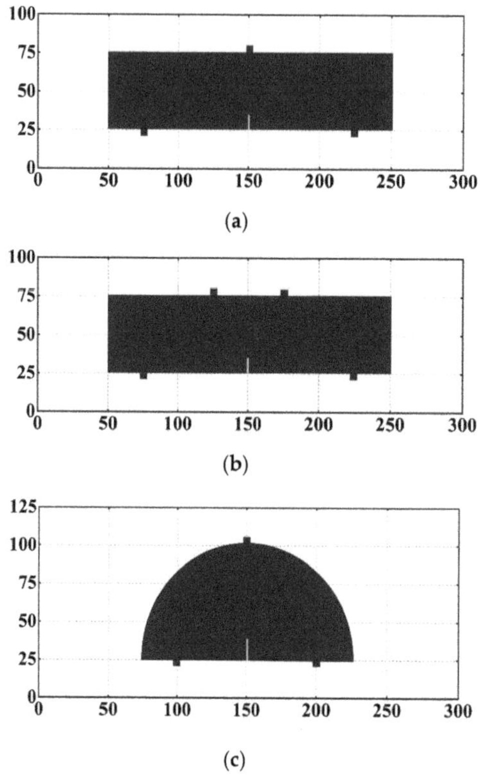

Figure 5. Virtual test schematic diagram: (**a**) Three-point bending virtual test; (**b**) Four-point bending virtual test; (**c**) Semi-circular bending virtual test.

4. Results and Discussion

This section constitutes the core of determining the optimal testing method. Firstly, the experimental results were presented and analyzed. Subsequently, by inputting the material parameters obtained in Section 3 into the PD numerical model, the simulation results for various virtual bending tests under different fracture energy inputs were obtained. Finally, three error metrics were selected to compare the test results and the simulation results. The optimal test was selected based on a comprehensive analysis of simulation error analyses.

4.1. Test Results

4.1.1. Load–Displacement Curves and Fracture Energy

Figure 6a–c show the three sets of load–displacement curves for the phosphogypsum-based specimens under three types of bending tests. The pre-peak curves of the Four-point bending test demonstrated a linear ascending trend. In contrast, the pre-peak curves of the Three-point bending test and the Semi-circular bending test showed an exponential uptrend. Specifically, the slope gradually increased as the curve progresses and approaches the peak point, indicating a gradual acceleration of the upward trend. For the three tests, the curve experienced a sudden decline after the peak point, which signifies the occurrence of fracture. This phenomenon verifies the brittle nature of the phosphogypsum materials and the assumption of the ideal elastic–brittle materials. With the increase in load, the stress concentration at the specimen crack leads to the appearance of small cracks, which rapidly propagate into through cracks. The sudden decline after the peak point indicated that this process was brief. The rapid transition from microcrack generation to through crack formation in brittle materials emphasizes the criticality of crack-resistant design for engineering brittle materials. The averaged results of the calculated and analyzed test outcomes, encompassing fracture energy, failure load, and failure opening displacement, are presented in Table 2. The fracture energy data revealed significant disparities of approximately 10–12% in the results across different tests, surpassing the variability within each test type. Among the three tests, the Semi-circular bending test resulted in highest value in fracture energy, followed by Four-point bending and Three-point bending. This fact is connected with the stress distribution and crack propagation paths, which are influenced by the shape and size of the specimen, as well as the loading method.

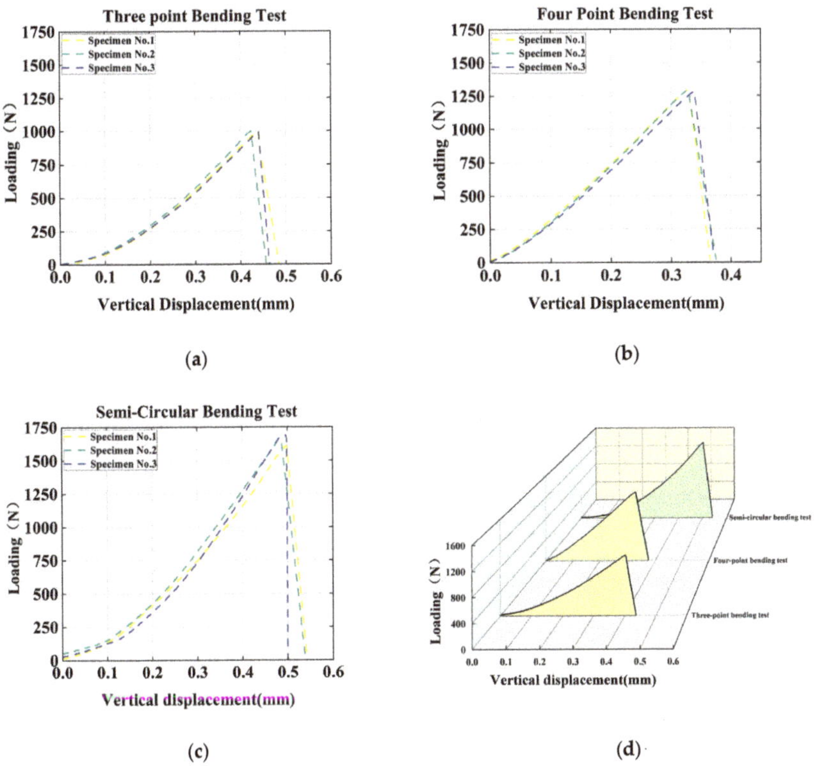

Figure 6. Bending tests results: (**a**) Load–displacement curve of Three-point bending test; (**b**) Load–displacement curve of Four-point bending test; (**c**) Load–displacement curve of Semi-circular bending test; (**d**) Load–displacement curve of three bending tests.

Table 2. Bending tests results.

Types	Peak Load/KN	Failure Displacement/mm	Failure CMOD/mm	Fracture Energy (J/m^2)
Three-point bending	1.009	0.4512	0.014	91.58
Four-point bending	1.286	0.3377	0.017	109.33
Semi-circular bending	1.691	0.5027	0.021	130.25

4.1.2. Displacement–CMOD Curves

Figure 7 summaries the displacement–CMOD curves obtained from the three types of bending tests. The displacement–CMOD curve of the Four-point bending test showed a linear rise at the beginning and kept stable after 0.017 mm. However, the pre-peak segments of the curves for the Three-point bending test and the Semi-circular bending test grew with a gradually decreasing slope. Moreover, the Semi-circular bending test led to deepest vertical displacement compared to the Four-point bending test and the Three-point bending test. These results are consistent with the results of load–displacement curves.

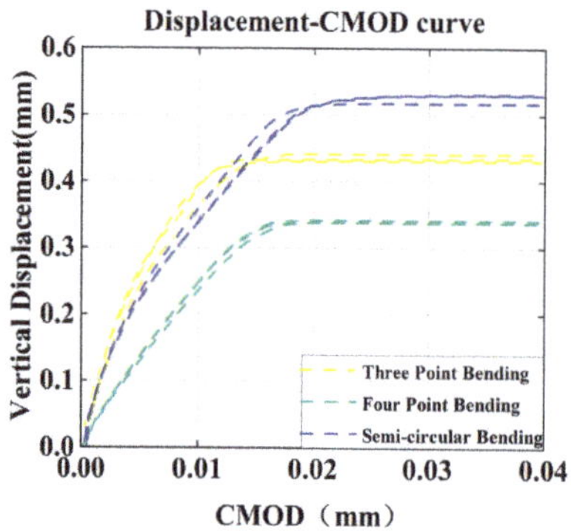

Figure 7. Displacement–CMOD curves.

4.1.3. Young's Modulus and Poisson's Ratio

Figure 8 depicts the load–displacement curve and bi-directional strain curve derived from the uniaxial compression test. Young's modulus and Poisson's ratio are individually calculated by analyzing the linear segments with curves. As a result, 3.3 GPa and 0.29 were calculated and used for the Young's modulus and Possion's ratio of phosphogypsum-based materials, respectively. The stress–strain curve exhibited a brief yielding phase, followed by a rapid decrease upon surpassing the elastic limit. This is consistent with the characteristics of elastic–brittle materials.

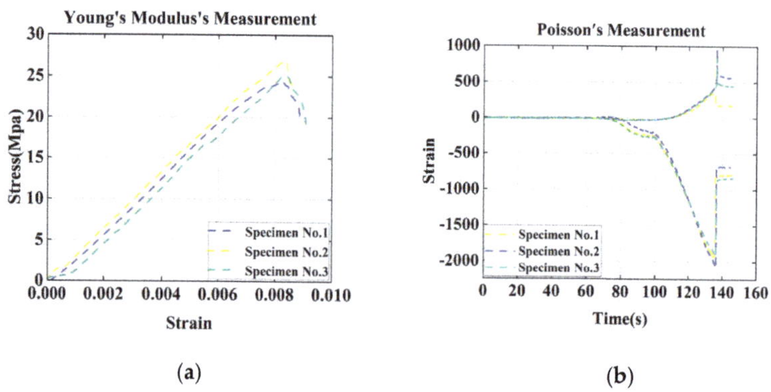

Figure 8. Uniaxial compression test results: (**a**) Load–displacement curve; (**b**) Biaxial strain curve.

4.2. Comparison between the Three Tests Based on PD Simulation

4.2.1. Numerical Results

Figure 9a–c illustrate the comparison between the simulated load–displacement curves under different fracture energy index inputs and the test results for various bending tests. The selected interval for the simulation results correspond to the range where the load demonstrates a stable increasing trend. Upon analyzing the simulated curves, it is shown that variations in fracture energy influenced the endpoint of the pre-peak curve, while the curve's trend was hardly affected by the changes in fracture energy. This result suggests that the characteristics of the pre-peak curve, particularly the linear segment, are primarily dependent on Young's modulus and Poisson's ratio for the same type of fracture test. Additionally, the Four-point bending test resulted in similar load–displacement curves between the simulated and test results among the three tests, followed by the Three-point bending test. However, the simulated results of the Semi-circular bending test were significantly different from the test results compared to that of the other tests, which indicates that the evolving traits of the Semi-circular bending test curve is partially characterized by the numerical model.

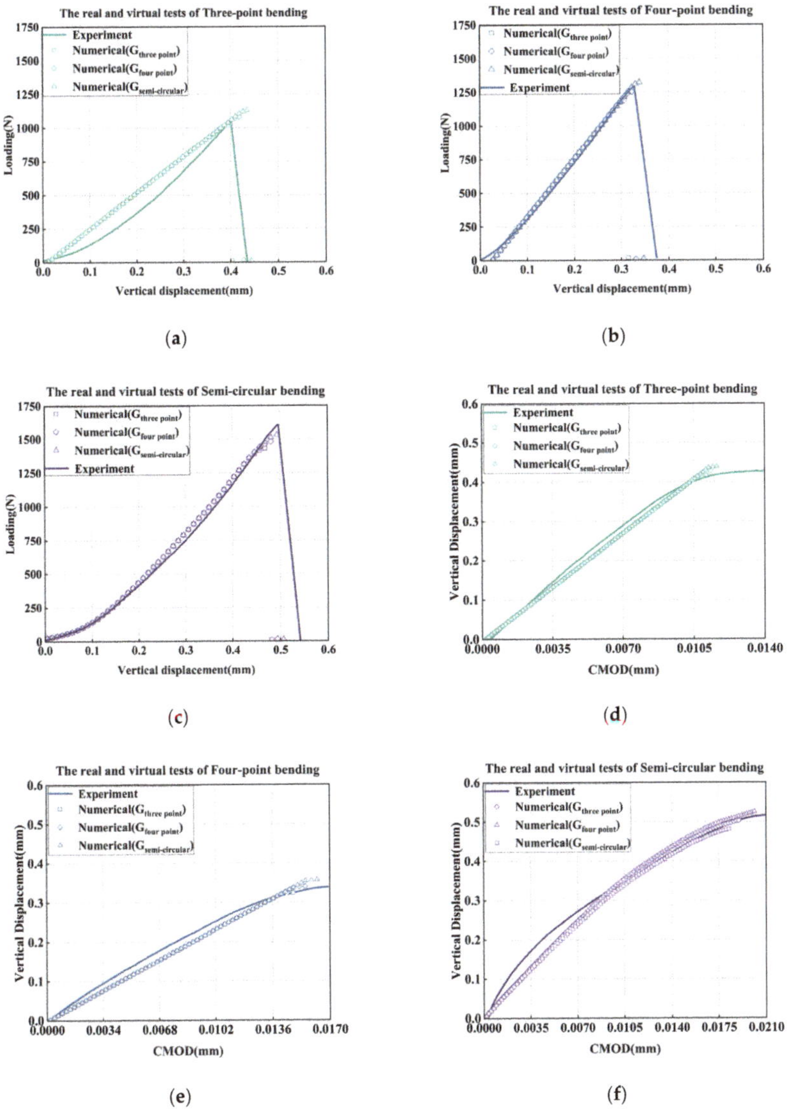

Figure 9. The comparison between simulation results and experimental results under different fracture toughness conditions: (**a**) Load–displacement curve of Three-point bending test; (**b**) Load–displacement curve of Four-point bending test; (**c**) Load–displacement curve of Semi-circular bending test; (**d**) Displacement–CMOD curve of Three-point bending test; (**e**) Displacement–CMOD curve of Four-point bending test; (**f**) Displacement–CMOD curve of Semi-circular bending test.

Figure 9d–f show the comparison between the simulated results and the actual experimental results of the vertical displacement–CMOD curve. The curves under different fracture energy input values exhibited a similar changing trend to the load–displacement curves, entering the failure stage successively after undergoing approximately the same trend. Among them, the simulated results of the Semi-circular bending model captured the trend of curvature change during the experimental process, corresponding to the simulated results of the load–displacement curve.

By comparing experimental and simulation results, it is observed that there are discrepancies in the rising stage between the experimental and simulated results for both the Three-point bending and Semi-circular bending tests according to the load–displacement curves or the displacement–CMOD curves. The main reason for the error is believed to be the difference between the selected constitutive model and the actual material characteristics. The ductility characteristics of the phosphogypsum-based material used in this study can be reflected in the descending section of Figure 8a. However, the bond-based brittle peridynamic constitutive model neglects the ductility of cementitious materials. Additionally, the consistency between the test results and the simulation results of the Four-point bending test suggests that the loading method of symmetrical double pressure heads can, to some extent, take the ductility characteristics of the material into consideration.

4.2.2. Quantitative Comparison between the Three Test Methods Based on Simulation Error

To quantitatively evaluate the fitting degree between the simulated curve and the actual experimental curve, this study introduces three parameters: Peak error, Difference error, and Absolute error [38]. Peak error reflects the accuracy of the value at the peak, that is, the accuracy of the value at the failure point. Difference error represents the error of the peak of the curve on the abscissa. Absolute error reflects the accuracy of the simulation results throughout the process. The results of the above three types of errors can be seen in Figure 10. The lower values of Peak error, Difference error and Absolute error indicate a better fitting degree between the simulated curve and the actual experimental curve, and vice versa.

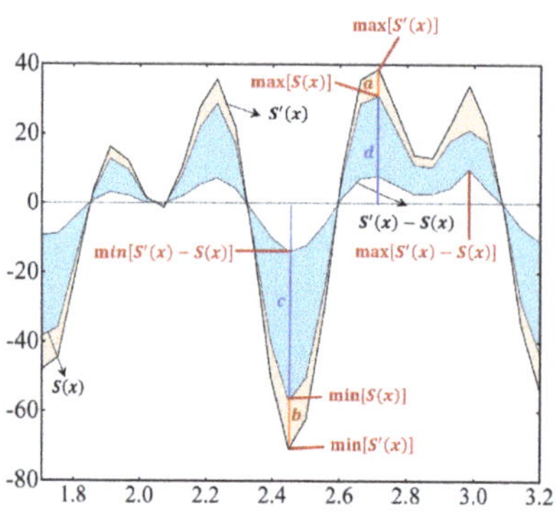

$$PeakError = \frac{a/d + b/c}{2} \qquad AbsoluteError = \frac{\square\, area}{\square + \square\, area}$$

$$DifferenceError = \frac{\max[S'(x) - S(x)]}{2d} + \frac{\min[S'(x) - S(x)]}{2c}$$

Figure 10. The schematic diagram of error definition.

Figure 11 displays the heatmap of the three parameters between the simulated results and the test results. The depth of color corresponds to the magnitude of the error value. In the coordinates, T, F, and S represent the Three-point, Four-point, and Semi-circular bending tests, respectively. For example, the coordinate (G(T), Accuracy(F)) means the error value between the simulated and experimental results of the Four-point bending test when using the fracture energy obtained from the Three-point bending test as the input into the numerical model. Most of the values on the main diagonal are the smallest in their respective rows, and the value at the center point is generally the smallest on the main diagonal. These results indicate that the fracture energy measured in each test ensures the accuracy when simulating the corresponding experiment, and the Four-point bending simulation has the highest precision. The two ends of the sub-diagonal have the largest values, implying that the fracture energy results obtained from the Three-point bending test and the Semi-circular bending test are not suitable for each other's numerical models for the materials. From the perspective of row distribution, the second row in each graph has relatively smaller values compared to the other rows, suggesting that the fracture energy measured from the Four-point bending test can achieve relatively high simulation accuracy for all test types. These facts preliminarily indicate that using the fracture energy obtained from the Four-point bending test as the input for the numerical models of different types of tests generally leads to relatively higher simulation accuracy. To make a more informed selection of the optimal experimental plan, this study supplemented three additional judgment criteria.

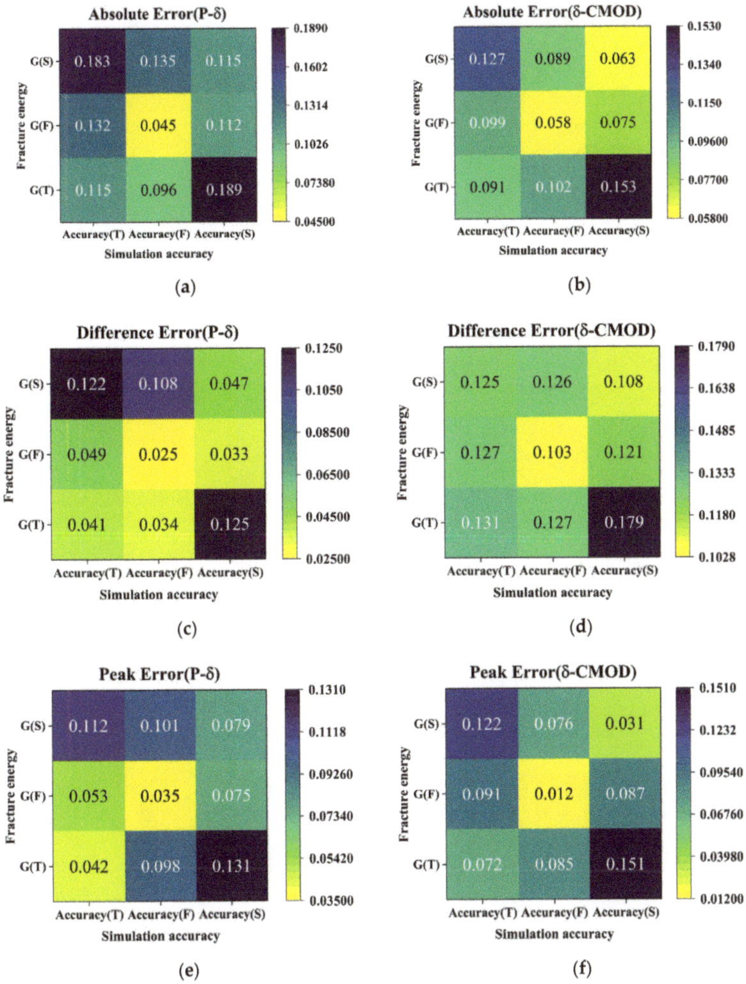

Figure 11. Error between simulated and experimental results: (**a**) Absolute Error of the P-δ Curve; (**b**) Absolute Error of the δ–CMOD Curve; (**c**) Difference Error of the P-δ Curve; (**d**) Difference Error of the δ–CMOD Curve; (**e**) Peak Error of the P-δ Curve; (**f**) Peak Error of the δ–CMOD Curve.

The rationale for selecting the optimal experimental plan for obtaining fracture energy is as follows: The fracture energy data should not only serve the numerical model of the bending test, from which the data are obtained effectively, but also be applicable to other bending test numerical models. Additionally, when this fracture energy data are used as input, the difference in simulation accuracy between the numerical models of different types of tests should not be too large. Otherwise, it can be inferred that the fracture energy data are not capable of serving numerical models of different types of bending tests. Based on this premise, this study relies on three indicators which were used for characterizing the simulation error to select the best experimental plan.

Indicator 1 represents the simulation error between the test results and the numerical model results, where the input fracture energy is obtained from the corresponding test. Indicator 2 indicates the average of two simulation errors between the test results and the numerical model results, where the input fracture energy is obtained from the other two tests. Indicator 3 represents the discrepancy of three simulation errors between the test results and the numerical model results, where the input fracture energy is obtained from the three tests.

Figure 12 summarizes the ranking sequences for all indicators. It is obviously observed that the fracture energy of the Four-point bending test generally resulted in the smallest values for indicators 1, 2, 3 of the three errors among the three tests. Additionally, the fracture energy of the Three-point bending test resulted in smallest value for Indicator 3 of Peak error. The above facts demonstrate that the Four-point bending test is the most suitable test for fracture energy characterization.

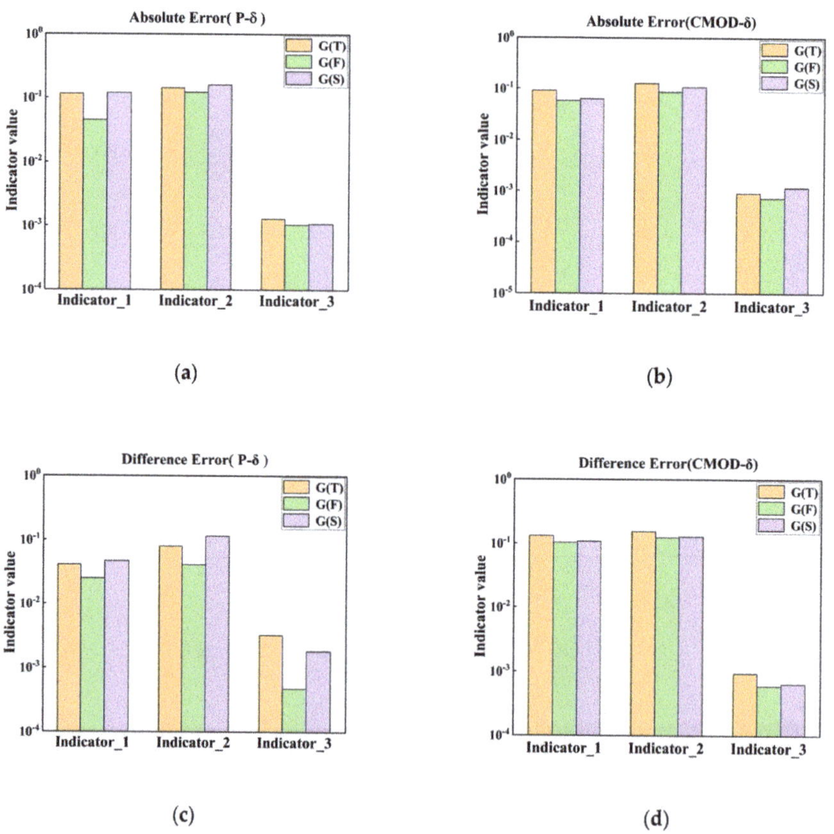

Figure 12. *Cont.*

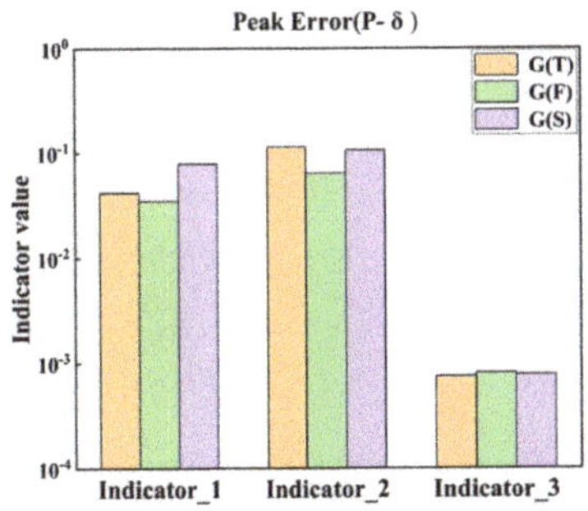

(e)

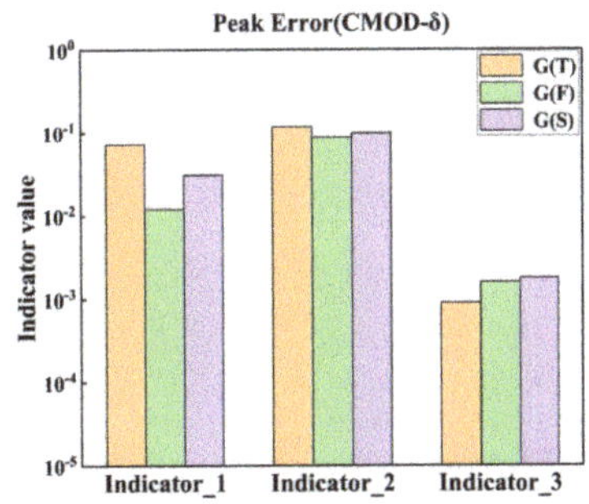

(f)

Figure 12. Summary of indicators: (**a**,**c**,**e**) Three Indicators calculated from the error values of the p-δ curves. (**b**,**d**,**f**) Three Indicators calculated from the error values of the CMOD-δ curves.

5. Conclusions

This study aims to analyze and compare the three bending tests based on the peridynamic theory and the numerical simulation method. According to the attained results, the conclusions are drawn as follows:

1. Regarding the phosphogypsum-based materials, there were significant differences in the load–displacement curves, displacement–CMOD curves and fracture energy between different bending tests, resulting in a variation of approximately 10–12% in the obtained material fracture energy.
2. The simulated load–displacement and displacement–CMOD values were consistent with the test results based on the elastic–brittle PD model. And the results of load–displacement and fracture energy of the three tests were in good agreement with that of the displacement–CMOD of the three tests.
3. The three error and indicator analyses indicate that the Four-point bending test is the best test method for fracture energy evaluation, followed by the Three-point bending test and the Semi-circular bending test.

Given that road base materials are commonly assumed to exhibit elastic–brittle behavior, this study adopts a peridynamic elastic–brittle constitutive model. It is worth noting that different conclusions may arise when employing alternative constitutive models. Moreover, peridynamic theory has also developed a criterion that controls the entire process of material damage based on tensile strength. Future research based on this criterion is recommended to investigate the optimal methods for tensile strength evaluation.

Author Contributions: Conceptualization, H.M.; methodology, H.M. and S.L.; formal analysis, H.M., S.L. and X.Z.; investigation, H.M. and K.Z.; resources, K.Z.; data curation, H.M.; writing—original draft preparation, H.M.; writing—review and editing, K.Z., S.L., J.D., X.F. and X.Z.; supervision, X.Z. All authors have read and agreed to the published version of the manuscript.

Funding: This research was funded by the National Key R&D Program of China (Grant No. 2022YFB2602603), the Science and Technology Project of Jiangxi Provincial Department of Transportation of China (Project No. 2021C0008) and the Gaopo Talents Support Program High-level and High-skill Leading Talents Cultivation Project.

Data Availability Statement: The original contributions presented in the study are included in the article. Further inquiries can be directed to the corresponding author.

Conflicts of Interest: Authors Kai Zhang and Xiangyang Fan were employed by the company Jiangxi Transportation Science Research Institute Co., Ltd. Author Sheng Liang was employed by Shenzhen Road & Bridge Group Co., Ltd. The remaining authors declare that the research was conducted in the absence of any commercial or financial relationships that could be construed as a potential conflict of interest.

References

1. Shen, W.; Zhou, M.; Zhao, Q. Study on lime–fly ash–phosphogypsum binder. *Constr. Build. Mater.* **2007**, *21*, 1480–1485. [CrossRef]
2. Liu, G.; He, M.; Chen, H.; Rashad, A.M.; Liang, Y. Study on the curing conditions on the physico-mechanical and environmental performance of phosphogypsum-based artificial aggregates. *Constr. Build. Mater.* **2024**, *415*, 135030. [CrossRef]
3. Amrani, M.; Taha, Y.; Kchikach, A.; Benzaazoua, M.; Hakkou, R. Phosphogypsum recycling: New horizons for a more sustainable road material application—ScienceDirect. *J. Build. Eng.* **2020**, *30*, 101267. [CrossRef]
4. Shen, W.; Zhou, M.; Ma, W.; Hu, J.; Cai, Z. Investigation on the application of steel slag–fly ash–phosphogypsum solidified material as road base material. *J. Hazard. Mater.* **2009**, *164*, 99–104. [CrossRef] [PubMed]
5. Liu, D.S.; Wang, C.Q.; Mei, X.D.; Zhang, C. An effective treatment method for phosphogypsum. *Mater. Rep.* **2019**, *26*, 30533–30539. [CrossRef] [PubMed]
6. Gui-jua, Z. Discussion on shrinkage crack and crack resistance of semi-rigid sub-base material. *Shanxi Archit.* **2007**, *6*, 169–170.
7. Li, Z.; Guo, T.; Chen, Y.; Wang, Y.; Chen, Y.; He, Q.; Yang, X.; Wang, J. Study on Performance of Retarded Composite Semi-Rigid Base Mixed with Rubber Powder. *Materials* **2022**, *15*, 4683. [CrossRef] [PubMed]
8. Khalilpour, S.; BaniAsad, E.; Dehestani, M. A review on concrete fracture energy and effective parameters. *Cem. Concr. Res.* **2019**, *120*, 294–321. [CrossRef]
9. Xu, M.; Wille, K. Fracture energy of UHP-FRC under direct tensile loading applied at low strain rates. *Compos. Part B Eng.* **2015**, *80*, 116–125. [CrossRef]
10. Kim, Y.R.; Wen, H. Fracture energy from indirect tension testing. *Asph. Paving Technol.* **2002**, *71*, 779–793.
11. Fei, C.; Linyan, Z.; Jiliang, F.; Yong, M.; Yanbin, Z. Research Progress on Test Methods of Asphalt Mixture's Low-temperature Anti-cracking Performance. *Mater. Rep.* **2021**, *35*, 127–137.
12. Guinea, G.; Planas, J.; Elices, M. Measurement of the fracture energy using three-point bend tests: Part 1—Influence of experimental procedures. *Mater. Struct.* **1992**, *25*, 212–218. [CrossRef]
13. Bencardino, F.; Rizzuti, L.; Spadea, G.; Swamy, R.N. Implications of test methodology on post-cracking and fracture behaviour of steel fibre reinforced concrete. *Compos. Part B Eng.* **2013**, *46*, 31–38. [CrossRef]
14. Cooper, G. Optimization of the three-point bend test for fracture energy measurement. *J. Mater. Sci.* **1977**, *12*, 277–289. [CrossRef]
15. Zhang, X. Investigating the mechanical properties of the fracture process zone of concrete using the energy method. *Eng. Mech.* **2008**, *25*, 18–23. (In Chinese)
16. Zhang, J. *Investigation on the Fracture Energy of Cement-Based Materials Based on the Analysis of Energy Dissipation D*; Institue of Engineering Mechanics, CEA: Beijing, China, 2022. [CrossRef]
17. Bermudo Gamboa, C.; Andersson, T.; Svensson, D.; Trujillo Vilches, F.J.; Martín-Béjar, S.; Sevilla Hurtado, L. Modeling of the fracture energy on the finite element simulation in Ti6Al4V alloy machining. *Sci. Rep.* **2021**, *11*, 18490. [CrossRef] [PubMed]
18. Khennane, A. *Introduction to Finite Element Analysis Using MATLAB and Abaqus*; CRC Press: Boca Raton, FL, USA, 2013.
19. Gu, Y.; Zhang, C.; Zhang, P.; Golub, M.V.; Yu, B. Enriched physics-informed neural networks for 2D in-plane crack analysis: Theory and MATLAB code. *Int. J. Solids Struct.* **2023**, *276*, 112321. [CrossRef]
20. Friedrich, L.F.; Colpo, A.B.; Kosteski, L.E.; Vantadori, S.; Iturrioz, I. A novel peridynamic approach for fracture analysis of quasi-brittle materials. *Int. J. Mech. Sci.* **2022**, *227*, 107445. [CrossRef]
21. Silling, S.A.; Epton, M.; Weckner, O.; Xu, J.; Askari, E. Peridynamic states and constitutive modeling. *J. Elast.* **2007**, *88*, 151–184. [CrossRef]
22. Mitchell, J.A. *A Non-Local, Ordinary-State-Based Viscoelasticity Model for Peridynamics*; Sandia National Laboratories (SNL): Albuquerque, NM, USA; Livermore, CA, USA, 2011.
23. Mitchell, J.A. *A Nonlocal, Ordinary, State-Based Plasticity Model for Peridynamics*; Sandia National Laboratories (SNL): Albuquerque, NM, USA; Livermore, CA, USA, 2011.
24. Sanfilippo, D.; Ghiassi, B.; Alexiadis, A. Peridynamic modelling and simulation of asphalt at low and high temperature. *Constr. Build. Mater.* **2023**, *367*, 130170. [CrossRef]
25. Huang, D.; Lu, G.; Liu, Y. Nonlocal Peridynamic Modeling and Simulation on Crack Propagation in Concrete Structures. *Math. Probl. Eng.* **2015**, *2015*, 858723. [CrossRef]
26. Keshavarzi, M.M.; Gilaki, M.; Sahraei, E. Characterization of in-situ material properties of pouch lithium-ion batteries in tension from three-point bending tests. *Int. J. Mech. Sci.* **2022**, *219*, 107090. [CrossRef]
27. Zhang, Y.; Zhang, S.; Deng, M. Four-point bending tests of ECC: Mechanical response and toughness evaluation. *Case Stud. Constr. Mater.* **2022**, *17*, e01573. [CrossRef]
28. Meng, Y.; Kong, W.; Gou, C.; Deng, S.; Hu, Y.; Chen, J.; Fan, L. A review on evaluation of crack resistance of asphalt mixture by semi-circular bending test. *J. Road Eng.* **2023**, *3*, 87–97. [CrossRef]
29. Gerstle, W.; Sau, N.; Aguilera, E. Micropolar peridynamic constitutive model for concrete. *J. Trans.* **2007**, 1–8.
30. Yaghoobi, A.; Chorzepa, M.G. Fracture analysis of fiber reinforced concrete structures in the micropolar peridynamic analysis framework. *Eng. Fract. Mech.* **2017**, *169*, 238–250. [CrossRef]

31. Wang, X.; Zhong, Y. Reflective crack in semi-rigid base asphalt pavement under temperature-traffic coupled dynamics using XFEM. *Constr. Build. Mater.* **2019**, *214*, 280–289. [CrossRef]
32. Huichao, Y. Research on Damage Mechanism and Its Identification Method of Crane Girder Based on Peridynamics. Ph.D. Thesis, School of Mechanical Engineering Southeast University, Nanjing, China, 2021. [CrossRef]
33. Silling, S.A.; Askari, E. A meshfree method based on the peridynamic model of solid mechanics. *Comput. Struct.* **2005**, *83*, 1526–1535. [CrossRef]
34. Lee, J.; Hong, J.W. Dynamic crack branching and curving in brittle polymers. *Int. J. Solids Struct.* **2016**, *100*, 332–340. [CrossRef]
35. Miura, D.; Ishida, Y.; Miyasaka, T.; Aoki, H.; Shinya, A. Reliability of different bending test methods for dental press ceramics. *Materials* **2020**, *13*, 5162. [CrossRef] [PubMed]
36. *JTG 3441-2024*; Test Methods of Materials Stabilized with Inorganic Binders for Highway Engineering. Ministry of Transport of the People's Republic of China: Beijing, China, 2024.
37. Zhao, G.; Yao, W.; Li, X.; Xu, Y.; Xia, K.; Chen, R. Influence of notch geometry on the rock fracture toughness measurement using the ISRM suggested semi-circular bend (SCB) method. *Rock Mech. Rock Eng.* **2022**, *55*, 2239–2253. [CrossRef]
38. Yingjie, Z. A Study on Integral Algorithm for Acceleration Test to Get Displacement and Application. Master's Thesis, Chongqing University, Chongqing, China, 2013.

buildings

MDPI

Article

Gypsum-Enhanced Red Mud Composites: A Study on Strength, Durability, and Leaching Characteristics

Shiying Yan, Yu Cheng, Wentong Wang *, Lu Jin and Ziyi Ding

College of Transportation, Shandong University of Science and Technology, Qingdao 266510, China; yanshiying@sdust.edu.cn (S.Y.); skd996718@sdust.edu.cn (Y.C.); skd991767@sdust.edu.cn (L.J.); dzy@sdust.edu.cn (Z.D.)
* Correspondence: wwt@sdust.edu.cn

Abstract: The strong alkalinity of red mud and the heavy metals it contains pose a serious threat to the environment. This study investigated the possibility of applying red mud as a solid waste material in road construction to mitigate the problem of red mud accumulation. Red mud was modified using titanium gypsum and phosphogypsum as curing agents. The effects of varying gypsum types and mixing ratios on the mechanical properties and heavy metal leaching of the resulting red mud-based materials were assessed using percussion tests, unconfined compressive strength measurements, scanning electron microscopy (SEM), and continuous heavy metal leaching tests. The results showed that the optimal moisture content for titanium gypsum–cement-stabilized red mud (RTC) exceeds that of phosphogypsum–cement-stabilized red mud (RPC), with RTC exhibiting a lower maximum dry density compared to RPC. When the gypsum admixture was within 10%, the strength of the RPC was higher than that of the RTC at the same and curing time. The reticulation in RPC-10 was denser. The cumulative heavy metal releases from both RTC and RPC were within the permissible limits for Class III groundwater discharge standards. Based on the comprehensive test results, RPC is identified as the superior modified red mud material, with an optimal mix ratio of red mud/phosphogypsum/cement of 87:5:8.

Keywords: red mud; gypsum; mechanical properties; microstructure; heavy metal leaching

1. Introduction

The rapid development of transportation construction in China has led to an increasing demand for roadbed filling materials in recent years. Since traditional roadbed materials consume a large amount of non-renewable soil resources, research on the use of solid wastes as roadbed fill materials has attracted widespread attention. China is a large alumina-producing country, and for every 1 ton of alumina produced, approximately 0.6 to 2.0 tons of red mud is generated [1]. The annual discharge of red mud is close to 150 million tons, and the cumulative discharge has exceeded 1.3 billion tons, making red mud the largest pollutant discharged from alumina plants. At present, the global red mud reserves are estimated to have exceeded 3 billion tons, the average utilization rate of red mud in the world is 15%, and the utilization rate of red mud in China is only 4% [2]. Red mud is highly productive, strongly alkaline, has a high risk of pollutant migration, and contains radioactive and toxic substances, and its massive accumulation not only occupies land resources but also leads to serious social and ecological environmental problems [3–5]. Therefore, there is an urgent need to find a suitable treatment for red mud, which can be solidified and used as roadbed filler, to realize the sustainable development of the alumina industry and reduce the impact of red mud on the socio-economic and ecological environment.

At present, domestic and foreign scientists' research on the resource utilization of red mud has three main directions: firstly, the mining and extraction of valuable metals; secondly, the production of building materials such as building bricks; and thirdly, the use of red mud as an adsorbent for wastewater treatment, especially for radioactive metal ions [6–9]. There are also some studies on the use of red mud as an adsorbent for heavy metal ions to treat soils contaminated with heavy metals. Xie et al. [10] investigated the stabilizing and curing effects of a red mud–fly ash composite curing agent on Cu^{2+} containing loess, and the results showed that this composite curing agent was able to rapidly adsorb Cu^{2+}, and at the same time, it enabled the Cu^{2+} to form $Cu(OH)_2$ precipitation in an alkaline environment, which effectively inhibited the migration of Cu^{2+}. Santona et al. [11] and Liang et al. [12] also stated in their studies that red mud can effectively reduce the solubility of heavy metals in contaminated soils. In addition, Sutar et al. [13] stated in their study that the application of red mud in construction materials, pollution control, and metal recycling all require quite high conditions and need to consider the risk of introducing new pollutants; therefore, more in-depth studies on red mud are needed, as well as a comprehensive assessment of the chemical and biological impacts it produces. Sarath et al. [14] replaced red mud with fly ash added gypsum to the mix and set up a mix ratio for strength and heavy metal leaching tests, which illustrated that fly ash and gypsum can significantly increase the strength of red mud and reduce the leaching of most heavy metals. Li et al. [15] stated that the degradation of bonding behavior due to the chloride-induced corrosion of steel strands is critical for the serviceability of prestressed concrete structures. Zhang et al. [16] studied the leaching of heavy metals from fly ash-based cementitious materials containing arsenic. Tests have shown that the solidification/stabilization technology for municipal solid waste incineration can effectively reduce the leaching of heavy metals from most fly ash-based cementitious materials containing arsenic.

The physical properties and composition of red mud are similar to those of soil, and the use of cured red mud as a roadbed filler can provide an effective way for the comprehensive utilization of red mud. Li et al. [17] investigated the synergistic effects of industrial solid wastes in ternary cementless composites based on red mud, ultra-fine fly ash, and ground granulated blast-furnace slag. The results showed that the addition of RM to the cementless composite led to reduced fluidity and compressive strength, as well as prolonged setting times. Deelwal [18] showed that red mud is denser and stronger compared to soil, and the cohesion of red mud is higher than that of conventional clay materials, so it can be used as a geotechnical material such as backfill, road base material, and embankment material, and it can be further stabilized by using lime, gypsum, and fly ash to improve its strength. Shi et al. [19] and Yu [20] verified the feasibility of cement–lime–phosphogypsum-cured red mud applied to road subgrade through indoor tests of road properties such as unconfined compressive strength and splitting resistance. Li [21] selected cement, lime, and phosphogypsum as the curing agents, using the indoor test and field test on the curing of red mud used as a roadbed filler. The results showed that the curing agent could improve the performance of red mud very well, but the improvement effects were greatly influenced by the admixture. The comprehensive use of cement, lime, and phosphogypsum curing of red mud is better than the curing of cement only, which can also cause excessive thermal fatigue [22]. Pu et al. [23] showed in their study that gypsum can improve the engineering utilization of soils and can be more effective when it is synergistic with cement. Li et al. [24] demonstrated experimentally that the Ca^{2+} in gypsum can react with silicate to form calcium silicate hydrate (C-S-H) or silica–aluminate groups, which improve the mechanical properties of terracotta-based grouting materials. Suo [25] and his workers used red mud and desulfurization gypsum to prepare cementitious materials, and their experiments proved that the addition of red mud and desulfurization gypsum can improve the performance of composite soils, and that the optimum ratio of desulfurization gypsum and red mud is 2:2. Zhang et al. [26] and Zhang et al. [27] investigated the mechanical properties of composite fillers using red mud and desulfurization gypsum as raw materials, and the results showed that red mud and desulfurization gypsum can promote the hydration reaction and improve the strength of the modified materials. In the above study, it can be seen that there is a great risk in using pure red mud as a roadbed filler, and because of the technology and cost, the utilization rate of red mud is very low. Therefore, there is a significant problem in terms of large-scale consumption.

In view of these, this paper selected titanium gypsum and phosphogypsum as curing agents, designed material mixing ratios, and investigated the effects of different gypsums and mixing ratios on the mechanical properties of red mud-based materials using indoor tests such as compaction tests, unconfined compressive strength, and SEM to explore the feasibility of using red mud-based materials for road filling materials. Meanwhile, since the geotechnical environmental evaluation of red mud is difficult and there is no mature research and relevant standard specification, this paper derives the data related to the leaching of heavy metal ions from red mud-based materials through the continuous heavy metal flume leaching test and evaluates the environmental safety of the groundwater of red mud used as a filling material based on the "Standard for groundwater quality" (GB/T 14848-2017) [28]. Finally, the test results were analyzed to propose the optimal material mix ratio that meets both the mechanical requirements of road subgrade materials and the requirements of groundwater standards. This provides a basis for applying red mud to road construction and solving the problem of red mud stockpiling.

2. Materials and Methods

2.1. Materials

2.1.1. Red Mud

The red mud used in the tests was selected from the Shandong Binzhou Beihai red mud dump. The original red mud was air-dried red mud, and its appearance was red lumps. Depending on the quality of the red mud required for the tests, the lumpy red mud was first air-dried, pulverized in a pulverizer, then dried at 105 °C to a constant weight and ground through a 0.075 mm square hole sieve. Its basic physical properties were investigated. The basic physical properties are shown in Table 1.

Table 1. Basic physical properties of red mud specimens.

Plastic Limit/%	Liquid Limit/%	The Natural Moisture Content/%	Optimum Moisture Content ω/%	Maximum Dry Density/$g{\cdot}cm^{-3}$	pH
27.4	17.21	35.9	24.1	1.82	10.49

The chemical composition and content in the red mud specimens from the Beihai were detected using X-ray fluorescence spectrometry, as shown in Table 2. The mineral composition of red mud mainly includes SiO_2, Fe_2O_3, $Al_2Si_2O_5(OH)_4$, $Mg_2CO_3(OH)_2(H_2O)_3$, and $AlO(OH)$.

Table 2. Main chemical composition and content of red mud.

Composition	SiO_2	Fe_2O_3	Al_2O_3	CaO	MgO	TiO_2	Na_2O	K_2O	Other
Content/wt%	28.09	25.54	19.81	1.71	0.29	2.15	10.17	0.31	11.43

2.1.2. Titanium Gypsum

The titanium gypsum required for the tests came from Jinan Yuxing Chemical Co., Ltd. (Jinan, China) and was sticky, reddish, and shaped like clay. The chemical composition and content in the titanium gypsum specimens was detected using an X-ray fluorescence spectrometer, as shown in Table 3. The mineral composition of the titanium gypsum mainly consisted of $CaSO_4$-$2H_2O$, Fe_2O_3, and SiO_2.

Table 3. Main chemical composition and content of titanium gypsum.

Composition	CaO	SO_3	Fe_2O_3	TiO_2	SiO_2	Na_2O	Al_2O_3	Other
Content/t%	27.43	34.42	9.28	5.49	4.34	1.13	1.37	0.786

2.1.3. Phosphogypsum

The phosphogypsum powder used for the tests was from Linyi Schkofen Chemical Co. Ltd., (Linyi, China) and its appearance was a light gray powder. The chemical composition and content in the phosphogypsum specimens were detected using X-ray fluorescence spectrometry, as shown in Table 4. The main mineral compositions of the phosphogypsum were $CaSO_4$-$2H_2O$, SiO_2, etc.

Table 4. Main chemical compositions and contents of phosphogypsum.

Composition	SO_3	P_2O_5	Al_2O_3	SiO_2	CaO	Other
Content/%	52.29	1.23	1.02	6.76	37.53	1.17

2.1.4. Cement

The ordinary silicate benchmark cement used in the tests was from Qufu Zhonglian Cement Co., Ltd. (Qufu, China) and had a dark gray appearance. The basic physical properties of the benchmark cement are shown in Table 5, and the chemical composition analysis results are shown in Table 6. The main mineral compositions of the cement were C_3S, C_2S, C_4AF, and CA.

Table 5. Physical properties of benchmark cement.

Fineness 0.08/% [1]	Specific Surface Area m^2/kg	Density g/cm^3	Solidification Time (min)		Flexural Strength (MPa)		Compressive Strength (MPa)	
			Initial Set	Final Set	3 days	7 days	3 days	7 days
0.4	345	3.12	95	156	6.3	7.8	28.0	39.5

[1] Percentage of sieve residue from a 0.08 mm square-hole sieve.

Table 6. Chemical composition and content of benchmark cement.

Composition	SO_2	Al_2O_3	Fe_2O_3	CaO	MgO	SO_3	Na_2O	f-CaO [1]	Cl^-	Other
Content/%	20.12	5.12	3.62	63.56	2.07	2.38	0.53	0.75	0.015	2.06

[1] Calcium oxide in a free state.

2.2. Specimen Preparation

A fixed cement admixture of 8%, along with titanium gypsum and phosphogypsum admixtures of 0%, 3%, 5%, 7%, and 10%, were used to prepare standard-sized specimens with a diameter of 39.1 mm and height of 80 mm at room temperature and pressure. The mass of the specimen was calculated according to the optimum moisture content and maximum dry density obtained from the compaction test results. The specimens were prepared according to 96% compaction. After the specimens were made, they were sealed in a Ziplock bag and placed in the curing room. The temperature of the curing room was controlled at 20 ± 2 °C, the relative humidity was controlled at 95%, and the curing ages were 7 d, 14 d, and 28 d. The specimen preparation process is shown in Figure 1. The titanium gypsum–cement-stabilized red mud matrix composites (RTC) and phosphogypsum–cement-stabilized red mud matrix composites (RPC) mix ratios are shown in Table 7.

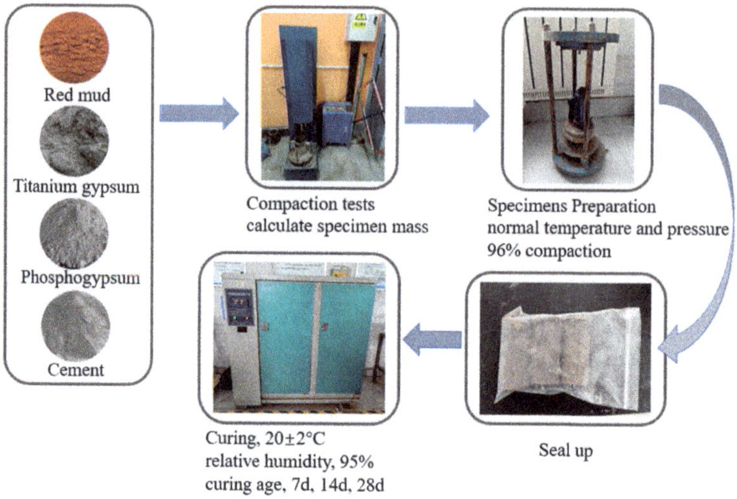

Figure 1. The specimen preparation process.

Table 7. RTC and RPC mix ratios.

Number	Admixture/%			Number	Admixture/%		
	Red Mud	Titanium Gypsum	Cement		Red Mud	Phospho Gypsum	Cement
RTC-0	92	0	8	RPC-0	92	0	8
RTC-3	89	3	8	RPC-3	89	3	8
RTC-5	87	5	8	RPC-5	87	5	8
RTC-7	85	7	8	RPC-7	85	7	8
RTC-10	82	10	8	RPC-10	82	10	8

2.3. *Experimental Methods*

2.3.1. Compaction Test

The compaction test was carried out using a standard (electric) compaction apparatus, and a compaction cylinder with an inner diameter of 10 cm and a height of 12.7 cm was selected according to the heavy compaction test method of soil in the "Test Methods of Soils for Highway Engineering" (JTG 3430-2020) [29]. The test was carried out three times, and the average value was taken as the test result.

2.3.2. Unconfined Compressive Strength

The unconfined compressive strength test was carried out based on "Test Methods of Soils for Highway Engineering" (JTG 3430-2020) [29]. The microcomputer-controlled electronic universal testing machine (WDW-S100 type) was used to control the loading speed of 1 mm/min. The specimens were prepared according to the compaction degree of 96%, and the unconfined compressive strength test was carried out when the curing age reached 7 d, 14 d, and 28 d, respectively. Three parallel specimens were set up for each group, and the average value was taken as the result of unconfined compressive strength.

2.3.3. Scanning Electron Microscopy (SEM)

The equipment used in this experiment was the ApreoSHiVac field emission scanning electron microscope. Before the experiment, the test block was knocked to the appropriate size, and a block sample with a relatively smooth surface was taken and dried to a constant weight in an oven at 105 °C. After removing the sample, it was placed in a drying tower until the sample was cooled to room temperature. The carbon conductive adhesive was first affixed to the sample stage, and then the sample was fixed to the conductive adhesive to ensure that the sample was firmly affixed to the sample stage. After fixing the sample, the sample was sprayed with gold, then the sample was observed and analyzed by selecting the appropriate magnification.

2.3.4. Continuous Flume Leaching Test

According to the block leaching test specification NEN 7375-2004 [30], a block sample of 4 cm × 4 cm × 16 cm was made using pure water in the laboratory as the leaching solution with a liquid–solid ratio of 10 L/kg. Eight consecutive cycles of leaching tests were conducted on the cement-stabilized red mud materials (RTC-0, RPC-0, and RC) and gypsum–cement-stabilized red mud matrix composites with different admixtures. At the end of each leaching stage, we measured the concentration of heavy metals in the leach solution. To minimize the experimental error, three groups of parallel tests were set up, and the test results were averaged to further analyze the leaching pattern of the heavy metals [31]. Figure 2 shows the continuous water tank leaching test setup in this paper, and Table 8 shows the leaching solution replacement time of the continuous water tank leaching test.

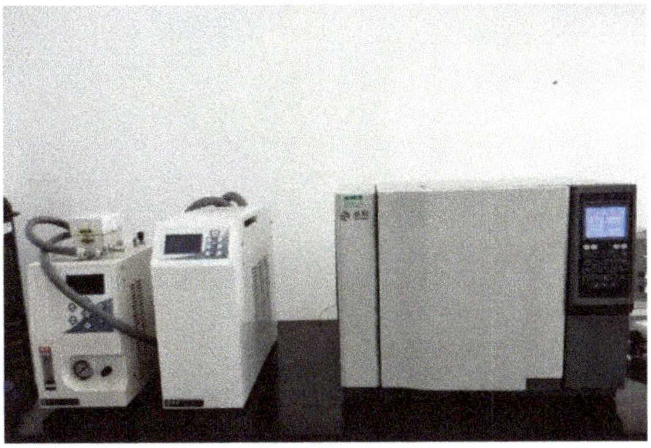

Figure 2. Continuous flume leaching test setup.

Table 8. Leach solution replacement time.

Leach Stage	Leach Solution Replacement Time	
	Cumulative Time/d	Interval Time/h
1	0.25	6
2	1	18
3	2.25	30
4	4	42
5	9	120
6	16	168
7	36	480
8	64	672

The following is the method of data processing used for the continuous flume leaching test NEN 7375-2004 [32]:

(1) The amount of a component leached E_i obtained for each stage i is shown in Equation (1):

$$E_i = \frac{V_i \times C_i}{m_d} \tag{1}$$

In the equation:

E_i: the amount of a component leached in the leaching stage i, mg/kg;
C_i: concentration of a component in the leach solution at leaching stage i, mg/L;
m_d: mass of sample dry matter in the column, kg.

(2) The leaching concentration E_i^* of a component obtained for each stage i is shown in Equation (2):

$$E_i^* = \frac{V_i \times C_i}{f \times A} \tag{2}$$

In the equation:

E_i^*: leaching concentration of a component at leaching stage i, mg/m^2;
V_i: volume of leach solution at leaching stage i, L;
C_i: concentration of a component in the leach solution at leaching stage i, μg/L;
f: a conversion factor, 1000 μg/mg;
A: surface area of the sample.

(3) The cumulative release of a component U_n is calculated as shown in Equation (3):

$$U_n = \sum_{i=1}^{8} E_i \tag{3}$$

In the equation:

U_n: cumulative leaching of a component in stages 1–8, mg/kg;
E_i: the amount of a component leached at the leaching stage i, mg/kg.

3. Results and Analysis

3.1. Comparative Analysis of Optimum Moisture Content and Maximum Dry Density

The RTC and RPC with different mixing ratios were subjected to compaction tests to obtain their optimum moisture contents and maximum dry densities. as shown in Table 9. Figure 3 shows the compaction characteristic curves of the cement-stabilized red mud-based materials with different gypsum types and different mixing ratios. Based on the figure, the optimal moisture content was less affected by the gypsum type and material ratio, and the optimal moisture content of the different gypsum-modified red mud composites was concentrated at 22~27%. In Figure 3a, the peak of the compaction curve gradually shifted to the lower right with the increase in the titanium gypsum admixture, the maximum dry density of RTC decreased with the increase in the titanium gypsum admixture by 0.11 g/cm^3, and the optimum moisture content increased with the increase in the titanium gypsum admixture by 0.87%. In Figure 3b, the peak of the compaction curve shifted irregularly with the increase in the phosphogypsum admixture. In this case, the optimum moisture content of RTC is slightly greater than that of RPC, and the maximum dry density of RTC is slightly less than that of RPC, which is caused by the fact that the water absorption of titanium gypsum is slightly greater than that of phosphogypsum.

Table 9. Compaction test results of composites with different mix ratios.

Number	Maximum Dry Density (g/cm^3)	Optimum Moisture Content/%	Number	Maximum Dry Density (g/cm^3)	Optimum Moisture Content/%
RTC-0	1.92	22.39	RPC-0	1.92	22.39
RTC-3	1.91	22.66	RPC-3	1.90	22.59
RTC-5	1.89	22.92	RPC-5	1.93	21.85
RTC-7	1.84	23.10	RPC-7	1.89	22.26
RTC-10	1.81	23.26	RPC-10	1.88	22.24

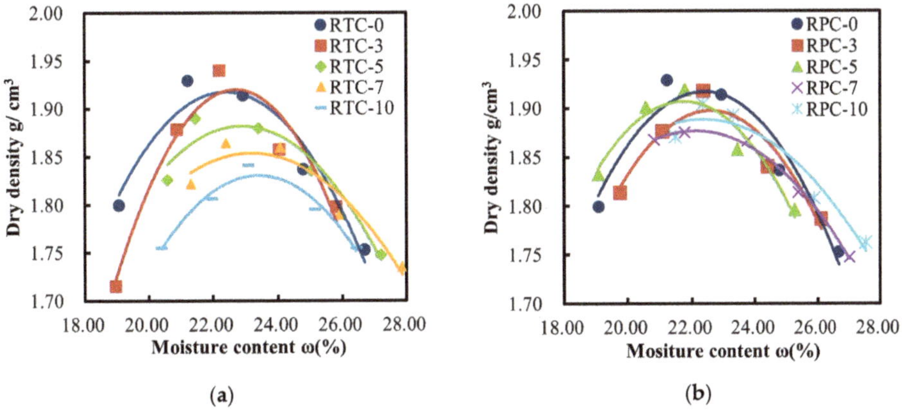

Figure 3. Compaction curves of samples with different gypsum and mix ratios. (**a**) RTC compaction curves with different mix ratios; (**b**) RPC compaction curves with different mix ratios.

3.2. Unconfined Compressive Strength Analysis

3.2.1. RTC Unconfined Compressive Strength Analysis

The results of the unconfined compressive strength tests of the RTC specimens are shown in Figure 4. The compressive strength of the RTC increased with increasing additions of titanium gypsum. When the titanium gypsum admixture was 0 wt%, there was no significant increase in the strength of the composites from 7 d to 28 d, and a decrease in strength was observed at 14 d. And, when the titanium gypsum admixture was 10 wt%, the 7 d strength reached 8.8 MPa, and the 28 d strength was able to reach 11.8 MPa. This indicates that the addition of titanium gypsum produces free Ca^{2+}, SO_4^{2-} in the mixture. The water in the red mud contained a large amount of OH^-, which can react with SiO_2 and Al_2O_3 in the red mud to generate reaction products such as calcium silicate hydrated, calcium aluminate hydrated, or calcium sulfoaluminate hydrated. It also wraps the red mud agglomerates with cement hydration products to form a more stable cemented whole [33]. The addition of titanium gypsum and cement caused the red mud-based composites to form a reticulated and stabilized structure with a combination of cementitious and crystalline associations through the processes of cement hydration and alkali excitation [34].

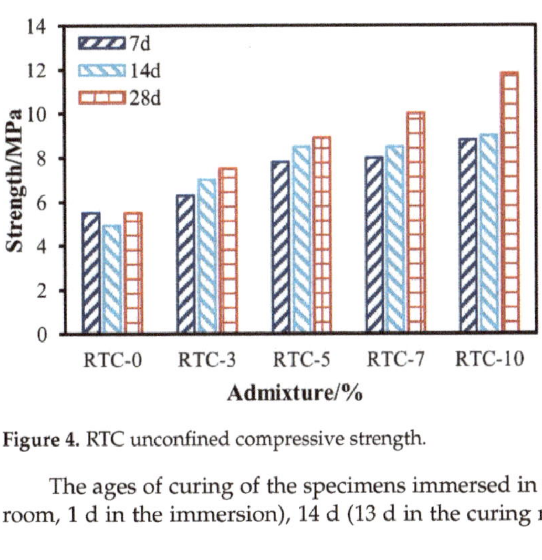

Figure 4. RTC unconfined compressive strength.

The ages of curing of the specimens immersed in water were 7 d (6 d in the curing room, 1 d in the immersion), 14 d (13 d in the curing room, 1 d in the immersion), and

28 d (27 d in the curing room, 1 d in the immersion) [35]. The unconfined compressive strength of the RTC after immersion is shown in Figure 5. A comparison of the unconfined compressive strength of the RTC after 7 d and 28 d of immersion was used to analyze the influence of the soaked conditions on the strength of the red mud-based composites. The results are shown in Figure 6.

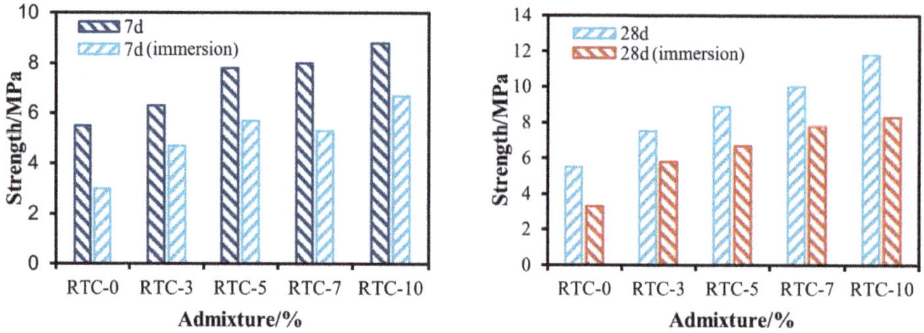

(**a**) (**b**)

Figure 5. Unconfined compressive strength after 7 d and 28 d of RTC curing. (**a**) Unconfined compressive strength after 7 d of RTC curing; (**b**) unconfined compressive strength after 28 d of RTC curing.

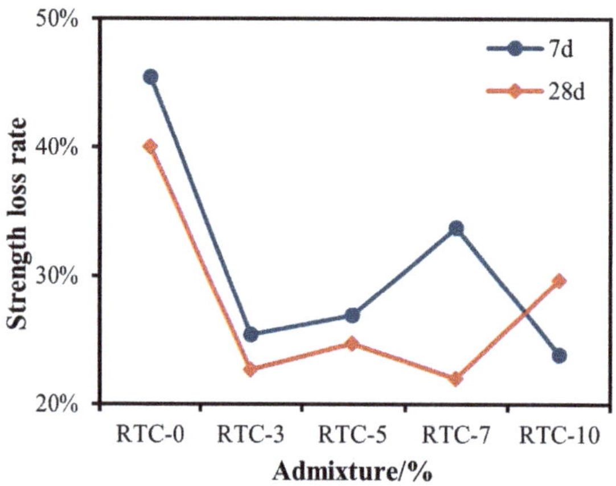

Figure 6. Rate of strength loss of RTC after immersion.

It can be found from the test results that when only cement was used to treat the red mud, the strength of the modified red mud was lost more after immersion, and the overall water stability was poor. The loss of strength reached 45% at 7 d and 40% at 28 d. In contrast, the red mud-based composites with added titanium gypsum had a relatively small loss of strength after immersion. The smallest loss was RTC-7, which was for 28 d, with a loss of 22%.

3.2.2. RPC Unconfined Compressive Strength Analysis

The results of the RPC unconfined compressive strength tests are shown in Figure 7. The admixture of phosphogypsum in the RPC had a significant effect on its strength. The strength of the modified red mud first increased and then decreased with phosphogypsum addition. The RPC-5 had the largest increase in unconfined compressive strength at 28 d, with a 125.5% increase in the 28 d strength compared to RPC-0. The modified red mud specimen with the 5 wt% phosphogypsum admixture had a maximum strength of 12.4 MPa at 28 d. This is because the Ca^{2+} and SO_4^{2-} dissolved in phosphogypsum accelerate the chemical reaction to produce a large number of ettringite with swelling properties. Ettringite fills the pores, and the mutual cross-linking between needle crystals binds many of the particles together to form a spatial structure along with hydrated calcium silicate and hydrated calcium silicate [36], which makes the structure denser and thus improves the strength of the RPC.

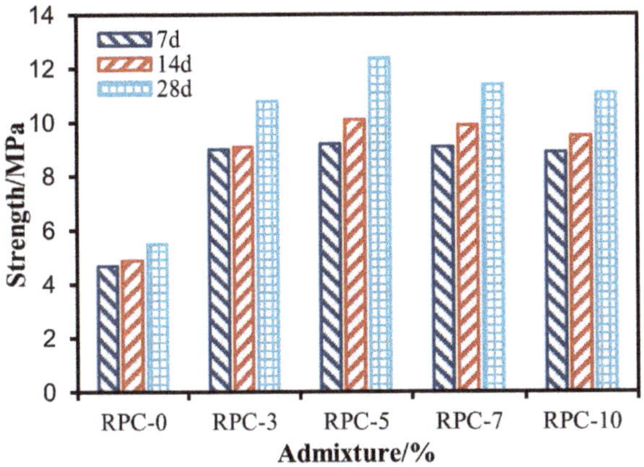

Figure 7. RPC unconfined compressive strength.

The immersion conditions for RPC were the same as those for RTC. The unconfined compressive strength of RPC after immersion is shown in Figure 8. A comparison of the unconfined compressive strength of RPC after 7 d and 28 d of immersion was used to analyze the influence of the soaked conditions on the strength of the red mud-based composites. The results are shown in Figure 9.

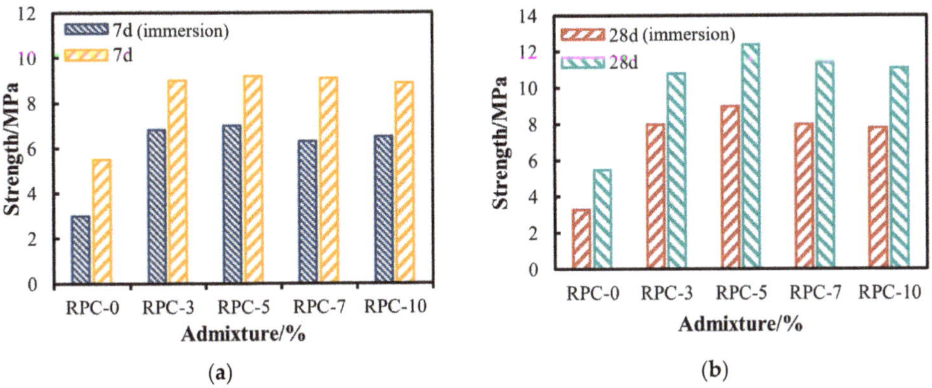

Figure 8. Unconfined compressive strength after 7 d and 28 d of RPC curing. (**a**) Unconfined compressive strength after 7 d of RPC curing; (**b**) unconfined compressive strength after 28 d of RPC curing.

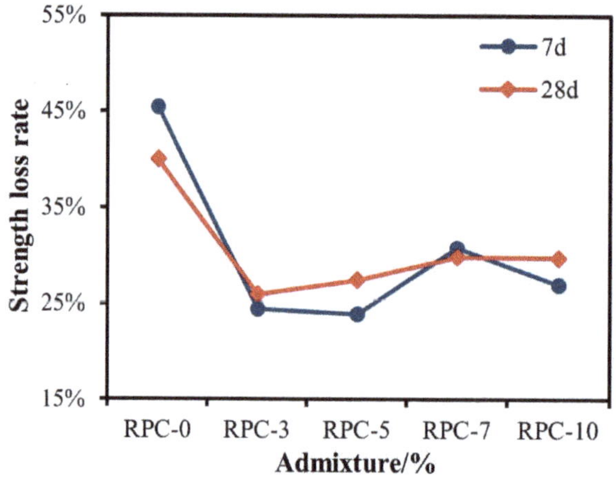

Figure 9. Rate of strength loss of RPC after immersion.

As can be seen from the figure, compared to RPC-0, the red mud-based composites with added phosphogypsum had a relatively small loss of strength after immersion. The smallest loss was RPC-5, which was for 7 d, with a loss of 24%.

3.2.3. Comparative Analysis of RTC and RPC Unconfined Compressive Strength

Figure 10 shows the variation curves of the unconfined compressive strength with the gypsum admixture after 7 d and 28 d for the modified red mud composites. The figure shows that the RTC 7 d and 28 d strengths increased. Figure 10a shows that the unconfined compressive strength of the RC was about 5.5 MPa after 7 d of curing, and the strength of the samples increased by 14.5%, 41.8%, 45.5%, and 61.6% as the titanium gypsum admixture was elevated to 3%, 5%, 7%, and 10%, respectively. When the phosphogypsum admixture was not more than 3%, the RPC strength increased significantly with the admixture, and the strength reached up to 9 MPa. The RPC strength changed slowly when the phosphogypsum admixture was more than 3%. Figure 10b shows that the RTC strength grew with the increases in the titanium gypsum admixture and reached 11.8 MPa when the admixture reached 10%, which was an increase of 114.5% compared to the RC strength. The RPC strength showed an increasing and then decreasing trend with the increase in phosphogypsum admixture, and the strength reached a maximum of 12.4 MPa when the phosphogypsum admixture was 5%, which was an increase of 125.5% compared to the RC strength. Based on the unconfined compressive strength after 7 d and 28 d, the strength of the phosphogypsum was higher than that of the titanium gypsum at the same admixture and same age, and the strength of the titanium gypsum-modified red mud exhibited an upward trend. However, when the gypsum admixture was controlled to be less than 10%, the RPC was superior to the RTC. This was due to the addition of gypsum, which promoted the hydration reaction within the matrix, generating a large number of ettringite (AFt) with calcium silicate hydrate (C-S-H) cross-growth, forming a dense net structure, and refining the internal pore size of the matrix [31]. It can be seen that the modified red mud composite with the best strength was RPC-5. The truss structure in the building was also formed in the RPC compared to the RTC, resulting in a higher strength of the specimen. This was consistent with the results of the microstructure analysis of the specimen.

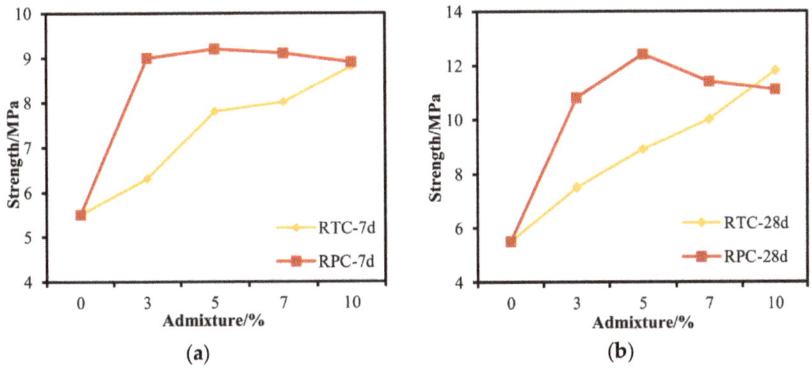

Figure 10. Variation curve of unconfined compressive strength of specimens with different gypsum admixtures. (**a**) Change curve of 7 d compressive strength with gypsum admixture; (**b**) change curve of 28 d compressive strength with gypsum admixture.

3.3. Microstructure Analysis

3.3.1. Comparative Analysis of RC and RTC Microstructures

Figure 11 shows the SEM micrographs of the specimen sections of RC specimens magnified by 2000 and 8000 times after 28 d of curing, respectively. The microstructures of the hydration products of the RC specimens at the age of 28 d consisted of part of a fine acicular material and amorphous gel, forming a spatial reticulation of the combination of colloidal–crystalline materials. However, the structure had more cracks and a certain number of unfilled pores. The fine acicular material was ettringite (AFt), and the amorphous gel was a calcium silicate hydrate (C-S-H). The C-S-H structure connected some fine particles, and the generation of these substances was the basis for the early strength development of the filling body. The overall structure of the filled body was denser, so the early strength was relatively high. With the increase in the age of curing, the hydration reaction of the cement continued, the activity of the red mud was stimulated, a small amount of AFt and a large amount of C-S-H were generated, and the crystals gradually developed [37].

Figure 11. SEM micrographs of RC at different magnifications. (**a**) SEM micrograph of RC specimen section at 2000×; (**b**) SEM micrograph of RC specimen section at 8000×.

Figure 12 shows the SEM micrographs of RTC specimens added with 5% titanium gypsum at a magnification of 2000 times and 8000 times after 28 d of curing. It can be seen that, with the addition of titanium gypsum, more particles with more complex shapes and different sizes appeared, and voids still existed between the particles. But compared with the specimens without titanium gypsum, the pores of the RTC-5 specimens became smaller at the age of 28 d, the number of cracks decreased, and finer acicular material AFt appeared. The fine acicular materials began to interweave with each other, the C-S-H gelation increased gradually, and the pores inside the specimens were gradually filled by the hydration products, which laid a structural foundation for enhancing the specimen's mechanical strength.

(a) (b)

Figure 12. SEM micrographs of RTC-5 at different magnifications. (**a**) SEM micrograph of RTC-5 specimen section at 2000×; (**b**) SEM micrograph of RTC-5 specimen section at 8000×.

Figure 13 shows the SEM micrographs of sections of the RTC specimens added with 10% titanium gypsum at magnifications of 2000 and 8000 times after 28 d of curing, respectively. As the admixture of titanium gypsum continued to increase, a coarser and wider AFt columnar structure was produced inside the matrix. The internal structure of the filled body was covered by a large amount of reticulated C-S-H, which further reduced the number of internal pores and cracks. This resulted in a dense structure, better integrity, and increased strength. This is because the addition of titanium gypsum produced a large number of free Ca^{2+} and $SO_4{}^{2-}$, and the water in the red mud contained a large number of OH^-, which could react chemically with the SiO_2 and Al_2O_3 in the red mud to generate reaction products such as calcium silicate hydrated, calcium aluminate hydrated, or calcium thioaluminate hydrated. This is the mechanism of action of the calcium-based material curing of the red mud-based composites. The reactions are as follows [33]:

$$x\mathrm{Ca(OH)_2} + \mathrm{SiO_2} + n\mathrm{H_2O} \rightarrow x\mathrm{CaO{\cdot}SiO_2}(n+1)\mathrm{H_2O} \tag{4}$$

$$x\mathrm{Ca(OH)_2} + \mathrm{Al_2O_3} + n\mathrm{H_2O} \rightarrow x\mathrm{CaO{\cdot}Al_2O_3}(n+1)\mathrm{H_2O} \tag{5}$$

This is also consistent with the strength test results.

(a) (b)

Figure 13. SEM micrographs of RTC-10 at different magnifications. (**a**) SEM micrograph of RTC-10 specimen section at 2000×; (**b**) SEM micrograph of RTC-10 specimen section at 15,000×.

3.3.2. RPC Microstructure Analysis

Figure 14 shows the SEM micrographs of sections of the RPC specimens with the 3% phosphogypsum admixture at magnifications of 4000 and 15,000 times after 28 d of curing, respectively. The figure shows that a large number of hedgehog-like crystals AFt appeared in the composites with 3% phosphogypsum added, while lumps were visible in the clusters of AFt crystals. The structures were gradually linked into a dense whole with fewer cracks under the connection of AFt. The addition of phosphogypsum produced free Ca^{2+} and $SO_4{}^{2-}$ ions in the composite material, while the moisture in the red mud contained a large amount of OH^-. With the increased pH in the moisture, they reacted chemically with the SiO_2 and Al_2O_3 in the red mud to generate reaction products such as calcium silicate hydrated, calcium aluminate hydrated, or calcium sulphoaluminate hydrated under the joint action. Together with the cement hydration products, they wrapped red mud aggregates, forming a more stable cemented whole [33].

(a) (b)

Figure 14. SEM micrographs of RPC-3 at different magnifications. (**a**) SEM micrograph of RPC-3 specimen section at 4000×; (**b**) SEM micrograph of RPC-3 specimen section at 15,000×.

Figure 15 shows the SEM micrographs of specimen sections of tbe RPC specimens with the 5% phosphogypsum admixture at magnifications of 4000 and 15,000 times after 28 d of curing. As can be seen from the figure, the increase in the phosphogypsum admixture to 5% further promoted the hydration reaction between the gelling material and the active substances in the red mud. A large amount of AFt and C-S-H grew interlaced and are tightly wrapped by the gel, forming a dense net structure, effectively refining the internal pore size and cracks of the matrix, and forming a truss structure like in the building, thus increasing the strength of the specimen. RPC-5 underwent cement hydration reaction and alkali excitation to form a stable mesh structure with a combination of cementitious and crystalline associations [31].

Figure 16 shows the SEM micrographs of specimen sections of RPC specimens with the 10% phosphogypsum admixture at magnifications of 4000 and 15,000 times after 28 d of curing. As can be seen from the figure, when the phosphogypsum admixture was increased to 10%, the pore size of the specimens increased, the number of cracks increased, and C-S-H did not form a mesh structure, which indicates that the addition of excess phosphogypsum affected the strength of the composites. This is because the mineral composition based on the Bayer method of the red mud was mainly silicoaluminate. As the reaction proceeded, a large amount of OH^- was dissolved. When the pH value rose to a certain level, the Ca^{2+} in the solution reacted chemically with active Si and Al_2O_3 to generate calcareous alumina, katenite, etc., and the Ca^{2+} concentration was gradually reduced. The concentrations of $SiO_4{}^{4-}$ and AlO^{2-} were due to the continuous depolymerization of the alumina–silicates in the raw material, which gradually increased, inhibiting the generation of AFt [38]. This corresponds to the results of the unconfined compressive strength tests.

(a) (b)

Figure 15. SEM micrographs of RPC-5 at different magnifications. (**a**) SEM micrograph of RPC-5 specimen section at 4000×; (**b**) SEM micrograph of RPC-5 specimen section at 15,000×.

(a) (b)

Figure 16. SEM micrographs of RPC-10 at different magnifications. (**a**) SEM micrograph of RPC-10 specimen section at 4000×; (**b**) SEM micrograph of RPC-10 specimen section at 15,000×.

3.4. Analysis of Heavy Metal Leaching Pattern

The results of the continuous flume leaching test after eight cycles are shown in Table 10 and Figure 17. The cumulative releases of the heavy metals As, Mo, Zn, and Cr were all larger, while the cumulative releases of Pb, Se, Ba, and Cu were smaller. However, the inhibition effects of phosphogypsum on As, Mo, and Cr leaching were better than that of the titanium gypsum, and the inhibition effects on Zn leaching were weaker than that of the titanium gypsum. When the gypsum admixture was the same at 5%, the cumulative release of As, Mo, and Cr in the RPC was 0.01655 mg/L, 0.02559 mg/L, and 0.00762 mg/L lower than that of RTC, respectively, and the cumulative release of Zn was 0.00719 mg/L higher than that of RTC. When the gypsum admixture was the same at 10%, the cumulative release of As, Mo, and Cr in the RPC was lower than that of RTC by 0.01078 mg/L, 0.05193 mg/L, and 0.01521 mg/L, respectively, and the cumulative release of Zn was higher than that of RTC by 0.01336 mg/L. The cumulative release of heavy metal elements through leaching in the RTC and RPC was found to meet the discharge standards for Class III groundwater. This is because the addition of titanium gypsum and phosphogypsum generates AFt, which is capable of interconnecting and filling the internal structure of the material so that the heavy metal elements in the material are encapsulated in the modified material in the form of complexes due to precipitation or other means [39]. However, there are long-term leaching behaviors of heavy metal elements in composites, such as Ni, Cr, Ba, As, Se, Zn, etc. The leaching behaviors of heavy metal elements can still be detected after several leaching cycles. This is because in the long-term leaching process, the gel structure is slowly eroded, and the heavy metal elements in the material diffuse into the leaching solution through the pore water. In contrast, the collodion minerals or nascent hydration products in the uneroded portion continue to encapsulate or adsorb the heavy metal elements, which continue to be present in the matrix and are not fully released [31]. This indicates that when the modified red mud material is applied to road sub-base materials, it is necessary to carry out a long-term test on the release of heavy metals to avoid the excessive release of heavy metals leading to environmental pollution.

Table 10. Results of continuous flume leaching tests.

	Group III Groundwater	RC	RTC-5	RTC-10	RPC-5	RPC-10
Be (mg/L)	≤0.0002	—	—	—	—	—
Cr (mg/L)	≤0.05	0.014419	0.02709	0.02572	0.01947	0.01051
Mn (mg/L)	≤0.1	0.00158	0.00014	—	—	—
Co (mg/L)	≤0.05	0.00020	0.00002	0.00003	0.00002	—
Ni (mg/L)	≤0.05	0.00034	0.00015	0.00039	0.00169	—
Cu (mg/L)	≤1.0	0.00273	0.00004	0.00085	0.00080	0.00044
Zn (mg/L)	≤1.0	0.02643	0.01017	0.0052	0.01736	0.01856
As (mg/L)	≤0.05	0.09978	0.04719	0.02151	0.03064	0.01073
Se (mg/L)	≤0.01	0.00078	0.00132	0.00141	0.00129	0.00133
Mo (mg/L)	≤0.1	0.05653	0.08971	0.08766	0.06412	0.03573
Cd (mg/L)	≤0.01	0.00013	—	—	—	—
Ba (mg/L)	≤1.0	0.00271	0.00144	0.00199	0.00204	0.00428
Hg (mg/L)	≤0.001	—	—	—	—	—
Pb (mg/L)	≤0.05	0.00767	—	0.00046	0.00013	0.00026

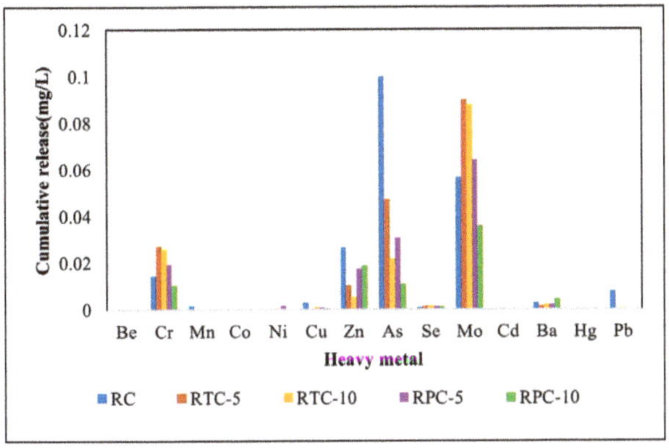

Figure 17. Cumulative heavy metal release results. Based on the above experimental results, the cumulative release pattern of the heavy metals As, Mo, and Ba from the modified red mud composites was analyzed.

3.4.1. Analysis of the Leaching Pattern of As

Figure 18 shows the cumulative release of As in the specimens with time and dosage for different gypsum admixtures. Based on the figure, the cumulative release of As reaches a steady state after the leaching time reaches 16 d. With the increase in titanium gypsum and phosphogypsum dosing, the total cumulative release of As exhibits a decreasing trend. For the same 5% admixture of titanium gypsum and phosphogypsum, the total leaching of As in the composites was 8.14 mg/m^2 and 5.07 mg/m^2, respectively, which was 8.88 mg/m^2 and 11.95 mg/m^2 less than that of the specimens without gypsum addition. For the same 10% admixture of titanium gypsum and phosphogypsum, the total leaching of As in the composites was 3.73 mg/m^2 and 1.82 mg/m^2, respectively, which was 13.29 mg/m^2 and 15.2 mg/m^2 less than that of the specimens without gypsum addition. It can be seen that phosphogypsum inhibits the leaching of As better than titanium gypsum with the same amount of gypsum admixture.

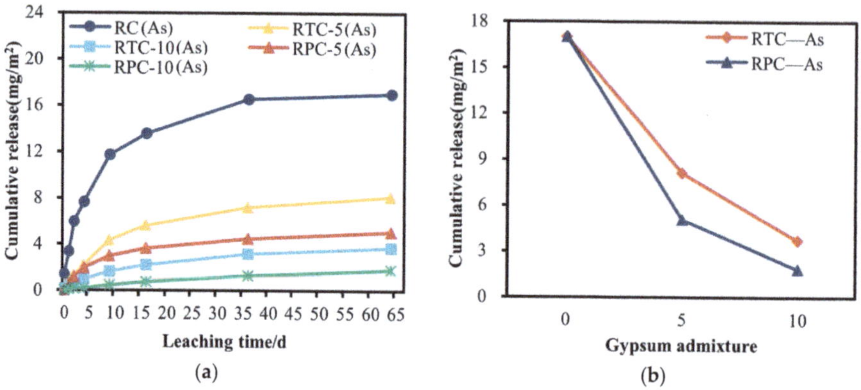

Figure 18. Variation curves of cumulative release of As with time and dosage in specimens with different gypsum admixtures. (**a**) Variations in the cumulative release of As with time; (**b**) variations in the cumulative release of As with admixture.

3.4.2. Analysis of the Leaching Pattern of Mo

Figure 19 shows the cumulative release of Mo in the specimens with time and dosage for different gypsum admixtures. Based on the figure, the cumulative release of Mo entered a slow growth phase after the leaching time reached 16 d. With increasing titanium gypsum and phosphogypsum incorporation, the overall total cumulative release of Mo both showed an increasing trend followed by a decreasing trend, but with differences. For the same 5% admixture of titanium gypsum and phosphogypsum, the total leaching of Mo in the composites was 15.74 mg/m^2 and 10.61 mg/m^2, respectively. The composites with titanium gypsum had a higher total amount of leaching than the composites with phosphogypsum by 5.13 mg/m^2. At the same 10% admixture of titanium gypsum and phosphogypsum, the total leaching of Mo in the composites was 15.20 mg/m^2 and 6.06 mg/m^2, respectively. The composites with titanium gypsum had a higher total amount of leaching than the composites with phosphogypsum by 9.14 mg/m^2. At the same dosage of 5%, both the titanium gypsum and phosphogypsum promoted the leaching of Mo in the composites. At the same dosage of 10%, the addition of titanium gypsum still promoted the leaching of Mo in the composites, while the addition of phosphogypsum inhibited the leaching of Mo in the composites.

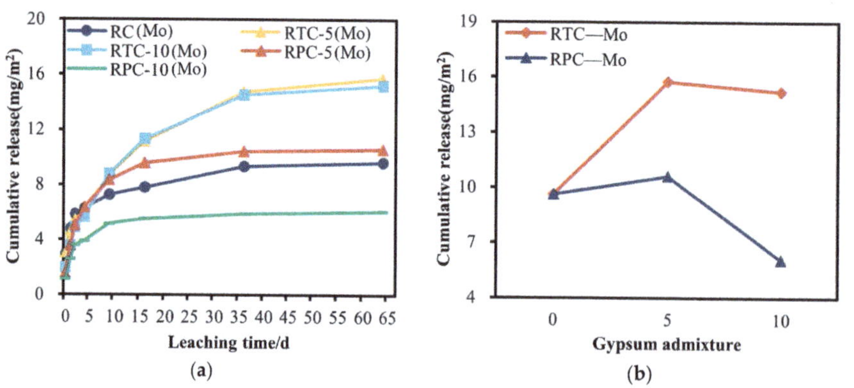

Figure 19. Variation curves of cumulative release of Mo with time and dosage in specimens with different gypsum admixtures. (**a**) Variations in the cumulative release of Mo with time; (**b**) variation in the cumulative release of Mo with admixture.

3.4.3. Analysis of the Leaching Pattern of Ba

Figure 20 shows the cumulative release of Ba in the specimens with time and dosage for different gypsum admixtures. Based on the figure, the cumulative release of Ba entered a slow growth phase after the leaching time reached 16 d. With increasing titanium gypsum and phosphogypsum incorporation, the overall total cumulative release of Ba both showed a decreasing trend followed by an increasing trend, but with differences. For the same 5% admixture of titanium gypsum and phosphogypsum, the total leaching of Ba in the composites was 0.25 mg/m^2 and 0.34 mg/m^2, respectively. The composites with titanium gypsum had a lower total amount of leaching than the composites with phosphogypsum by 0.09 mg/m^2. At the same, for the 10% admixture of titanium gypsum and phosphogypsum, the total leaching of Ba in the composites was 0.34 mg/m^2 and 0.73 mg/m^2, respectively. The composites with titanium gypsum had a lower total amount of leaching than the composites with phosphogypsum by 0.39 mg/m^2. At the same dosage of 5%, both titanium gypsum and phosphogypsum inhibited the leaching of Ba in the composites. At the same dosage of 10%, the addition of titanium gypsum still inhibited the leaching of Mo in the composites, while the addition of phosphogypsum promoted the leaching of Mo in the composites.

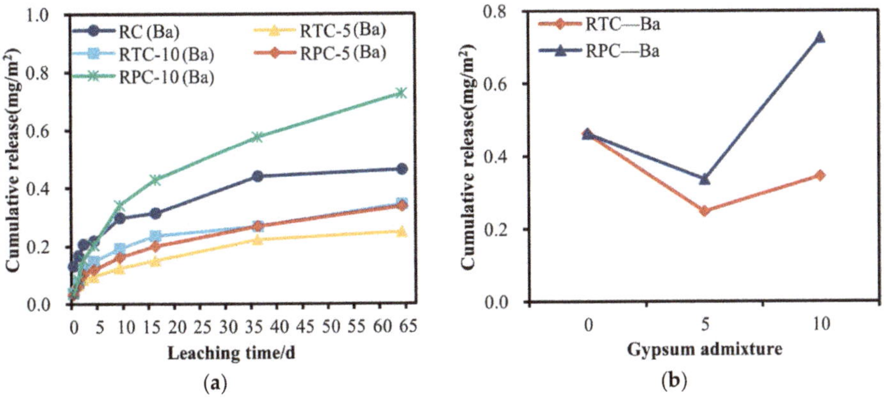

Figure 20. Variation curves of cumulative release of Ba with time and dosage in specimens with different gypsum admixtures. (**a**) Variations in the cumulative release of Ba with time; (**b**) variations in the cumulative release of Ba with admixture.

4. Conclusions and Prospects

4.1. Conclusions

In this paper, the compaction characteristics, unconfined compressive strength, microstructure, and heavy metal leaching patterns of RTC and RPC with different gypsum admixtures were investigated, and the optimal ratio of gypsum–cement-stabilized red mud composites was selected. The main conclusions are as follows:

(1) The results of the compaction test showed that there were minor differences in the optimal moisture content between RTC and RPC, while the maximum dry densities of the two were very close to each other, at 1.91 g/cm^3 and 1.93 g/cm^3, respectively.
(2) Based on the unconfined compressive strength at 7 d and 28 d, when the admixture of gypsum was controlled to be less than 10%, the RPC was better than the RTC. When the phosphogypsum admixture was 5%, the strength reached a maximum of 12.4 MPa. RPC-5 showed a 125.5% increase in strength compared to RC.
(3) Based on the 28 d microstructure of the specimens, phosphogypsum cured better at the optimum mixing ratio. Among them, RPC-5 had the best curing effect, with a denser structure with fewer pores.

(4) The cumulative releases of As, Mo, Zn, and Cr from the 8% cement-stabilized red mud material in the continuous flume leaching experiments were larger. The cumulative releases were 0.09978 mg/L, 0.05653 mg/L, 0.02643 mg/L, and 0.014419 mg/L, respectively. The two gypsum additions had the same effect on the leaching of the As and different effects on the leaching of Mo and Ba.
(5) Summarizing all the test results, the optimal modified red mud material that met both the mechanical requirements of road subgrade materials and the requirements of groundwater standards was the RPC, and the optimal mix ratio was red mud/phosphogypsum/cement = 87:5:8.

4.2. Prospects and Limitations

(1) In this paper, only the effects of two gypsums on the mechanical properties and heavy metal leaching of red mud-based composites were investigated when the cement was 8%. The effects of the two gypsums on the composites were not investigated when the cement dosage was less than 8% or more than 8%. In addition, only 96% compaction was used to prepare the specimens in this paper. The effects of other compaction levels on the material performance characteristics were not investigated.
(2) The research in this paper only carried out indoor tests, and no work was carried out on the experimental sections of the modified red mud. The large-scale application of modified red mud for road construction requires long-term stability testing of gypsum–cement-stabilized composites. The results of the heavy metal leaching tests likewise indicate the need to analyze groundwater and soil hazards through long-term monitoring of the cumulative releases of heavy metals from the materials.

Author Contributions: Conceptualization, S.Y. and Y.C.; methodology, S.Y.; software, W.W.; validation, S.Y., Y.C. and W.W.; formal analysis, S.Y.; investigation, S.Y., L.J. and Z.D.; resources, Y.C.; data curation, Y.C.; writing—original draft preparation, S.Y.; writing—review and editing, S.Y. and W.W.; visualization, W.W.; supervision, L.J. and Z.D.; project administration, L.J.; funding acquisition, Y.C. All authors have read and agreed to the published version of the manuscript.

Funding: This research was funded by the project on Green Low Carbon Design and High Resource Utilization of Concrete Materials (Funder: Y.C. Funding number: SKDHKQ20240166) and the Research on Utilization Technology of Resourceful Highway Engineering of Stone Industry Waste Sludge (Sawdust) (Funder: Y.C. Funding number: JS-22-1378).

Data Availability Statement: The raw data supporting the conclusions of this article will be made available by the authors on request.

Conflicts of Interest: The authors declare no conflicts of interest.

References

1. Editorial Board of the Dictionary of Environmental Science. *Dictionary of Environmental Science*, 2nd ed.; China Environmental Science Press: Beijing, China, 2008; pp. 60–61.
2. Zhang, Y.; Wang, Y.; Wang, K.; Dou, Z.; Liu, Y.; Zhao, Q.; Niu, L.; Zhang, Z.; Han, J.; Guo, J. A Method of Iron Extraction and Direct Cementation of High Iron Red Mud. China Patent No. 201910291219.4, 12 October 2021.
3. Liang, G.; Chen, W.; Nguyen, A.; Nguyen, T. Red mud carbonation using carbon dioxide: Effects of carbonate and calcium ions on goethite surface properties and settling. *J. Colloid Interface Sci.* **2018**, *517*, 230–238. [CrossRef] [PubMed]
4. Khairul, M.; Zanganeh, J.; Moghtaderi, B. The composition, recycling and utilisation of Bayer red mud. *Resour. Conserv. Recycl.* **2019**, *141*, 483–498. [CrossRef]
5. Wang, Y. Strength Characteristics and Micro-Mechanism of Red Mud-Based Modified Silt Loam. Master's Thesis, Zhongyuan University of Technology, Zhengzhou, China, 2022.
6. Xie, W.; Zhang, N.; Li, J.; Zhou, F.; Ma, X.; Gu, G.; Zhang, W. Optimization of condition for extraction of aluminum and iron from red mud by hydrochloric acid leaching. *Chin. J. Environ. Eng.* **2017**, *11*, 5677–5682. [CrossRef]
7. Ning, L.; He, D.; Chen, W.; Zhou, K.; Peng, C.; Zhang, X.; Teng, C. Sulfuric Acid Leaching and Kinetics Study for Separation of Iron and Scandium from Red Mud. *Min. Metall. Eng.* **2019**, *39*, 81–84+88. [CrossRef]
8. Luo, S.; Liu, M.; Yang, L.; Chang, J.; Yang, W.; Yan, X.; Yu, H.; Shen, Y. Utilization of waste from alumina industry to produce sustainable cement-based materials. *Constr. Build. Mater.* **2019**, *229*, 116795. [CrossRef]

9. Bai, B.; Bai, F.; Li, X.; Nie, Q.; Jia, X.; Wu, H. The remediation efficiency of heavy metal pollutants in water by industrial red mud particle waste. *Environ. Technol. Innov.* **2022**, *28*, 102944. [CrossRef]
10. Xie, J.; He, B.; Chen, T.; Zhang, N.; Peng, X.; Nie, X.; Ma, F. Characterization of red mud and fly ash based for co-curing of Cu^{2+}-contaminated soil. *Case Stud. Constr. Mater.* **2024**, *20*, e02769. [CrossRef]
11. Santona, L.; Castaldi, P.; Melis, P. Evaluation of the interaction mechanisms between red muds and heavy metals. *J. Hazard. Mater.* **2006**, *136*, 324–329. [CrossRef] [PubMed]
12. Liang, Z.; Peng, X.; Luan, Z. Immobilization of Cd, Zn and Pb in sewage sludge using red mud. *Environ. Earth Sci.* **2012**, *66*, 1321–1328. [CrossRef]
13. Sutar, H.; Mishra, S.C.; Sahoo, S.K.; Chakraverty, A.P.; Maharana, H.S. Progress of Red Mud Utilization: An Overview. *Am. Chem. Sci. J.* **2014**, *4*, 255–279. [CrossRef]
14. Sarath Chandra, K.; Krishnaiah, S. Strength and leaching characteristics of red mud (bauxite residue) as a geomaterial in synergy with fly ash and gypsum. *Transp. Res. Interdiscip. Perspect.* **2022**, *13*, 100566. [CrossRef]
15. Li, H.; Yang, Y.; Wang, X.; Tang, H. Effects of the position and chloride-induced corrosion of strand on bonding behavior between the steel strand and concrete. *Structures* **2023**, *58*, 105500. [CrossRef]
16. Zhang, M.; Wu, Y.; Li, Y.; Zhou, R.; Yu, H.; Zhu, X.; Quan, H.; Li, Y. Risk assessment for the long-term stability of fly ash-based cementitious material containing arsenic: Dynamic and semidynamic leaching. *Environ. Pollut.* **2024**, *345*, 123361. [CrossRef]
17. Li, Z.; Gao, M.; Lei, Z.; Tong, L.; Sun, J.; Wang, Y.; Wang, X.; Jiang, X. Ternary cementless composite based on red mud, ultra-fine fly ash, and GGBS: Synergistic utilization and geopolymerization mechanism. *Case Stud. Constr. Mater.* **2023**, *19 Pt A*, e02410. [CrossRef]
18. Deelwal, K. Evaluation of characteristic properties of red mud for possible use as a geotechnical material in civil construction. *Int. J. Adv. Eng. Technol.* **2014**, *7*, 1053–1059.
19. Shi, M.; Tian, X.; Wang, H.; Yu, C.; Du, X.; Zhang, R. Performance of Cement-Lime-Phosphogypsum Solidified Red Mud as Road Base Material. *J. Changjiang River Sci. Res. Inst.* **2024**, *41*, 114–120. [CrossRef]
20. Yu, C. Feasibility Study on Phosphogypsum Modified Cement-Lime Stabilized Red Cement Road Base Material. Master's Thesis, Southeast University, Nanjing, China, 2021.
21. Li, C. Study on the Basic Performance and Engineering Application of Bayer Red Mud Used in Roadbed Filling Applications. Master's Thesis, Shandong Jianzhu University, Jinan, China, 2018.
22. Zeng, H.; Jin, M.; Li, W.; Gao, C.; Ma, Y.; Guan, Q.; Liu, J. Performance evolution of low heat cement under thermal cycling fatigue: A comparative study with moderate heat cement and ordinary Portland cement. *Constr. Build. Mater.* **2024**, *412*, 134863. [CrossRef]
23. Pu, S.; Zhu, Z.; Huo, W. Evaluation of engineering properties and environmental effect of recycled gypsum stabilized soil in geotechnical engineering: A comprehensive review. *Resour. Conserv. Recycl.* **2021**, *174*, 105780. [CrossRef]
24. Li, Z.; Zhang, J.; Li, S.; Gao, Y.; Liu, C.; Qi, Y. Effect of different gypsums on the workability and mechanical properties of red mud-slag based grouting materials. *J. Clean. Prod.* **2020**, *245*, 118759. [CrossRef]
25. Suo, C.; Wen, H.; Cao, J.; Dong, X. Performance of a Composite Soil Prepared with Red Mud and Desulfurized Gypsum. *Ksce J. Civ. Eng.* **2022**, *26*, 47–56. [CrossRef]
26. Zhang, S.; Ding, J.; Lai, Z.; Guo, Q.; Wan, X. Mechanical and microstructural properties of silt roadbed filling improved with cement, red mud and desulfurization gypsum. *Eur. J. Environ. Civ. Eng.* **2024**, *28*, 176–196. [CrossRef]
27. Zhang, J.; Yan, G.; Bai, X.; Kong, S.; Li, J.; Li, G.; Ge, Z.; Huang, J. Resource Utilization Potential of Red Mud: A Study on the Micro-Mechanism of the Synergistic Effect of Multiple Solid Waste Filling Materials. *Sustainability* **2023**, *15*, 15532. [CrossRef]
28. *GB/T 14848-2017*; Standard for Groundwater Quality. 2nd ed. Standards Press of China: Beijing, China, 2017.
29. *JTG 3430-2020*; Test Methods of Soils for Highway Engineering. 1st ed. China Communications Press: Beijing, China, 2020; 122–127, 213–217.
30. *NEN 7375-2004*; Leaching Characteristics—Determination of the Leaching of Inorganic Components from Moulded or Monolitic Materials with a Diffusion Test—Solid Earthy and Stony Materials. Milieukwaliteit: Amsterdam, The Netherlands, 2004.
31. Huang, Z.; Su, X.; Zhang, J.; Luo, D.; Chen, Y.; Li, H. Study on leaching of heavy metals from red mud-phosphogypsum composite materials. *Inorg. Chem. Ind.* **2022**, *54*, 133–140. [CrossRef]
32. Rong, S. Study on the Leaching Behavior of Rare Metals in Red Mud under pH-Influenced Conditions. Master's Thesis, Southwest Jiaotong University, Chengdu, China, 2020.
33. Sun, Z. Study on the Engineering Technology and Environmental Impacts of the Bayer Method of Filling Roadbeds with Red Mud. Master's Thesis, Shandong University, Jinan, China, 2017.
34. Li, Y. Research on Road Use Technology of Industrial Waste Red Mud. Master's Thesis, Chongqing Jiaotong University, Chongqing, China, 2021.
35. Zhang, X.; Fu, L. Application of cement stabilized aggregates in municipal projects. *Heilongjiang Sci. Technol. Inf.* **2008**, *10*, 190.
36. Zhao, Y. Experimental Study on Regeneration of Subgrade Phosphogypsum Lime Stabilized Soil. Master's Thesis, Jiangsu University, Zhenjiang, China, 2008.
37. Lu, H. Development of New Soft Soil Curing Agent and Research on Its Reaction Mechanism. Master's Thesis, Jinan University, Guangzhou, China, 2006.

38. Zhang, J.; Huang, Z.; Su, X.; Chen, Y.; Li, H.; Li, W. Effect of phosphogypsum addition on properties of red mud-phosphogypsum composites. *China Nonferrous Metall.* **2023**, *52*, 121–130. [CrossRef]
39. Cho, J.H.; Eom, Y.K.; Lee, T.G. Stabilization/solidification of mercury-contaminated waste ash using calcium sodium phosphate (CNP) and magnesium potassium phosphate (MKP) processes. *J. Hazard. Mater.* **2014**, *278*, 474–482. [CrossRef]

buildings

MDPI

Article

Investigation into the Characteristics of Expansion and Compression Deformation of Interbedded Weak Expansive Rocks in Water Immersion

Yaning Wang [1], Yuchen Li [2], Haoyu Qin [2], Yangui Zhu [1], Yibo Yao [1], Jin Jin [1], Tao Zheng [1], Qingting Qian [2] and De Chen [2,3,*]

1 China Railway First Survey and Design Institute Group Co., Ltd., Xi'an 710043, China
2 School of Civil Engineering, Southwest Jiaotong University, Chengdu 610031, China
3 Key Laboratory of High-Speed Railway Engineering of Ministry of Education, Southwest Jiaotong University, Chengdu 610031, China
* Correspondence: chende@swjtu.edu.cn

Abstract: In order to investigate the deformation characteristics of interbedded weak expansive rocks in water immersion, the sandstone–mudstone interbedded structures were taken as the object of this study. A total of 27 sets of indoor immersion tests were designed with three influencing factors, namely, the layer thickness ratios of sandstone and mudstone (1:1, 2:1, 3:1), the occurrence of the rock layers (flat, oblique, and vertical), and the overburden loadings (0 kPa, 12.5 kPa, and 25 kPa). Tests were conducted to obtain the deformation time series data of the samples during the immersion loading process. Based on this, the influence pattern of each influencing factor on the sample deformation was analyzed individually. The results show that with the increase in overburden loading and rock inclination angle, the sample develops from expansion deformation to compression deformation. Changes in the layer thickness ratio will not change the deformation trend of the sample, and the decrease in the relative mudstone content will only reduce the absolute value of the sample's expansion and compression deformation. The deformation stability rate of the sample under load is 5~7 times that under no load. The increase in layer thickness ratio and rock inclination angle will lead to different degrees of attenuation of sample expansion force in the range of 8.91~38.68% and 51.00~58.83%, respectively. The research results of this paper can provide a meaningful reference for the design and maintenance of a high-speed railway subgrade in a weak expansive rock area with an interbedded structure.

Keywords: subgrade engineering; weak expansive rocks; interbedded structure; indoor water immersion test; expansion and compression deformation

1. Introduction

Layered rock mass has obvious layered structures such as a stratification plane and a schistosity plane. As one of the special structures, interbedded rocks are characterized by the repeated alternation of two or more lithology strata due to the continuous change in the sedimentary environment, which is usually manifested as the difference in rock hardness [1]. Compared to a single homogeneous rock mass, interbedded rock mass exhibits a high degree of anisotropy [2]. Due to the change in boundary conditions such as a complex in situ stress environment, excavation, and support, and the influence of environmental factors such as rainfall, evaporation, and the freeze–thaw cycle, its deformation and failure mechanism is very complex. Geotechnical structures over the long-term serving in such complex environments face severe safety challenges. Therefore, it is necessary to explore the basic mechanical behavior and deformation failure mechanism of the interbedded rock mass. This is of great practical significance to ensure the rational design and safe service of geotechnical structures in such areas.

Scholars first focused on the study of mechanical properties and deformation mechanisms of interbedded rock mass. They are accustomed to conducting numerical simulation inversion analyses of the structural damage based on experimental data, and then introducing a structural response into the calculation model, which is compared with the field monitoring data to verify the validity of the model [3,4]. Maheshwari [5] used the elastic–viscoelastic correspondence principle to explain the long-term viscoelastic response caused by underground excavations in the layered rock mass, and proposed a simple calculation method for the deformation of the layered rock mass. Wu et al. [6] conducted different times of wet–dry cycle tests on 116 silty mudstone and silty sandstone samples. The test results reveal the strength-weakening characteristics and mechanisms of these two types of rocks. Through the established strength-weakening model, they discussed the influence of the wet–dry cycle on the strength of soft–hard interbedded rock mass. In recent years, many new techniques have been applied to the preparation and failure characterization of interbedded rock samples. Tian et al. [7] used 3D-printing sand core technology to prepare soft–hard interbedded rock mass samples with different inclination angles. The deformation difference between soft and hard rocks is verified by the captured digital images, and the anisotropic failure evolution mode of soft–hard interbedded rock mass is revealed. Luo et al. [8] explored the strain evolution and crack propagation process of soft–hard interbedded composite rock-like samples by digital image correlation (DIC) technology. A simplified mechanical model of interbedded rock mass under uniaxial compression has also been established.

The stability problem of interbedded rock slopes is very prominent. Frequent geological disasters pose a great threat to the safety of engineering and personnel [9–11]. Some scholars have obtained the engineering characteristics of the slope through on-site geological surveys. By combining indoor tests and numerical simulations, the deformation and failure mechanism of an interbedded structure slope has been explored. On this basis, the failure evolution process of the slope is divided into stages with different characteristics [12–15]. Safety evaluation is also the research focus in the field of slope engineering [16]. Wyllie and Wood [17,18] verified the feasibility of the limit equilibrium method to analyze toppling deformation by several examples of anti-dip rock slope engineering. Liu et al. [19] established an independent cantilever beam model based on the theory of fracture mechanics, and then proposed an improved calculation method for the stability of soft and hard interbedded anti-dip rock slopes. Based on the deformation compatibility of rock mass, Wang et al. [20] proposed a theoretical method for evaluating the stability of soft and hard interbedded anti-dip rock slopes, namely, DCM.

The upper and lower rock masses of the interbedded surrounding rock are uneven in hardness and poor in self-stability. The surrounding rock is easy to slide or bend along the inclined rock surface due to excavation disturbance, resulting in a bias phenomenon. Local stress concentration and excessive deformation will promote the initiation and propagation of cracks. When this is serious, it will lead to shear failure of the tunnel structure [21,22]. Previous studies have mainly focused on exploring the stress distribution characteristics and failure evolution law of tunnels in a single-lithologic layered rock mass. But a unified understanding has not yet been formed. In recent years, research results on the large deformation mechanism of soft and hard interbedded surrounding rock tunnels have been emerging. Zhao et al. [23] took the Zhengzhou–Wanzhou high-speed railway mudstone-interbedded–sandstone-bedding rock tunnel as the research object. The effects of burial depth, bedding dip angle and spacing, and groundwater on the pressure and deformation characteristics of the tunnel surrounding rock were investigated. They also revealed the failure mechanism of large-span bedding tunnel lining. Based on the field monitoring data, Liu and Chen et al. [24,25] analyzed the deformation characteristics and failure modes of the Muzhailing tunnel. They divided the evolution process of tunnel large deformation into five stages: premise, gestation, development, occurrence, and treatment.

The sandstone–mudstone interbedded structure is widely distributed all over the world, especially in the southwest, northwest, central, and southern regions of China [26]. A large number of studies have shown that mudstone and sandstone have weak expansibility and will expand and soften in water [27]. With the continuous growth of high-speed railway construction mileage in the world, the problem of subgrade uplift deformation caused by expansive rock is becoming more and more prominent. The strict requirements of track smoothness limit the popularization and application of the ballastless track in such areas. In recent years, some scholars have made useful explorations on the above issues. Dai et al. [28] obtained the expansive characteristics of red-bed mudstone by taking the subgrade of Neijiang North Railway Station as an example. They proposed a decoupling analysis method of seepage and swelling to invert and predict the uplift deformation of the subgrade. According to the field deformation monitoring results of the subgrade, Wang et al. [4] analyzed the influence of the expansion rate of the subgrade filler and mudstone foundation on the uplift response of the ballastless track. Then, they gave the prediction method of uplift based on the numerical calculation results. Some scholars have used a physical model test, an in situ test, and other methods in practice. On this basis, they explored the mechanical behavior and deformation characteristics of the subgrade in an expansive rock area under water immersion conditions [29–35].

Although there are a considerable number of research results on interbedded rock mass, they are mainly concentrated in the fields of slope and tunnel engineering. There are relatively few studies on the deformation response of a sandstone–mudstone interbedded structure, which is often encountered in the practice of high-speed railway subgrade engineering. In addition, the existing research is mostly limited to the influence of the engineering characteristics of a single expansive rock on the deformation of the subgrade [36–39]. Few studies have been reported on weak expansive rocks, which consider the complex deformation and damage mechanisms. In addition, in order to better control the parameters such as layer thickness ratio, inclination angle, and loading, most of the existing studies used similar materials to prepare samples [40,41]. Some scholars used numerical simulation or other methods [42] to analyze the deformation characteristics of interbedded rock mass. However, the results obtained by these methods are difficult to truly reflect the deformation characteristics of weak expansive rocks with an interbedded structure in a natural state.

Based on the above problems, this study took the sandstone–mudstone interbedded structures as the research object. This research utilized a self-designed expansion and compression deformation characterization test equipment to design 27 sets of indoor immersion tests. These were based on three influencing factors: the thickness ratio of sandstone and mudstone, the occurrence of rock mass, and the overburden loading. Tests were conducted to obtain the deformation time series data of the samples during the immersion loading process. Based on this, the influence pattern of each influencing factor on the sample deformation was analyzed individually. The research results are expected to provide a valuable reference for the rational design, operation, and maintenance of a ballastless track subgrade of a high-speed railway in a weak expansive rock area with an interbedded structure.

2. Experimental Scheme Design

2.1. Test Equipment

The research team developed a specialized testing apparatus to investigate the expansion and compression deformation characteristics of weak expansive rocks with a typical sandstone intercalated with mudstone interbedded structure when immersed in water. This apparatus, depicted in Figure 1, is designed to measure the expansion and compression deformation characteristics of weak expansive rocks under applied loads. The primary components of the equipment include a sample model box, a hydraulic loading system, a force sensor, and an array of displacement sensors.

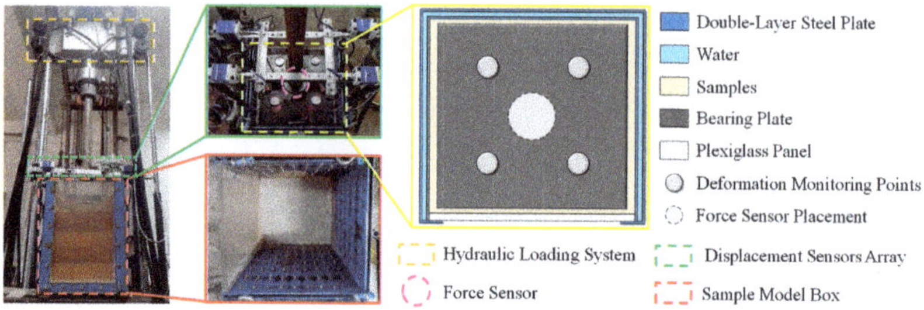

Figure 1. Expansion and compression deformation characterization test equipment.

The sample model box, constructed with reference to the dimensions of a large-scale dynamic triaxial test (300 mm × 600 mm), features an internal loading space measuring 475 mm in length, 440 mm in width, and 600 mm in height. The box is fabricated from a double-layer steel plate, welded and fitted with a high-strength transparent plexiglass panel bolted to the front side. This design allows for easy observation and documentation of the layer thickness ratio, occurrence, and deformation of sandstone and mudstone interbedded structure samples during the loading process. A 10 mm gap between the double-layer steel plates accommodates even distribution of water permeation holes on the inner steel plate's surface, promoting rapid immersion and saturation of the samples in a tri-directional water environment. Any excess water can be efficiently drained through a water valve located at the bottom rear of the sample model box.

The hydraulic loading system is an electro-hydraulic servo with a single-channel pseudo-dynamic loading capability. The actuator has a vertical loading capacity of 250 kN and a stroke of ±200 mm, enabling both static and dynamic loading. A force sensor, fabricated from alloy steel and positioned at the actuator's extremity, accurately measures the load applied. Additionally, a bearing plate, sized to match the sample model box, is placed on the sample surface to transmit and distribute the vertical load. Test monitoring data primarily include the vertical displacement of the sample surface. A 2 × 2 array of displacement sensors is utilized, with four corresponding deformation monitoring points located at the bearing plate's corners. The bearing plate's monitored vertical displacement indirectly indicates the sample's overall expansion and compression deformation values. The parameters for the laser displacement sensors are detailed in Table 1.

Table 1. Laser displacement sensors parameters.

Model Number	Measuring Center Distance	Measuring Range	Repeatability	Diameter of Light Speed	Straightness	Supply Voltage	Installation Mode
HG-C1200	200 mm	±80 mm	70 μm	About 50 μm	±0.1% F.S	24 V	Support

2.2. Experiment Method

The sandstone and mudstone samples utilized in the experiment were sourced from the section of a high-speed railway in Guangxi, which is characterized by weak expansive rock subgrades. The samples, extracted on-site, underwent processes of hammering, air-drying, crushing, and sieving before being set aside for use. Their fundamental physical properties were ascertained through laboratory testing, as detailed in Table 2.

This study focuses on the interbedded structure of sandstone and mudstone as the research subject, examining the impact of three factors: the ratio of sandstone to mudstone layer thickness, the occurrence of rock strata, and the applied overlying load. These factors are considered in relation to the expansion and compression deformation characteristics of interbedded weak expansive rocks when immersed in water. To investigate each factor, three distinct design schemes are implemented, resulting in a total of 27 experimental groups. The specific parameters for these experiments are detailed in Table 3.

Table 2. Fundamental physical properties of rock samples for testing.

Rocks Samples	Natural Water Content (%)	Natural Density (g/cm³)	Dry Density (g/cm³)	Particle Density (g/cm³)	Optimal Water Content (%)	Single-Axis Compressive Strength (MPa)		Expansion Ratio (%)	Expansion Force (MPa)	Cohesive Force (MPa)
						Natural	Drying			
Mudstone	19.17	2.07	2.47	2.77	21	0.964	20.4	1.3	0.055	11.3
Sandstone	13.19	2.30	2.60	2.69	13	0.748	6.98	0.1	0.015	30.0

Table 3. Values of experiment parameters.

Level	Investigation Factors		
	Thickness Ratio (Sandstone: Mudstone)	Occurrence	Loading
1	1:1	Flat Layer (the inclination angle is 0°)	0 kPa
2	2:1	Oblique Layer (the inclination angle is 30°)	12.5 kPa
3	3:1	Vertical Layer (the inclination angle is 90°)	25 kPa

According to the geological survey report of the sampling site area, only the rock mass in the range of atmospheric and immersion influence depth was considered, and the thickness ratio of sandstone layer to mudstone layer fluctuates within the range of 0.74:1~4.56:1. In order to facilitate the experimental design and make the test results representative [8,40,41], three levels of 1:1, 2:1 and 3:1 were selected in this study. On this basis, the influence of layer thickness ratio on the expansion and compression deformation of weak expansive rock with interbedded structure was investigated.

Referring to 'TB 10102-2023 Specification for Soil Test of Railway Engineering' [43], sandstone and mudstone samples were individually stirred and wetted at their respective optimal water contents during the experiment. The prepared loose rock samples were then transformed into interbedded structure samples using static pressure method, with the sandstone samples exhibiting a brown color and the mudstone samples a reddish-brown hue. The concrete sample filling scheme is illustrated in Figure 2.

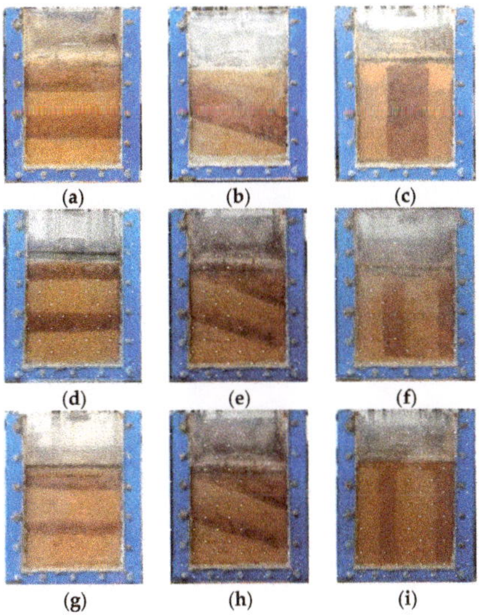

Figure 2. Sample filling scheme diagram: (a) Thickness Ratio 1:1–Flat Layer; (b) Thickness Ratio 1:1–Oblique Layer; (c) Thickness Ratio 1:1–Vertical Layer; (d) Thickness Ratio 2:1–Flat Layer; (e) Thickness Ratio 2:1–Oblique Layer; (f) Thickness Ratio 2:1–Vertical Layer; (g) Thickness Ratio 3:1–Flat Layer; (h) Thickness Ratio 3:1–Oblique Layer; (i) Thickness Ratio 3:1–Vertical Layer.

Once the sample filling is complete, water is gradually introduced into the sample model box through the water valve until the bearing plate's surface is submerged. Throughout the loading process, a thin layer of water consistently covers the surface of the bearing plate. The displacement sensor array is employed to capture the time history data of the bearing plate's vertical displacement changes under the influence of various factors. These data serve as an indicator to indirectly measure the overall expansion and compression deformation of the sample. Once the expansion and compression deformation stabilize, flooding and loading can cease. Subsequently, the sample model box's contents, including the samples and water, are removed. The next cycle of flooding and loading can then commence after the samples have been replenished according to the subsequent filling scheme.

3. Analysis of Experiment Results

3.1. Orthogonal Experimental Analysis

In light of the extensive workload and the considerable time required for comprehensive experiments, an orthogonal experiment was employed in this study to perform a preliminary analysis of the factors that influence the expansion and compression deformation characteristics of interbedded weak expansive rocks when immersed in water. An $L_9(3^4)$ orthogonal array, encompassing four factors at three levels each, was designed to acquire the expansion and compression deformation data of the sample upon reaching stability under the condition of a combination of all factors. Subsequently, a range analysis of the test data was conducted, as presented in Table 4.

Table 4. Orthogonal experiment table.

Test Number	A (Thickness Ratio)	B (Occurrence)	C (Loading)	D (Blank Column)	Expansion and Compression Deformation Value (mm)
1	1	1	1	1	2.28
2	1	2	3	2	−0.61
3	1	3	2	3	−0.22
4	2	1	3	3	−0.06
5	2	2	2	1	−0.19
6	2	3	1	2	1.12
7	3	1	2	2	0.10
8	3	2	1	3	0.99
9	3	3	3	1	−0.20
K_1	1.450	2.320	4.390	1.890	
K_2	0.870	0.190	−0.310	0.610	
K_3	0.890	0.700	−0.870	0.710	
k_1	0.483	0.773	1.463	0.630	
k_2	0.290	0.063	−0.103	0.203	
k_3	0.297	0.233	−0.290	0.237	
R	0.193	0.710	1.753	0.427	

The effect curve, as depicted in Figure 3, is derived from the data presented in Table 4.

Figure 3 clearly illustrates that the loading is the predominant factor influencing the expansion and compression deformation of sandstone intercalated with mudstone. This is followed by the occurrence and the thickness ratio. As these factors increase, the samples tend to undergo compression deformation. It is evident that as the ratio of sandstone to mudstone layer thickness increases, the mudstone content diminishes progressively. Consequently, the interior content of the hydrophilic clay minerals also decreases, which is manifested in the reduced expansion deformation of the water-saturated sample. The overlying load significantly inhibits the water-immersed expansion deformation of weak expansive rocks. Thus, with an increase in the loading, the expansion deformation of the weak expansive rocks with an interbedded structure also gradually diminishes. When the loading value escalates to match the expansive force of the sample, the expansion deformation reaches zero; if the loading continues to rise, the sample undergoes compressive deformation. Among these nine sets of experiments, the minimum value for deformation is −0.61 mm (indicating compression deformation), while the maximum value is 2.28 mm (indicating expansion deformation), with a minimum absolute value of −0.06 mm. The factor combinations corresponding to these three groups are A1B2C3, A1B1C1, and A2B1C3, respectively. Based solely on the result data, it is challenging to precisely determine the impact of rock occurrence on the water-immersed expansion and compression deformation of sandstone intercalated mudstone samples. However, it can be tentatively deduced that an increase in the rock inclination angle restricts the expansion deformation of weak expansive rocks, a hypothesis that requires further validation through subsequent experiments.

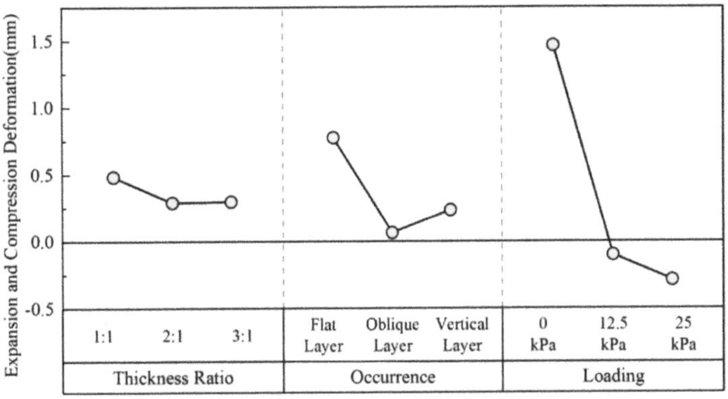

Figure 3. Orthogonal experiment effect curve.

3.2. *Expansion and Compression Deformation Trend Analysis*

To elucidate the impact of diverse factors on the expansion and compression deformation behaviors of sandstone intercalated with mudstone interbedded structure samples, three parameters—thickness ratio, occurrence, and loading—were designated as control variables. The stable expansion and compression deformation of the samples under uniform control conditions were portrayed through a 3D mapping surface, generated by the integration of these factors. This approach allowed for a visual representation of the sample's expansion and compression deformation characteristics. The ensuing analysis outcomes are delineated in the subsequent sections.

3.2.1. Thickness Ratio as Control Condition

The 3D mapping surfaces in Figure 4 illustrate the stable expansion and compression deformation of samples under combinations of factors at various thickness ratios. From this figure, it is evident that for different layer thickness ratios, the stability value of the water-immersed expansion and compression deformation of weak expansive rocks with sandstone intercalated with mudstone shows a consistent trend with changes in occurrence and loading level. This trend is characterized by significant expansion deformation when the flat layer is unloaded and substantial compression deformation when the vertical layer is subjected to 25 kPa. As the rock inclination angle increases gradually along with the loading grade, the expansion deformation of the sample decreases progressively, transitioning from expansion to compression deformation.

Overall, as the layer thickness ratio continues to rise, the absolute value of the sample's expansion and compression deformation also diminishes gradually, which is represented by a lighter color on the 3D mapping surface. Analysis indicates that an increased layer thickness ratio leads to a reduction in mudstone content within the sample. This limits the expansion deformation caused by the water absorption of hydrophilic mineral components, allowing the accumulated expansion deformation to easily dissipate along the sandstone layer. During compression deformation, the decreased mudstone content also results in fewer internal pores. Consequently, the dynamic expansion of mudstone rapidly fills these pores, hindering particle sliding and compaction. Moreover, since the sandstone is relatively dense, the overall compressible space within the sample of expansible rock with a weak interbedded structure is limited, resulting in a smaller absolute value of compression deformation.

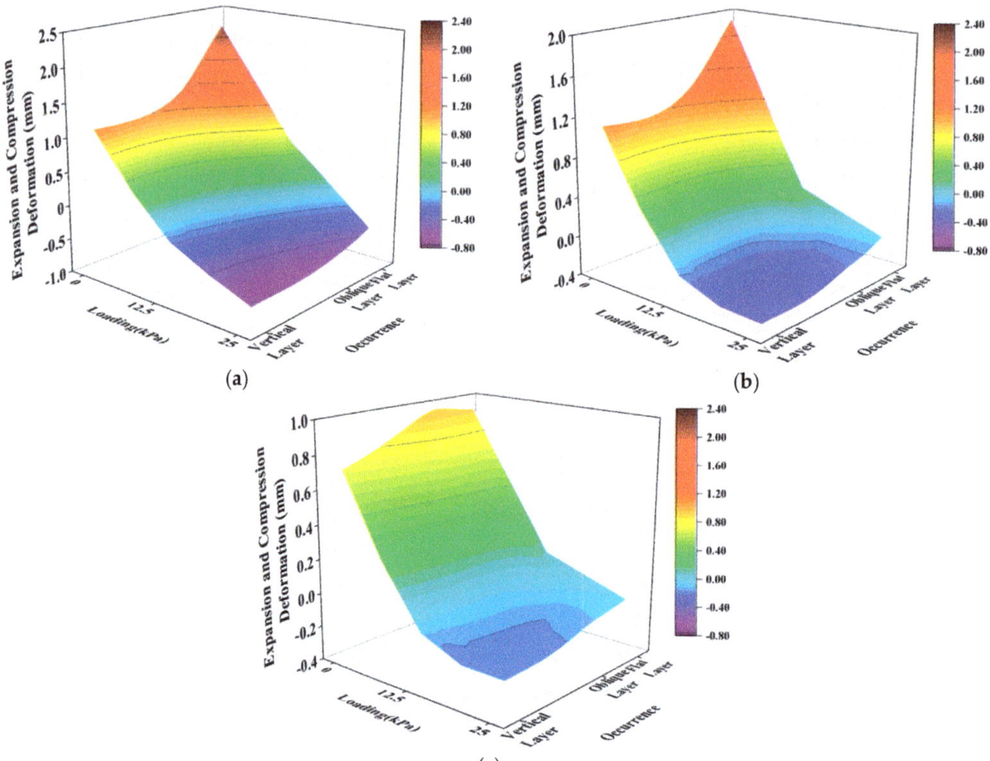

(**a**)

(**b**)

(**c**)

Figure 4. Three-dimensional mapping surfaces at different thickness ratio levels: (**a**) Thickness Ratio 1:1; (**b**) Thickness Ratio 2:1; (**c**) Thickness Ratio 3:1.

3.2.2. Occurrence as Control Condition

The 3D mapping surfaces illustrating the stable expansion and compression deformation of samples, as influenced by the combinations of various factors at different occurrence levels, are depicted in Figure 5.

Figure 5 illustrates that the alteration in rock occurrence does not influence the pattern of expansion and compression deformation characteristics for weak expansive rocks with a sandstone intercalated mudstone interbedded structure when immersed in water. In the expansive deformation segment, an increase in the layer thickness ratio results in a decrease in the value of expansion deformation. Conversely, in the compressive deformation segment, the absolute value of compression deformation for weak expansive rock samples with an interbedded structure diminishes progressively as the thickness ratio increases. The impact of the loading on sample expansion and compression deformation remains predominant, with peak values of expansion and compression deformation occurring at thickness ratios of 1:1 under 0 kPa and 1:1 under 25 kPa, respectively. The cause analysis aligns with previous discussions.

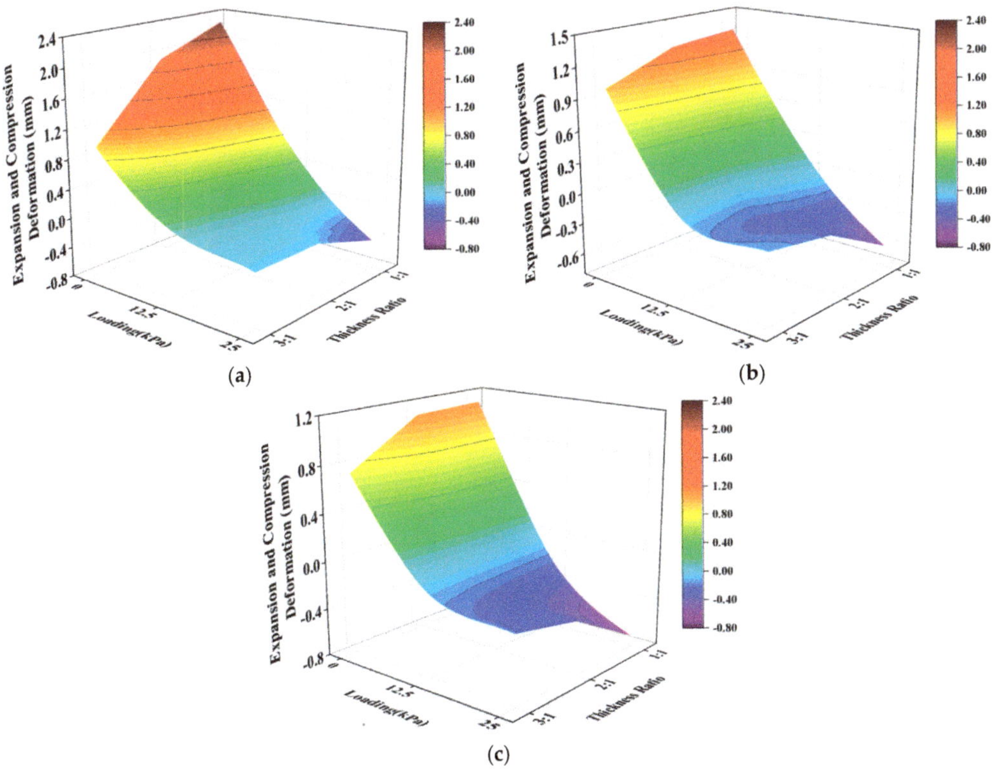

Figure 5. Three-dimensional mapping surfaces at different occurrence levels: (**a**) Flat Layer (0°); (**b**) Oblique Layer (30°); (**c**) Vertical Layer (90°).

As the inclination angle of the sandstone and mudstone interbedded structure sample increases, the overall expansion deformation of the sample progressively diminishes, while the compression deformation augments. In other words, the sample's overall deformation evolves from expansion to compression, particularly evident when the layer thickness ratio is 1:1–25 kPa. Analysis indicates that the experiment primarily gauges the surface expansion and compression deformation of the sample, constrained by lateral deformation as a whole. Given a constant layer thickness ratio—that is, constant mudstone content—the vertical deformation of the sample is dictated by the horizontal area of the sample, which correlates with the rock layer's inclination angle. A greater inclination angle results in a reduced effective expansion area for the weak expansive rock in the horizontal direction. Consequently, the cumulative amount of mudstone's expansion deformation decreases, and with the same mudstone content, the entire sample exhibits equivalent potential for compression deformation. The dynamic equilibrium between expansion and compression deformation within the microscopic mudstone sample continues to favor compression deformation, leading to the transformation from expansion to compression deformation after the sample is fully submerged in water and stabilized. Whether this deformation trend can be quantitatively computed using theoretical formulas requires further investigation.

3.2.3. Loading as Control Condition

The 3D mapping surfaces in Figure 6 illustrate the stable expansion and compression deformation of the samples under various loading levels and combinations of different factors. From Figure 6, it is evident that when the loading is considered as a control condition, it becomes challenging to uniformly describe the deformation behavior of samples in conditions of water immersion, expansion, and compression. Nevertheless, on an overall basis, there is a consistent trend observed: as the loading level increases, irrespective of the layer thickness ratio or the level of rock occurrence, the weak expansive rock samples of sandstone intercalated with mudstone tend to shift from exhibiting expansion deformation to exhibiting compression deformation. The loading stands out as the most influential factor affecting the expansion and compression deformation of the sample.

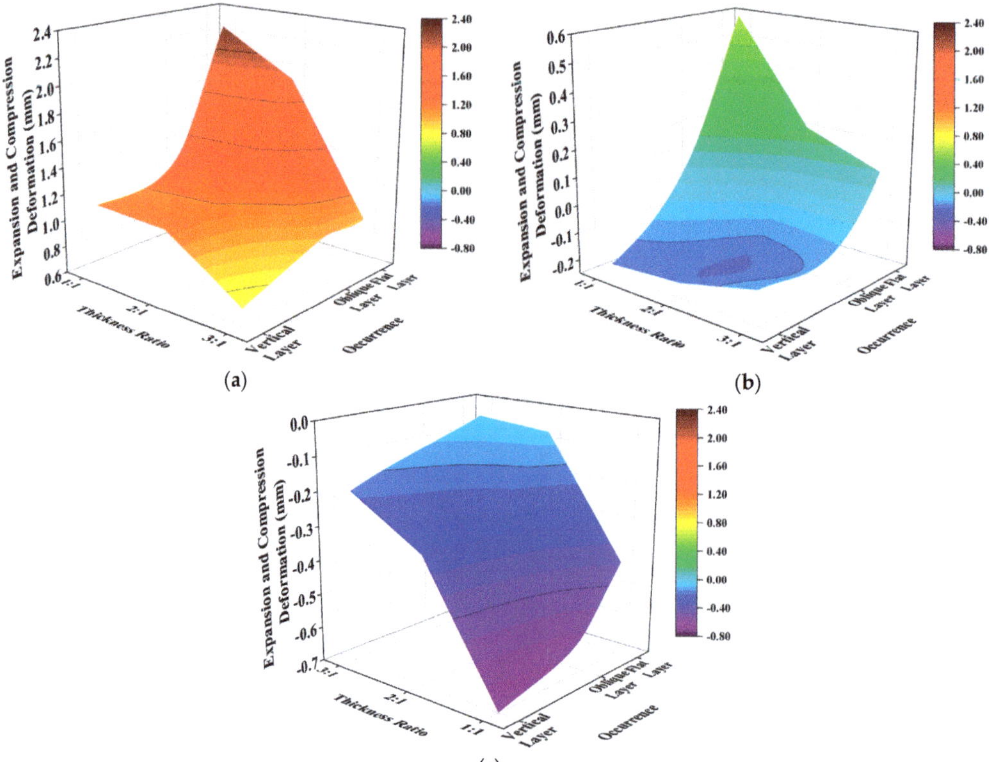

Figure 6. Three-dimensional mapping surfaces at different loading levels: (**a**) 0 kPa; (**b**) 12.5 kPa; (**c**) 25 kPa.

Under 0 kPa conditions, as depicted in Figure 6a, the sample undergoes expansion deformation for all combinations of a variety of factors. The maximum and minimum expansions are observed at 2.28 mm and 0.72 mm, respectively, notably evident when the layer thickness ratio is either 1:1 (flat layer) or 3:1 (vertical layer). It is apparent that the expansion deformation diminishes with an increase in both the layer thickness ratio and the rock inclination angle.

When subjected to a loading of 12.5 kPa, as illustrated in Figure 6b, the sample consistently demonstrates expansion deformation in flat layer structures and compression deformation in vertical layer structures, irrespective of the layer thickness ratio. This behavior signifies that the overall expansive force within the sample diminishes progressively with an increment in the rock inclination angle. When the rock layer's inclination ranges between 0° and 90°, the total expansive force of the sample precisely counterbalances the overburden load of 12.5 kPa.

As the loading increases to 25 kPa, as depicted in Figure 6c, the entire sample exhibits compression deformation. In this instance, the impact of the layer thickness ratio on the sample's expansion and compression behavior in water immersion significantly outweighs the occurrence of rock strata. Furthermore, when the layer thickness ratio is at least 2:1, the overall expansion and compression behavior of the sample remains largely unaltered. If the foundation contains a relatively small amount of weak expansive rocks, it is unnecessary to consider the influence of rock occurrence. Instead, it suffices to manage the overlying load value of the weak expansive rocks through embankment filling reinforcement design, ensuring that expansion deformation does not occur. This approach minimizes the need for extensive land replacement, thereby reducing the overall cost of the project.

3.3. *Time-History Analysis of Expansion and Compression Deformation*

The prior discussion primarily concentrates on the stability value associated with the deformation of a sandstone intercalated with mudstone interbedded structure under conditions of water immersion. However, in practical engineering applications, it is imprudent to concentrate solely on this particular value. The rate of deformation is an additional parameter that necessitates monitoring.

Considering the scarcity of engineering cases involving vertical layer (90° inclination angle) weak expansive rock formations with interbedded structures, this study focuses on analyzing test results for flat layer (0° inclination angle) and oblique layer (30° inclination angle). The time-history curves depicting the expansion and compression deformation of both the flat and oblique layer samples were plotted. These curves were subsequently fitted with hyperbolic functions, with the specific results being displayed in Figure 7.

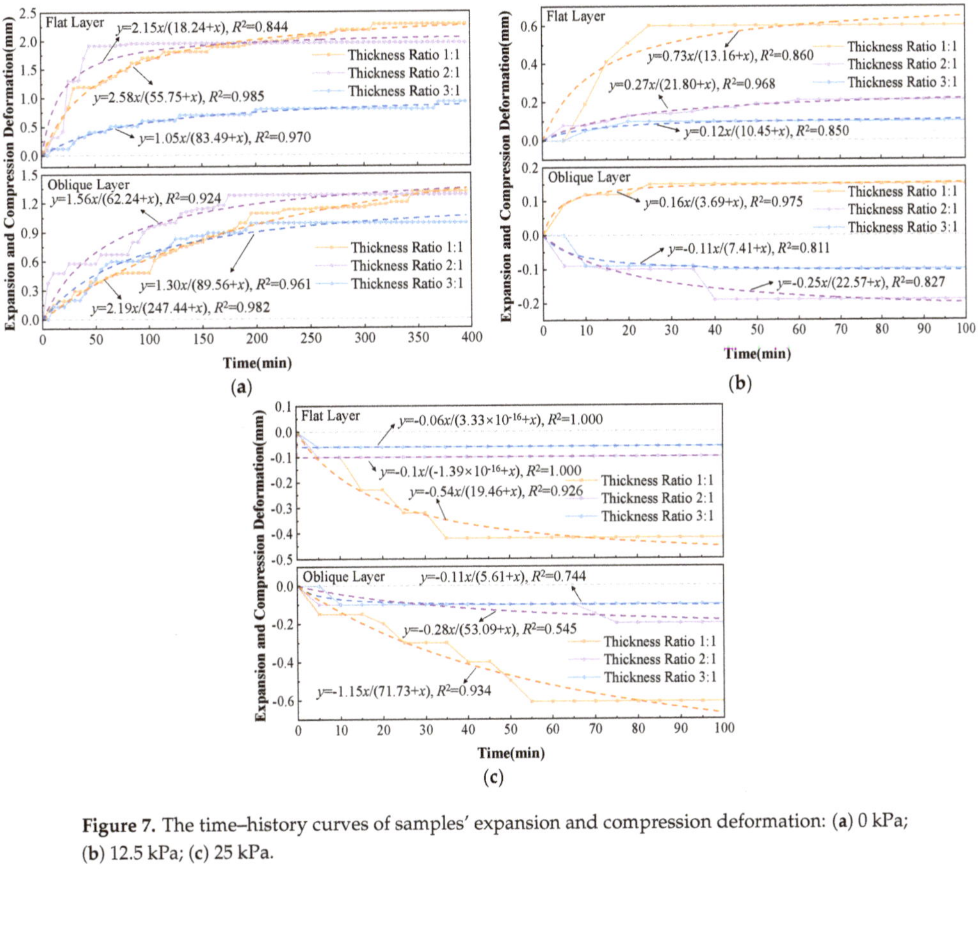

Figure 7. The time–history curves of samples' expansion and compression deformation: (**a**) 0 kPa; (**b**) 12.5 kPa; (**c**) 25 kPa.

As depicted in Figure 7, for the flat layer structure, the samples' expansion forces of different layer thickness ratios oscillate between 12.5 and 25 kPa. Regarding the oblique layer structure, when the layer thickness ratio is 1:1, the expansion force ranges from 12.5 to 25 kPa; however, when the ratio exceeds 1:1, the expansion force diminishes to a span of 0 to 12.5 kPa. Moreover, the rate at which the sample's expansion and compression deformation stabilizes under a load is substantially faster compared to its unloaded state. Specifically, in the absence of a load, the sample requires at least 5~7 h to achieve a stable deformation state, whereas under a load, this stabilization occurs within approximately 1 h. Additionally, when the applied load approaches or surpasses the sample's expansion force, deformation can rapidly progress. There appears to be no substantial disparity in the duration of the stability period. It can thus be inferred that in the context of a subgrade construction within regions characterized by weak expansive rocks with an interbedded structure, the compactness of the site can potentially be enhanced through precompression, tamping, and other such techniques. Consequently, maintaining a relatively high level of overlying load on the interbedded structure could significantly abbreviate the time required for the foundation to reach a stable deformation state post-construction.

The time-history curves of expansion and compression deformation that demonstrated a superior hyperbolic fitting effect were selected. Subsequently, the first-order derivatives of these fitting curves were calculated, respectively. This enabled the plotting of the sample's deformation rate over time under the influence of various factors, as depicted in Figure 8.

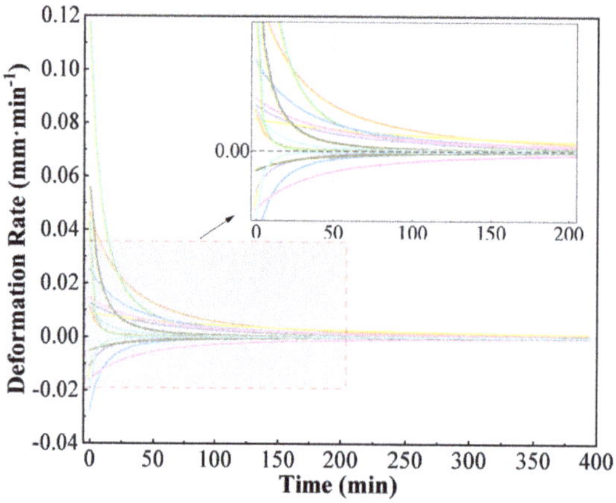

Figure 8. Deformation rate—time relation curves.

An analysis of Figure 8 reveals that irrespective of whether the sample is undergoing expansion or compression deformation, the deformation rate remains relatively high between 0 and 50 min after immersion. During this period, the deformation progresses rapidly, yet the rate of deformation diminishes quickly. From 50 to 200 min, the samples' expansion and compression deformation rates enter a phase of gradual decline, ultimately approaching zero, indicating a slow progression toward a stable state. Post 200 min, the rates of expansion and compression deformation are significantly reduced, suggesting that the samples have achieved a stable condition in terms of their deformation development.

The expansion and compression deformation process of a weak expansive rock sample with sandstone intercalated with mudstone interbedded structure can be categorized into three distinct stages: rapid deformation, slow deformation, and completed development. In the rapid deformation stage, the samples' expansion and compression deformation develops rapidly, potentially reaching 90% of the final stable state under loading. In the subsequent slow deformation stage, the rate of deformation development decreases, nearing completion, with minimal incremental changes in the expansion and compression values. Finally, in the completed development stage, any further expansion and compression deformation is virtually halted, rendering its impact on the superstructure in practical engineering applications negligible.

3.4. Expansion Force Fitting

The analysis presented above indicates that the expansion and compression deformation characteristics of weak expansive rocks with an interbedded structure, when immersed in water, are most significantly influenced by the overlying load. The expansion and compression deformation of such rock can be effectively managed by regulating the value of the overlying load. Consequently, determining the expansion force of weak expansive rock with an interbedded structure is crucial for constructing a high-speed railway subgrade in areas characterized by this type of rock.

Previous research indicates that the expansion deformation of expansive rock (or soil) exhibits an inverse correlation with the overlying load. The vertical axis represents the expansion and compression deformation of the sample, while the horizontal axis denotes the corresponding load values. A relationship curve between the expansion and compression deformation and the overlying load is plotted, which is subject to nonlinear fitting. The x-coordinate value at the intersection of the fitted curve with y = 0 represents the expansion force of the sample under the current factor level. The fitting curve for the sample, considering the combined effect of various factors on the expansion and compression deformation versus the loading, is illustrated in Figure 9.

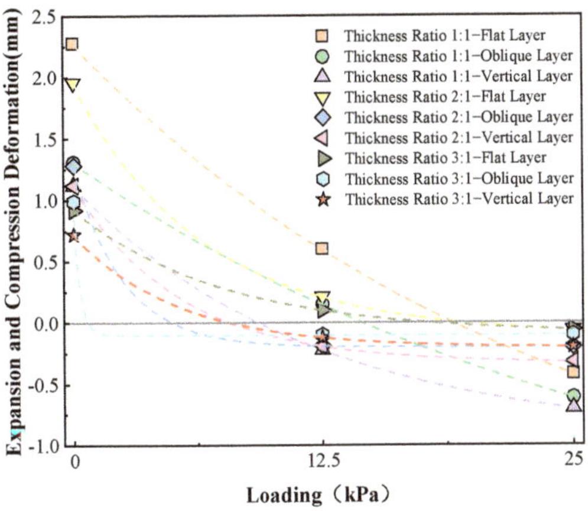

Figure 9. Expansion and compression deformation—Loading fitting curve.

The x-coordinate values of each fitted curve at the intersection with y = 0 are presented as follows.

The data presented in Table 5 are graphically represented on a 3D mapping surface, as depicted in Figure 10.

From Figure 10, it is evident that the expansion force of the weak expansive rock sample, characterized by an interbedded structure of sandstone and mudstone, diminishes as the layer thickness ratio and rock inclination angle increase. The calculated reductions are within the ranges of 8.91% to 38.68% and 51.00% to 58.83%, respectively. The data indicate that the expansion force is more significantly influenced by the rock occurrence level than by the layer thickness ratio.

For projects involving underlying foundations with thin, steeply inclined weak expansive rock layers, the overlying load can be determined through geological surveys and laboratory testing. If the expansion force is within acceptable limits, no additional treatment measures are necessary. In cases where the weak expansive rock layer is shallow, thick, and gently inclined, the expansion force value should be calculated based on the specific project requirements. The embankment surface load level can be maintained by either designing an appropriate embankment fill height or by excavating the weak expansive rock layer to a certain depth for partial filling. Post-construction, the expansion and compression deformation of the weak expansive rock with the underlying interbedded structure remains at a low level, allowing for rapid and stable development.

Table 5. The expansion force of the sample of each factor level combination.

Thickness Ratio	Occurrence	Expansion Force (kPa)
Thickness Ratio 1:1	Flat Layer	19.08
	Oblique Layer	14.53
	Vertical Layer	9.35
Thickness Ratio 2:1	Flat Layer	19.36
	Oblique Layer	9.09 *
	Vertical Layer	7.97
Thickness Ratio 3:1	Flat Layer	17.38
	Oblique Layer	8.91 *
	Vertical Layer	7.88

* The data were modified by manual fitting.

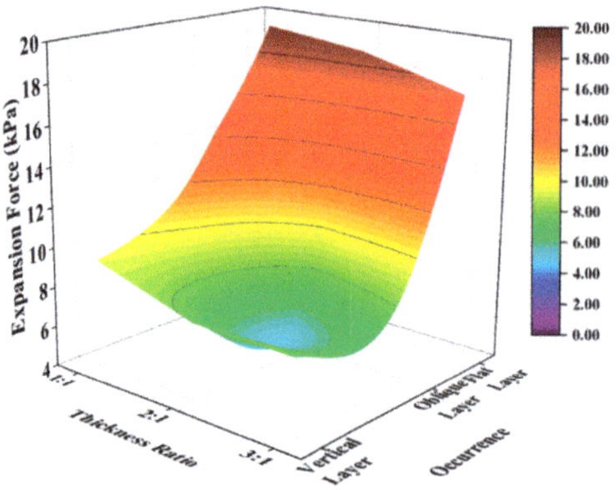

Figure 10. Three-dimensional mapping surface of sample expansion force.

4. Conclusions

Two typical weak expansive rocks, sandstone and mudstone, were selected in this study. Nine sets of weak expansive rock samples with different forms of interbedded structures were prepared, using the layer thickness ratios (1:1, 2:1, and 3:1) and the rock occurrence (flat, oblique, and vertical) as the controlling factors. Then, a total of 27 sets of indoor immersion deformation characteristic tests were designed based on different overburden loadings (0 kPa, 12.5 kPa, 25 kPa). The influence laws of three influencing factors on the expansion and compression deformation of weak expansive rocks with interbedded structures were comprehensively analyzed. The main conclusions are as follows:

1. The overburden loading has the most significant effect on the expansion and compression deformation of sandstone–mudstone interbedded structure samples. Its orthogonal test effect curve showed a clear monotonic decreasing trend and the decrease was the largest. The increase in overburden loading and rock inclination angle will cause the sample to change from expansion to compression deformation. The change in layer thickness ratio will only affect the absolute value of deformation, but will not change the deformation trend of the sample.
2. The expansion and compression deformation of sandstone interbedded with mudstone samples has obvious three-stage characteristics, 0~50 min for the rapid deformation stage, 50~200 min for the slow deformation stage, and 200 min later for the completed development stage. In addition, the deformation stability rate of

the sample under load is obviously faster, which is about 5~7 times that under no load. However, when the overburden loading value exceeds the expansion force of the sample, the difference in the deformation stability period of the sample is not significant.

3. The expansion force of the sample decreases with the increase in the layer thickness ratio level and the inclination angle of the rock layer, and the reduction ranges are 8.91~38.68% and 51.00~58.83%, respectively. Therefore, the influence of rock occurrence on the expansion force of the sample is more significant.

Author Contributions: Conceptualization, Y.W. and Y.L.; methodology, H.Q. and Q.Q.; formal analysis, Y.L., Y.Z., Y.Y., J.J. and T.Z.; investigation, H.Q.; data curation, Y.W., Y.Z. and D.C.; writing—original draft preparation, Y.L. and D.C.; writing—review and editing, Y.W., Y.Y. and J.J.; visualization, Y.L., T.Z. and Q.Q.; supervision, Y.W. and D.C.; project administration, Y.W., Y.Z., Y.Y. and D.C. All authors have read and agreed to the published version of the manuscript.

Funding: This research was sponsored by the Research Fund of the National Natural Science Foundation of China (Grant No. 51808462), the Applied Basic Research Project of Sichuan Science and Technology Department (general project) (Grant No. 2023NSFSC0346), the Inner Mongolia Autonomous Region science and technology planning project (2021GG0038), the Major science and technology project of Hohhot (2021-Key-Social-3), the Science and Technology project of Inner Mongolia Transportation Department (NJ-2022-14), the interdisciplinary research project of the Southwest Jiaotong University (A0920502052301-404), the new interdisciplinary cultivation fund of the Southwest Jiaotong University (YH1500112432284), the China Postdoctoral Science Foundation (Grant No. 2018 M643520), the Applied Basic Research Project of Sichuan Science and Technology Department (Free Exploration Type) (Grant No. 2020YJ0039), and the Key R & D Support Plan of Chengdu Science and Technology Project—Technology Innovation R & D Project (Grant No. 2019-YF05-00002-SN).

Data Availability Statement: The raw data supporting the conclusions of this article will be made available by the authors on request.

Conflicts of Interest: Authors Yaning Wang, Yangui Zhu, Yibo Yao, Jin Jin and Tao Zheng were employed by the company China Railway First Survey and Design Institute Group Co., Ltd. The remaining authors declare that the research was conducted in the absence of any commercial or financial relationships that could be construed as a potential conflict of interest.

References

1. Liu, X.; Chen, X.; Song, X. Physical simulation test on deformation and failure mechanism of soft and hard interbedded rock masses. *Chin. J. Rock Mech. Eng.* **2023**, *42*, 3980–3995. [CrossRef]
2. Liu, Z.; Zhang, R.; Lan, T.; Zhou, Y.; Huang, C. Laboratory test and constitutive model for quantifying the anisotropic swelling behavior of expansive soils. *Appl. Sci.* **2024**, *14*, 2255. [CrossRef]
3. Li, H.; Wang, J.; Li, H.; Wei, S.; Li, X. Experimental study on deformation and strength characteristics of interbedded sandstone with different interlayer thickness under uniaxial and triaxial compression. *Processes* **2022**, *10*, 285. [CrossRef]
4. Wang, R.; Cheng, J.; Gao, L.; Li, Z.; Qi, Y.; Wang, M.; Ding, B. Research on the swelling mechanism of high-speed railway subgrade and the induced railway heave of ballastless tracks. *Transp. Geotech.* **2021**, *27*, 100470. [CrossRef]
5. Maheshwari, P. Analysis of deformation of linear viscoelastic two layered laminated rocks. *Int. J. Rock Mech. Min. Sci.* **2021**, *141*, 104681. [CrossRef]
6. Wu, Q.; Liu, Y.; Tang, H.; Kang, J.; Wang, L.; Li, C.; Wang, D.; Liu, Z. Experimental study of the influence of wetting and drying cycles on the strength of intact rock samples from a red stratum in the Three Gorges Reservoir area. *Eng. Geol.* **2023**, *314*, 107013. [CrossRef]
7. Tian, Y.; Wu, F.; He, L.; Tian, H.; Huang, M.; Chen, W. Investigation of 3D printing sand core technology on the mechanical behaviors of soft–hard interbedded rock masses. *J. Mater. Civ. Eng.* **2023**, *35*, 04023440. [CrossRef]
8. Luo, P.; Li, D.; Ma, J.; Zhou, A.; Zhang, C. Experimental investigation on mechanical properties and deformation mechanism of soft-hard interbedded rock-like material based on digital image correlation. *J. Mater. Res. Technol.* **2023**, *24*, 1922–1938. [CrossRef]
9. Qin, Y.; Yang, G.; Liu, B.; Xu, J. Study on deformation and failure mechanism of low-dip red bed slope with soft-hard interbedded structure: A case study of Chishui, China. *Nat. Hazards* **2024**, 1–19. [CrossRef]
10. Chen, G.; Cai, P.; Zhan, J.; Yang, Y.; Yao, Z.; Yu, Z. Deformation patterns and failure mechanisms of soft-hard-interbedded anti-inclined layered rock slope in Wolong open-pit coal mine. *Appl. Sci.* **2024**, *14*, 3082. [CrossRef]

11. Dai, Z.; Huang, K.; Chi, Z.; Chen, S. Model test study on the deformation and stability of rainfall-induced expansive soil slope with weak interlayer. *Bull. Eng. Geol. Environ.* **2024**, *83*, 76. [CrossRef]
12. Goodman, R.; Bray, J. Toppling of rock slopes. proceedings, specialty conference on rock engineering for foundations and slopes. *Boulder* **1976**, *2*, 739–760.
13. Zhang, B.; Tang, H.; Fang, K.; Ding, B.; Gong, Q. Experimental study on deformation and failure characteristics of interbedded anti-inclined rock slopes induced by rainfall. *Rock Mech. Rock Eng.* **2024**, *57*, 2933–2960. [CrossRef]
14. Zhang, B.; Tang, H.; Sumi, S.; Ding, B.; Zhang, L.; Ning, Y. Exploring the deformation and failure characteristics of interbedded anti-inclined rock slopes: Insights from physical modelling tests. *Rock Mech. Rock Eng.* **2024**, *57*, 1603–1628. [CrossRef]
15. Dong, M.; Zhang, F.; Lv, J.; Hu, M.; Li, Z. Study on deformation and failure law of soft-hard rock interbedding toppling slope base on similar test. *Bull. Eng. Geol. Environ.* **2020**, *79*, 4625–4637. [CrossRef]
16. Ye, W.; Gao, C.; Liu, Z.; Wang, Q.; Su, W. A Fuzzy-AHP-based variable weight safety evaluation model for expansive soil slope. *Nat. Hazards* **2023**, *119*, 559–581. [CrossRef]
17. Wyllie, D.; Wood, D. Stabilization of toppling rock slope failures. In Proceedings of the 33rd Highway Geology Symposium, Engineering Geology and Environmental Constraints in Vail, Colorado, 1982; Colorado Geological Survey: Denver, CO, USA, 1983.
18. Wyllie, D. Toppling rock slope failures-examples of analysis and stabilization. *Rock Mech.* **1980**, *13*, 89–98. [CrossRef]
19. Liu, Y.; Huang, D.; Peng, J.; Guo, Y. Analysis of the effect of rock layer structure on the toppling failure evolution of soft-hard interbedded anti-dip slopes. *Eng. Fail. Anal.* **2023**, *145*, 107005. [CrossRef]
20. Wang, R.; Zheng, Y.; Chen, C.; Zhang, W. Theoretical and numerical analysis of flexural toppling failure in soft-hard interbedded anti-dip rock slopes. *Eng. Geol.* **2023**, *312*, 106923. [CrossRef]
21. Yu, J. Study on reasonable construction method for tunnels in contact zone of red clay and sandstone interbedded with mudstone. *Mod. Tunn. Technol.* **2021**, *58*, 174–181. [CrossRef]
22. Zhao, D.; Ji, Q.; Wang, G. Research on mechanical characteristics of large-span tunnels in bedding mudstone with sandstone strata. *Tunn. Constr.* **2021**, *41*, 740–748. [CrossRef]
23. Zhao, D.; He, Q.; Ji, Q.; Wang, F.; Tu, H.; Shen, Z. Similar model test of a mudstone-interbedded–sandstone-bedding rock tunnel. *Tunn. Undergr. Space Technol.* **2023**, *140*, 105299. [CrossRef]
24. Liu, W.; Chen, J.; Chen, L.; Luo, Y.; Shi, Z.; Wu, Y. Deformation evolution and failure mechanism of monoclinic and soft-hard interbedded strata: Study of muzhailing tunnel. *J. Perform. Constr. Facil.* **2021**, *35*, 04021042. [CrossRef]
25. Chen, J.; Liu, W.; Chen, L.; Luo, Y.; Li, Y.; Gao, H.; Zhong, D. Failure mechanisms and modes of tunnels in monoclinic and soft-hard interbedded rocks: A case study. *KSCE J. Civ. Eng.* **2020**, *24*, 1357–1373. [CrossRef]
26. Zhang, Z.; Wang, T. Failure modes of weak interlayers with different dip angles in red mudstone strata, Northwest China. *Bull. Eng. Geol. Environ.* **2023**, *82*, 156. [CrossRef]
27. Wang, T.; Yan, C. Investigating the influence of water on swelling deformation and mechanical behavior of mudstone considering water softening effect. *Eng. Geol.* **2023**, *318*, 107102. [CrossRef]
28. Dai, Z.; Guo, J.; Yu, F.; Zhou, Z.; Li, J.; Chen, S. Long-term uplift of high-speed railway subgrade caused by swelling effect of red-bed mudstone: Case study in Southwest China. *Bull. Eng. Geol. Environ.* **2021**, *80*, 4855–4869. [CrossRef]
29. Wang, B.; Wang, Q.; Zhang, R.; Wang, T.; Zhang, T.; Liang, K. Large scale model test of water swelling deformation of mudstone under lateral confinement. *J. Disaster Prev. Mitig. Eng.* **2020**, *40*, 337–342. [CrossRef]
30. Xue, Y.; Wang, Q.; Ma, L.; Wang, B.; Zhang, T. Experimental study on in-situ flooding response characteristics of mudstone foundation for high-speed railway. *J. China Railw. Soc.* **2020**, *42*, 144–150. [CrossRef]
31. Yang, G.; Ding, J. Model test on expansion and shrinkage deformation in expansive soil roadbed. *China J. Highw. Transp.* **2006**, *19*, 23–29. [CrossRef]
32. Duan, J.; Yang, G.; Liu, Y.; Kan, J.; Qiu, M. Deformation characteristics of double-line ballastless track subgrade under water immersion of expansive rock foundation. *J. Cent. South Univ. Sci. Technol.* **2020**, *51*, 156–164. [CrossRef]
33. Cui, X.; Wang, Q.; Zhang, R.; Wang, C.; Xue, Y. Expansive deformation test on ballastless track of expansive mudstone. *J. Railw. Sci. Eng.* **2017**, *14*, 2339–2344. [CrossRef]
34. Cao, Z. Field soaking tests on the moderate-strong expansive soil ground of the high-speed railway subgrade. *J. Railw. Sci. Eng.* **2022**, *19*, 3467–3476. [CrossRef]
35. Duan, J.; Yang, G.; Liu, Y.; Kan, J.; Zhang, L. Influence of water immersion on mechanical and deformation behaviors of expansive soil subgrade for ballastless track. *J. China Railw. Soc.* **2022**, *44*, 119–126. [CrossRef]
36. Zhao, S.; Zheng, J.; Yang, J. Stability analysis of embankment slope considering water absorption and softening of subgrade expansive soil. *Water* **2022**, *14*, 3528. [CrossRef]
37. Lan, T.; Zhang, R.; Yang, B.; Meng, X. Influence of swelling on shear strength of expansive soil and slope stability. *Front. Earth Sci.* **2022**, *10*, 849046. [CrossRef]
38. Li, J.; Wang, Q.; Zhang, R.; Zhang, T.; Wang, T.; Liang, K. Experimental study on swelling law of expansive soil during humidification. *Hydro-Sci. Eng.* **2018**, *3*, 86–94. [CrossRef]
39. Zuo, Q.; Li, P.; Li, X.; Chen, F. Swelling damage evolution of argillaceous slate in a water-rich environment. *Q. J. Eng. Geol. Hydrogeol.* **2023**, *56*, qjegh2022-128. [CrossRef]
40. Kang, Y.; Gu, J.; Wei, M. Mechanical properties and acoustic emission characteristics of soft-hard interbedded rocks under different loading rates. *J. Northeast. Univ. (Nat. Sci.)* **2023**, *44*, 399–407. [CrossRef]

41. Huang, F.; Zhou, Y.; Li, T.; Hu, X. Laboratory experimental study on mechanical properties and failure modes of soft and hard interbedded rock mass. *J. China Coal Soc.* **2020**, *45*, 230–238. [CrossRef]
42. Han, B.; Fu, Q.; Wang, C. Numerical simulation for creep characteristics of soft and hard interphase rock mass. *J. Phys. Conf. Ser.* **2020**, *1600*, 012075. [CrossRef]
43. *TB 10102-2023*; Specification for Soil Test of Railway Engineering. China Railway Publishing House: Beijing, China, 2023.

buildings

MDPI

Article

Interface Interaction of Waste Rubber–Asphalt System

Jinfei Su [1,2], Peilong Li [2,3,*], Guangxin Zhu [1], Xiaoxu Wang [1] and Shihao Dong [1,2,*]

1 College of Transportation, Shandong University of Science and Technology, Qingdao 266590, China; jinfeisu@sdust.edu.cn (J.S.); 202111140330@sdust.edu.cn (G.Z.); xiaoxuwang@sdust.edu.cn (X.W.)
2 Key Laboratory of Intelligent Construction and Maintenance of Civil Aviation Airport, Chang'an University, Xi'an 710064, China
3 Highway School, Chang'an University, Xi'an 710064, China
* Correspondence: lipeilong@chd.edu.cn (P.L.); dongshihao@sdust.edu.cn (S.D.)

Abstract: Asphalt pavement construction is a large-volume project, with the ability to recycle the industrial waste and reduce carbon emissions. Rubber-modified asphalt is a carbon-neutralized asphalt-based material, facilitating the recycling of waste rubber materials and improving the road performance of the asphalt mixture. To evaluate the interface interaction of the rubber–asphalt system and its effect on the viscosity characteristics of rubber-modified asphalt, the contact properties of rubber particles in asphalt were analyzed on a microscopic level. Rubber swelling tests and solvent elution tests were conducted on the rubber–asphalt system under different preparation conditions. The swelling ratio, degradation ratio, and swelling–degradation ratio were proposed to evaluate the interface interaction. The results show that the interface interaction of the rubber–asphalt system can be divided into the following three stages: swelling, effective degradation, and over-degradation. The degree of swelling is mainly affected by the content and size of the rubber particles and it is physically condensed, while the degradation is mainly affected by the preparation temperature and preparation time. The effective interface interaction greatly affects the viscosity with the building of the stable three-dimensional network structure. The stronger the interface interaction, the greater the viscosity of the rubber-modified asphalt, except for the 25% content of rubber particles. The gel film will be generated on the surface of the rubber particles throughout the swelling and effective degradation, increasing the viscosity of the rubber-modified asphalt.

Keywords: rubber–asphalt system; interfacial interaction; particle contact properties; three-dimensional net microstructure; viscosity characteristics

1. Introduction

Energy shortages and environmental pollution are prominent issues of the 21st century [1]. The recycling of waste materials represents an effective approach to address both of these challenges and to reduce carbon emissions. With the advancement of the automotive industry, the rapid accumulation of waste rubber tires has become a global issue, commonly referred to as "black pollution". Additionally, the road performance of asphalt mixtures needs to be improved [2,3]. The utilization of waste rubber powder as a modifier to prepare rubber-modified asphalt, which is a carbon-neutralized asphalt-based material, harnessing the elastic and flexible properties of rubber to enhance asphalt performance, has emerged as a carbon emission reduction and resource-efficient solution [4]. This waste rubber recycling technology has found increasing applications in the field of road construction [5]. Ilyin et al. [6] analyzed the effect of rubber particles on the rheological properties of asphalt. The study found that the addition of rubber particles improved the rutting resistance of road binders. Xiao et al. [7] reported that the addition of rubber particles can increase the Voids in Mineral Aggregates (VMAs) and improve the resistance to rutting in asphalt mixtures. Mashaan et al. [8] focused on the influence of tire rubber particles on the performance of Stone Mastic Asphalt (SMA) mixtures. Mull et al. [9] found that the crack resistance of rubber-modified asphalt pavements surpasses that of base pavements. Chen and Li [10] employed the Discrete Element Method (DEM) for a microstructural analysis of rubber particle-modified asphalt mixtures. Tan et al. [11] explored the production process of rubber-modified asphalt concrete. Huang et al. [12] investigated the fatigue properties of rubber-modified asphalt mixtures. Cao et al. [13] analyzed the road performance of rubber-modified asphalt mixtures utilizing various activation and modification methods. Yu and Wu [14] studied the design method of rubber-modified asphalt mixtures. Wang [15] analyzed the influence of particle size and the content of rubber powder on the high-temperature rheological properties of asphalt mixtures. Tang et al. [16] examined the impact of mineral powder on the road performance of rubber-modified asphalt mixtures.

Due to the steric effect of rubber particles and the interface adhesive effect between rubber particles and asphalt, rubber-modified asphalt exhibits a higher viscosity and more complex viscoelastic, lubrication, and rheological properties [17]. Investigating the interactions between rubber particles and asphalt aids in understanding the modification mechanisms of rubber-modified asphalt, facilitating the development of asphalt with an improved performance and, in turn, enhancing the service performance of road surfaces. Guo et al. [18,19], through the establishment of molecular models for the components of rubber particles and asphalt, demonstrated the compatibility between rubber and asphalt. The study found that the rubber particles' content significantly influenced the storage stability of rubber-modified asphalt. Liseane et al. [20] proposed that the digestion time determined the elastic properties of rubber particles and its solubility in asphalt. Dong et al. [21] simulated the swelling process of rubber powder in asphalt and discovered that rubber particles only absorb the lighter components of asphalt, with swelling rates initially increasing rapidly and then stabilizing. Frantzis et al. [22] analyzed the impact of particle swelling on the storage stability of rubber-modified asphalt and proposed asphalt diffusion coefficients and dissolution parameters. Ni et al. [23] examined the relationship between rubber particles' swelling degree and asphalt components, discussing the devulcanization during the rubber swelling process. Bradley et al. [24] explored the interaction of rubber particles with asphalt and the particle effects. Wang et al. [25] investigated the microstructure of rubber particles and the interface properties between rubber particles and the base asphalt. Frédérique et al. [26] employed a wetting dynamics approach to calculate the surface energies between rubber particles and asphalt, explaining the compatibility between rubber particles and asphalt.

The rubber particles absorb the saturated and aromatic components of asphalt, leading to an increase in volume during the preparation process of rubber modified asphalt. This absorption triggers the degradation of rubber powder, releasing small molecules such as styrene and butadiene, thus facilitating the exchange of substances between rubber particles and asphalt [27]. Based on these observations, this investigation designed rubber powder swelling tests and solvent elution experiments and proposed swelling rate and degradation rate as parameters to evaluate the physicochemical behavior of rubber powder during the preparation of rubber-modified asphalt.

In this investigation, the rubber-modified asphalt was considered as a rubber–asphalt system. The research delved into the movement characteristics of rubber particles at a microscopic level and introduced swelling and degradation rates to characterize the interface interactions within the rubber–asphalt system. These findings will contribute to the exploration of the relation of the interface interaction and viscosity of rubber-modified asphalt, providing a theoretical foundation for the development of the sustainable rubber-based asphalt material and an analysis of the composition mechanism of rubber-modified asphalt.

2. Materials and Preparation Methods of Rubber-Modified Asphalt

2.1. Base Asphalt

SK 90# asphalt was selected as the base asphalt, and the relevant testing was conducted in accordance with the requirements of the Standard Test Method of Bitumen and Bituminous Mixtures for Highway Engineering (JTG-E20 2011) [28]. The property parameters for the asphalt are presented in Table 1.

Table 1. The properties of asphalt.

Properties		Standard Requirement	Test Results	Method
Penetration (25 °C, 5 s, 100 g)/0.1 mm		80~100	88.6	T0604
Penetration index, PI		−1.0~+1.0	−0.6	T0604
Ductility (5 cm/min, 10 °C)/cm		≥25	79.5	T0605
Ductility (5 cm/min,15 °C)/cm		≥100	>100	T0605
Softening point (ring and ball method)/°C		≥45	46	T0606
Flash point (open bottle method)/°C		≥245	292	T0611
Solubility (trichloroethylene)/%		≥99.5	99.88	T0607
Density (15 °C) g/cm^3		≥1.01	1.034	T0603
RTFOT(163 °C, 85 min)	Mass change/%, no more than	±0.8	−0.065	T0609
	Residual penetration ratio (25 °C)/%	≥57	61.2	T0604
	Residual ductility (10 °C)/cm	≥8	10	T0605
	Residual ductility (15 °C)/cm	≥8	47.3	T0605

2.2. Rubber Powder

The rubber powder utilized in the experiments was a product of Huayi Rubber Company, located in Dujiangyan. This rubber powder was available in various particle sizes, specifically 20 mesh, 40 mesh, 60 mesh, 80 mesh, and 100 mesh. It exhibited a black color and did not contain any visible particles perceptible to the naked eye, as shown in Figure 1.

Figure 1. The rubber powder before mixing with asphalt.

According to the relevant requirements of the Ground Vulcanized Rubber of Scrap Tires for Highway Engineering (JT/J 797-2011) [29], the physicochemical properties of the rubber powder were tested with the test indicators presented in Table 2.

Table 2. The physicochemical properties of rubber.

Properties	Test Results	Standard Requirement
Density/(g·cm^{-3})	1.19	1.10~1.30
Moisture content/%	0.47	<1
Metal content/%	0.028	<0.05
Fiber content/%	0.49	<1
Ash content/%	5.3	≤8
Acetone extract/%	14.7	≤22
Carbon black content/%	31	≥28

2.3. Preparation Method

Preliminary research in the project indicated that the optimal conditions for preparing rubber-modified asphalt using a stirring method at a rotational speed of 500 r/min were as follows: a rubber size of 20 mesh, a rubber content of 20% asphalt mass, a preparation temperature of 180 °C, and a preparation time of 90 min [30]. To investigate the interface interactions within the rubber–asphalt system and their mechanistic impact on viscosity, this investigation adopted a control variate method. The preparation parameters for rubber-modified asphalt are presented in Table 3.

Table 3. Preparation parameters of rubber-modified asphalt.

Preparation Conditions	Values
Rubber size/Mesh number	20, 40, 60, 80
Rubber content/%	5, 10, 15, 20, 25
Preparation temperature/°C	150, 160, 170, 180, 190
Preparation time/min	30, 60, 90, 120, 150

Firstly, the SK90# base asphalt was heated in an oven to 160 °C and was then transferred into an oil bath. The temperature was rapidly increased on a heating plate to reach the designated preparation temperature. To mitigate the adverse effects of asphalt aging, it is important that the heating temperature in the oven does not exceed 160 °C, and the heating time in the oil bath does not exceed 20 min. Secondly, the stirring machine was activated, and the rotational speed was set at 500 r/min. Finally, the pre-weighed rubber powder was added in five equal portions to the base asphalt, and rubber-modified asphalt was prepared under the specified conditions.

3. Experimental Method

The interface interaction between rubber particles and asphalt is a process involving the swelling, devulcanization, and degradation of rubber powder, and the combined effects of these processes impacts the properties of rubber-modified asphalt. The degree of swelling and degradation is related to the rubber content, rubber size, preparation temperature, and preparation time. Rubber powder absorbs the light components of asphalt, leading to an initial mild swelling effect. However, as swelling reaches a certain level, the degradation process accelerates, resulting in the disintegration of rubber particles and the breaking of molecular chains. The development of swelling and degradation determines the extent to which the interface interaction between rubber particles and asphalt affects viscosity. To quantitatively characterize the interface interaction between rubber particles and asphalt, the swelling tests and degradation tests designed are employed for this investigation [31].

3.1. Swelling Test of Rubber Powder

In preparation for the experiment, a sieve with a mesh number of 120 was selected along with a corresponding support base. Approximately 100 g of the prepared rubber-modified asphalt was poured onto the sieve. The sieve and its support base were then placed inside an oven preheated to 160 °C, where they were maintained for approximately one hour. The rubber-modified asphalt exhibited a liquid state under high-temperature environments, with the base asphalt flowing into the support base, while rubber particles were retained on the sieve.

A small quantity of rubber particles from the sieve was placed onto glass slides to prepare the test specimens. These specimens were then subjected to observation under an optical microscope to measure the size of the rubber particles. The optical microscope used is the ZOOM-2860 microscope produced by Shanghai Yuguang Instrument Company, the magnification can be expanded from 2× to 180×, and the maximum distance can reach 280 mm. The separation process of the rubber particles and asphalt is illustrated in Figure 2.

To reduce the experimental error, the size of fifty rubber particles was measured for each set of specimens, and the average size was calculated.

(a) (b)

Figure 2. The separation of rubber and modified asphalt. (**a**) Before separation. (**b**) After separation.

3.2. Solvent Elution Test

Firstly, the rubber-modified asphalt with a weight of 4 g was placed within a pre-folded filter paper. Subsequently, the filter paper was folded to seal the top end, ensuring that the rubber powder would not leak from the filter paper. The folded filter paper was then inserted into a prepared Soxhlet extractor. In a 250 mL volumetric flask, trichloroethylene was added to approximately two-thirds of the volume. Finally, the flask was heated and the extraction process was repeated until the liquid in the extractor turned colorless. To minimize error, five parallel experiments were conducted. Rubber is an inert material, which is difficult to solubilize in trichloroethylene. To verify the accuracy of the research, three blank samples were prepared to test the solubility of rubber particles in the trichloroethylene, and only 1 g rubber particles were placed in the blank sample. The filter paper that wraps the rubber particles was placed in trichloroethylene, and was then weighed and dried. The experimental process and the rubber powder after elution are depicted in Figures 3 and 4, respectively.

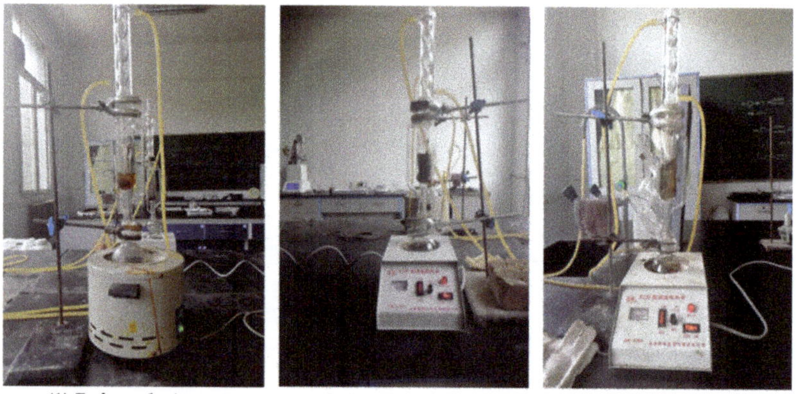

(1) Before elution (2) During elution (3) After elution

Figure 3. Process of elution.

Figure 4. The rubber after elution.

3.3. Evaluation Indices

The swelling ratio V_R was proposed to characterize the increased extent of volume in the rubber particles. Assuming that the rubber particles are approximate spheres, the calculation method for the swelling ratio is expressed in Equation (1):

$$V_R = \frac{R_1^3 - R_2^3}{R_2^3} \quad (1)$$

where R_1 is the average diameter of rubber after swelling, in um; R_2 is the average diameter of rubber before swelling, in um.

The degree of the degradation rate, T, was proposed to evaluate the reduction degree of the mass of the rubber powder, expressed in Equation (2):

$$T = \frac{M{\cdot}a - m}{M{\cdot}a} \quad (2)$$

where M is mass of the rubber-modified asphalt before elution, in g; m is mass of the rubber powder after elution, in g; and a is the content of the rubber powder.

The volume expansion of rubber particles occurs due to their absorption of the light components from the asphalt, leading to an increase in the steric effect within these particles. Throughout the degradation process, there is a continual precipitation of smaller molecular substances, enhancing the connections between rubber particles and asphalt, as well as among different rubber particles. However, as the extent of degradation becomes more substantial, the rubber particles undergo disintegration, nearly dissolving within the asphalt. To quantitatively evaluate the interface interaction within the rubber–asphalt system, the swelling–degradation rate φ was proposed, as expressed in Equation (3):

$$\varphi = V_R + T \quad (3)$$

4. Interface Interaction of Rubber–Asphalt System

As depicted in Figure 5, the swelling ratio initially showed an increase in preparation time and then decreased, reaching its maximum value at 90 min. Rubber powder, being an inert high-molecular-weight material, finds it challenging to achieve complete molecular integration with asphalt to form a stable modified asphalt. Consequently, numerous free-standing rubber particles disperse within the base asphalt. Upon the addition of rubber powder to asphalt and subsequent heating and stirring, the lighter components within the asphalt permeate and diffuse into the rubber particles, leading to the swelling of these rubber particles. Concurrently, chemical changes such as the degradation of rubber polymer chains occur.

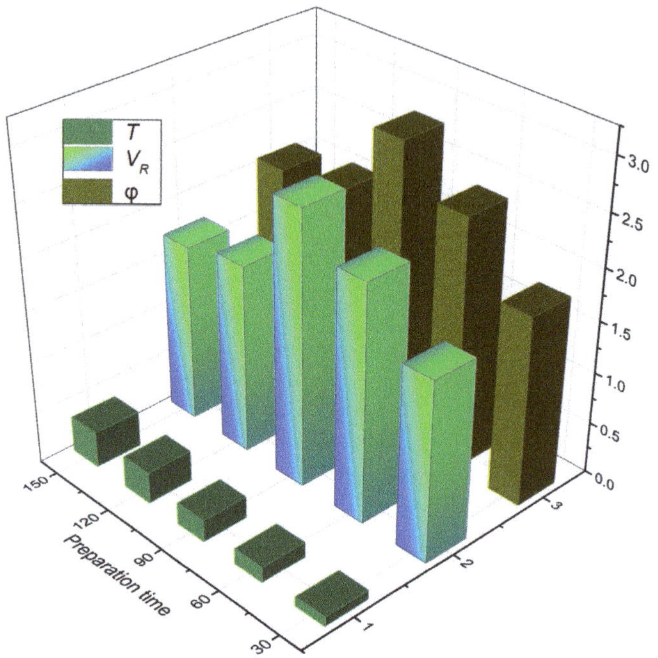

Figure 5. The change of interface interaction parameters with preparation time.

With the increase in preparation time, the volume of rubber particles continues to increase due to the swelling behaviors. However, when the preparation time exceeds 90 min, the swelling ratio notably decreases. The degradation of rubber powder not only diminishes the mass of the rubber particles, but also contributes to a decrease in their volume. During the initial stages of preparation, the degradation behavior of the rubber particles is relatively insignificant, exerting a slight influence on volume. From Figure 5, when the preparation time is prolonged, the extent of degradation of rubber powder significantly increases, promoting the generation of small molecules, ultimately leading to a reduction in the volume of rubber particles.

Rubber-modified asphalt is a multiphase dispersion rubber–asphalt system resulting from complex physicochemical interactions between rubber particles and asphalt. The rubber–asphalt system comprises rubber particles, an interface viscous phase, and the liquid asphalt phase. The swelling and degradation behaviors of the rubber particles within the base asphalt are the sources of the interface interactions of the rubber–asphalt system. When rubber particles were enveloped by the asphalt, the lighter components were absorbed, causing an expansion in the volume of the rubber particles. Simultaneously, as the volume expands, the polymer chains experience fragmentation, giving rise to the generation of small molecular substances that dissolve into the asphalt. As Figure 6 illustrates, compared with the original rubber powder, the rubber particles undergo swelling and degradation, transferring and exchanging substances within the base asphalt. Consequently, this process leads to an interconnection among rubber–asphalt–rubber and the formation of a three-dimensional network structure.

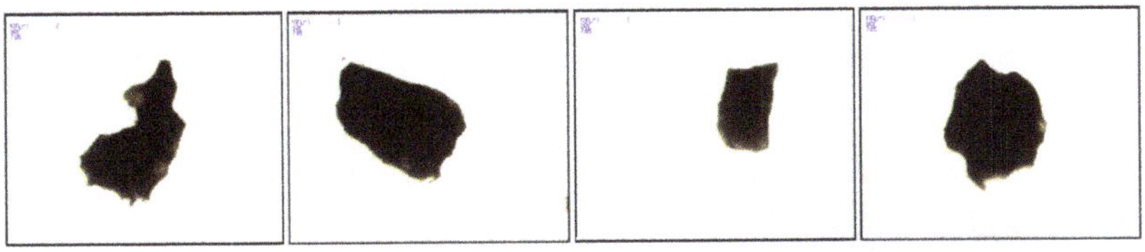

(a)

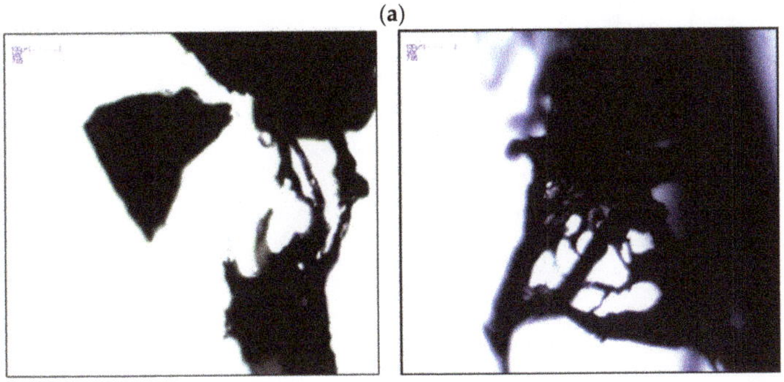

(b)

Figure 6. Microscopic characteristic at the optimal preparation conditions. (**a**) Original rubber particles [27]; (**b**) Rubber–asphalt system.

The pattern of change in the swelling–degradation rate is consistent with that of the swelling rate, indicating that the interface interactions in the rubber–asphalt system are the strongest at a preparation time of 90 min.

As demonstrated in Figure 7, it can be observed that the swelling rate of rubber-modified asphalt increases first and then decreases as the size of the rubber powder particles increases, reaching its maximum value at a size of 60 mesh. On the one hand, as the size of rubber powder particles increases, there are fewer pores and cracks on the surface of the rubber powder [32], making it more difficult for the light components in asphalt to penetrate the interior of the rubber powder. On the other hand, larger particle sizes result in a lower quantity of rubber powder, with a rubber powder content of 20%. During the preparation process, larger rubber particles have fewer opportunities for contact and collision, weakening its ability to absorb the light components from the asphalt.

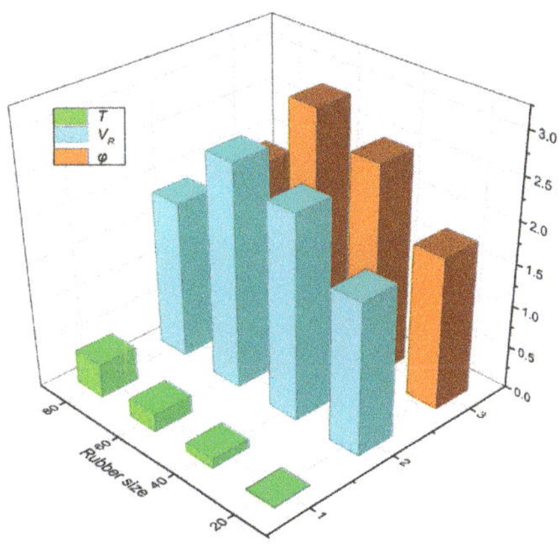

Figure 7. The change of interface interaction parameters with size of rubber.

Based on Figure 7, it is evident that the degradation rate of rubber particles increases with an increase in particle size. When the preparation time and preparation temperature remain constant, the extent of the degradation of the rubber powder is contingent upon both the mass of the rubber powder and the degree of individual rubber particle swelling. A smaller rubber powder particle size yields a more porous structure and a larger specific surface area, thereby leading to a greater degree of degradation at the same preparation temperature and preparation time. As the lightweight components are absorbed by rubber particles, the volume of rubber particles increases and degradation gradually initiates during the swelling process, indicating that degradation is contingent upon the swelling process.

At the point where the particle size reaches 80 mesh, a reduction in swelling rate becomes evident. Normally, decreasing the particle size of the rubber powder can enhance its capability to absorb lightweight components. Hence, it can be deduced that this phenomenon is a result of the degradation behavior of rubber powder, resulting in a more significant reduction in its volume compared to the increase in volume caused by the absorption of lightweight components. This phenomenon is herein defined as "excessive degradation".

The swelling–degradation rate of rubber powder follows the same trend as the swelling rate, indicating that the interface interaction within the rubber–asphalt system is at its strongest intensity when the rubber powder particle size is 60 mesh.

5. The Effect of Interface Interaction on the Viscosity of Rubber-Modified Asphalt

A comprehensive analysis of Figures 8 and 9 elucidates that the viscosity of rubber-modified asphalt exhibits a consistent trend with the degradation rate of rubber powder, and this relationship follows a highly significant quadratic function with a correlation coefficient as high as 0.998. This indicates that the interface interactions of the rubber–asphalt system constitute the primary source of variation in viscosity of rubber-modified asphalt. Notably, the interface interactions between rubber particles and asphalt are at their strongest at a temperature of 180 °C, leading to the formation of the most stable three-dimensional network structure and consequently yielding the highest viscosity.

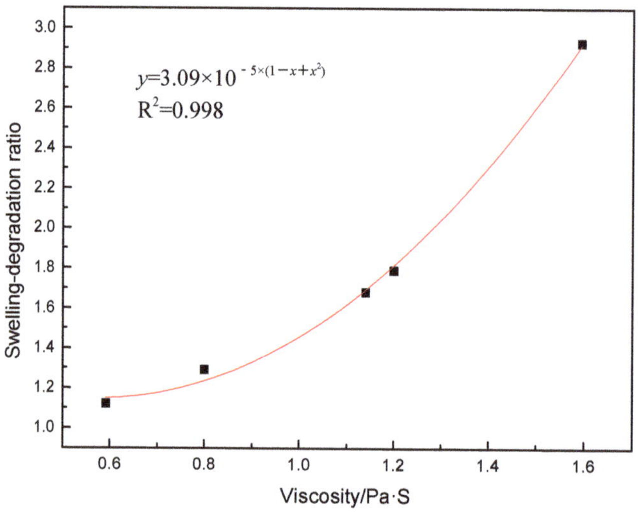

Figure 8. Correlation between φ and viscosity at different preparation temperatures.

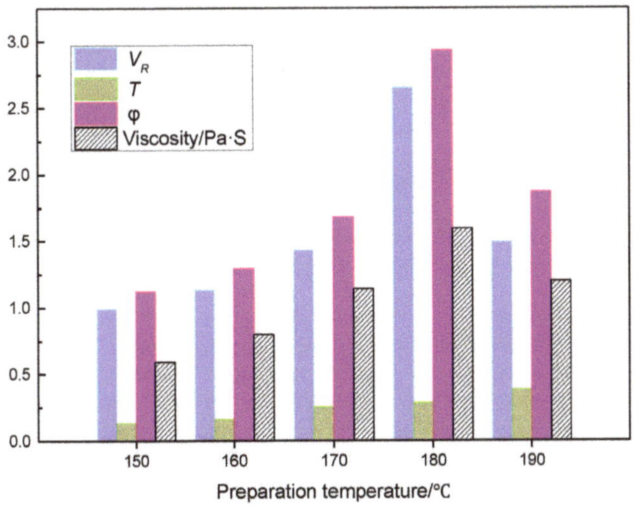

Figure 9. Comparison of viscosity and interaction parameters at different reaction temperatures.

At temperatures below 180 °C, both the swelling rate and degradation rate of the rubber powder increase with rising temperature. This phenomenon can be attributed to the poor flowability of asphalt at lower temperatures. On the one hand, the movement of rubber particles is impeded, and the time required for their uniform dispersion increases compared to higher temperature conditions, consequently reducing the time for the swelling of the rubber particles. On the other hand, the ability of the lighter components in asphalt to infiltrate decreases, and lower temperatures adversely affect the reactivity of rubber particles, resulting in a reduced capacity to absorb lighter components. When subjected to external forces, the intrinsic movement behaviors of the rubber particles are notably influenced by temperature, with higher temperatures intensifying the extent of their motion and augmenting the interaction forces when they collide with other rubber particles. As a result, individual rubber particles are more likely to generate the forces that differ from the movement direction of asphalt, thereby increasing their chances of contacting the light components within the asphalt [33]. As the temperature rises, the capacity of rubber particles to absorb lighter components steadily expands, yet concurrently, the extent of degradation increases. When the temperature surpasses 180 °C, the swelling rate of the rubber powder experiences a dramatic reduction. This phenomenon signifies an excessive level of degradation of the rubber powder, enhancing its capability to absorb lightweight components, ultimately leading to a reduction in the volume of the rubber particles. Therefore, the degradation of the rubber powders enhances the swelling, while leading to a decrease in the swelling rate. Consequently, achieving optimal interface interactions at lower temperatures necessitates the extension of the mixing duration, while temperatures exceeding 180 °C call for a reduction in the mixing time.

Based on an exploration of the interface interactions of the rubber–asphalt system, while keeping other preparation conditions constant, this investigation demonstrates the mechanism behind the variation in the viscosity of rubber-modified asphalt by changing the preparation time. Rubber particles exhibit an increased volume of light component absorption, and as degradation progresses, the rubber–asphalt system gradually forms a three-dimensional network structure, leading to an increase in the viscosity of rubber-modified asphalt [34,35]. However, when the temperature exceeds 180 °C, with a preparation time of 90 min, rubber particles undergo further degradation, generating more small molecular substances. While this enhances the bonding capacity among rubber particles, it results in a reduced individual rubber particle elasticity and a decrease in the stability of the three-dimensional network structure, consequently leading to a decrease in the viscosity of rubber-modified asphalt. The physicochemical behavior of the rubber–asphalt system, which gradually enhances the stability of the three-dimensional network structure, is termed an effective interface interaction, and the corresponding degradation behavior is referred to as an effective degradation.

Furthermore, previous research has revealed that when the preparation temperature is excessively elevated, and the preparation time is unduly prolonged, the degrading rubber particles can undergo polymerization reactions or react with asphalt to form polar compounds [30]. Additionally, asphalt is susceptible to aging with higher temperatures, leading to an increase in its viscosity. These phenomena show that the viscosity reduction caused by excessive degradation is greater than the viscosity increase caused by the aging of asphalt.

As depicted in Figure 10, all other concentrations exhibit a prominent linear relationship between the swelling–degradation rate and viscosity, except for at a rubber content of 25%. When the rubber content increases to 25%, the total surface area for individual rubber particles to interact with light components decreases throughout the preparation process. Consequently, the interface interaction between rubber particles and asphalt weakens, and the interconnection among rubber particles is less robust. As a result, it becomes challenging to form a stable three-dimensional network structure within the rubber–asphalt system. However, the condition of a rubber content of 25% leads to the generation of a significant quantity of insufficiently swollen free rubber powder, augmenting the steric hindrance effect on rubber particles. Thus, the rubber-modified asphalt prepared with a rubber content of 25% exhibits the highest viscosity.

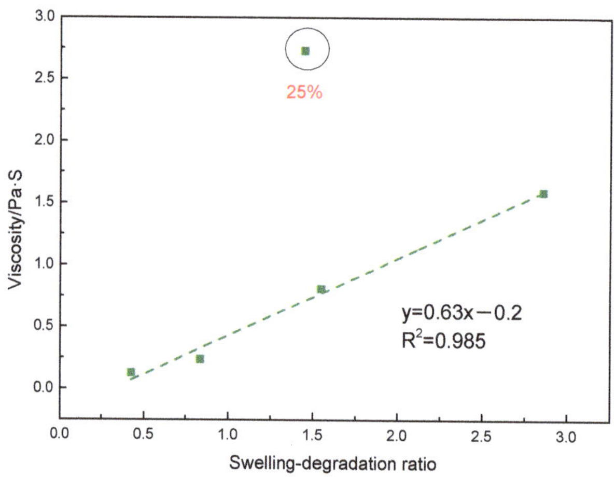

Figure 10. Correlation between φ and viscosity under different rubber contents.

As Figure 11 illustrates, the viscosity of rubber-modified asphalt rises with the increase in the degradation rate of swelling, except for at a rubber content of 25%. When the rubber content is relatively low, the probability of contact and collision between rubber particles diminishes. On the one hand, it hinders the capacity of individual rubber particles to interact further with other regions of the asphalt, once they have absorbed light components from the local asphalt region and undergone swelling and degradation. On the other hand, it allows individual rubber particles to be readily encapsulated by asphalt, effectively obstructing the passage for mass transfer between the rubber particles and asphalt, thereby diminishing the degree of interface interaction within the entire rubber–asphalt system. These factors collectively render the formation and stability of a three-dimensional network structure more challenging.

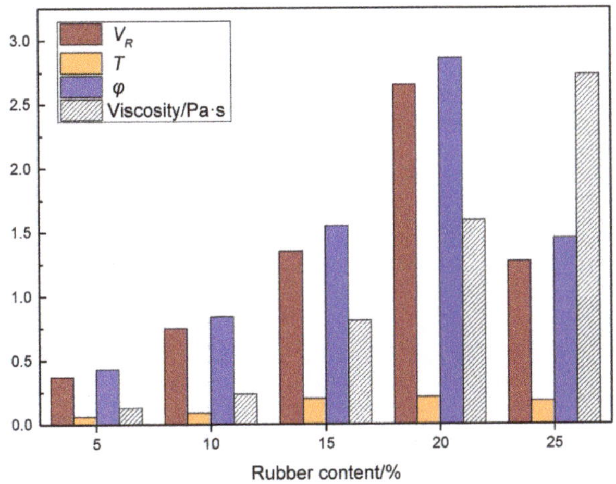

Figure 11. Comparison of viscosity and interaction parameters under different rubber contents.

However, under the combined influence of swelling and degradation, a gel-like film forms on the surface of the rubber particles, intensifying the adhesive effects between rubber and asphalt. As the content of rubber powder increases, this adhesive effect becomes more pronounced, thereby augmenting the steric hindrance effect on rubber particles, resulting in a higher viscosity of the rubber-modified asphalt. With the increasing rubber content, particles become less prone to immobilization by asphalt due to collision and friction, thereby increasing the probability of contact with lighter components from other regions and an enhancement in swelling and degradation capabilities. When the rubber content reaches 20%, both swelling and degradation rates peak. The swelling augments the volume of rubber particles, while degradation enhances the inter-particle connectivity, significantly improving the stability of the three-dimensional network structure. Moreover, it amplifies the contact area between the gel-like film and asphalt, resulting in an increase in the viscosity of rubber-modified asphalt.

6. Conclusions

This investigation conducted swelling tests and solvent elution tests to evaluate the swelling and degradation behaviors of rubber powder in asphalt under various preparation conditions. The swelling rate, degradation rate, and the combined swelling–degradation rate were proposed to characterize the interface interactions of the rubber–asphalt system. Furthermore, the influence mechanisms of interface interactions on the viscosity of rubber-modified asphalt were revealed. The main conclusions are as follows:

(1) The degradation of rubber powder is reliant on swelling, where swelling causes an increase in the volume of rubber particles, consequently enhancing the steric hindrance effect of individual particles, while degradation intensifies the connectivity among rubber particles. The interface interaction between rubber particles and asphalt facilitates the transfer of substances and promotes the formation of a three-dimensional network structure. The interface interactions within the rubber–asphalt system can be categorized into the following three stages: swelling, effective degradation, and excessive degradation.

(2) The stability of the three-dimensional network structure is a critical determinant of the viscosity of rubber-modified asphalt. The rubber size, the rubber content, and the preparation time and preparation temperature influence the inter-particle collision probability, individual particle activity, and the flow properties of asphalt, thereby impacting the interface interactions of the rubber–asphalt system. Stronger interface interactions lead to a higher viscosity in rubber-modified asphalt. The interface interactions of the

rubber–asphalt system are strongest under the optimal preparation conditions, all of which correlate with the highest viscosity of rubber-modified asphalt.

In the following research work, based on the interface interactions of the rubber–asphalt system, the contact slip test device independently developed will be adopted, taking into consideration aggregate factors, to explore the bonding/lubrication interface transition behavior of rubber-modified asphalt. This will provide a theoretical basis and new direction for determining the optimal mixing and compaction temperatures.

Author Contributions: Project administration, P.L.; funding acquisition: P.L. and J.S.; writing—original draft preparation: J.S.; investigation: G.Z.; methodology: S.D.; formal analysis: X.W.; validation: S.D. and X.W.; software: G.Z. All authors have read and agreed to the published version of the manuscript.

Funding: This research was funded by the National Natural Science Foundation of China [grant number 52278426], the Applied Basic Research Project of the Ministry of Transport of China [grant number 2014319812151], the Natural Science Basic Research Plan in Shaanxi Province of China [grant number 2019JM195], the Natural Science Basic Research Plan in Shandong Province [grant number ZR2023QE122], the Fundamental Research Funds for the Central Universities [grant number 300102213530, 300102212206], and the Open Research Project of Shandong Key Laboratory of Highway Technology and Safety Assessment [grant number SH202302].

Data Availability Statement: Data are contained within the article.

Conflicts of Interest: The authors declare no conflicts of interest.

References

1. Yuan, D.; Jiang, W.; Sha, A.; Xiao, J.; Wu, W.; Wang, T. Technology method and functional characteristics of road thermoelectric generator system based on Seebeck effect. *Appl. Energy* **2023**, *331*, 120459. [CrossRef]
2. Jiang, W.; Yuan, D.; Shan, J.; Ye, W.; Lu, H.; Sha, A. Experimental study of the performance of porous ultra-thin asphalt overlay. *Int. J. Pavement Eng.* **2022**, *23*, 2049–2061. [CrossRef]
3. Li, H.; Ling, X.; Zhao, J.; Li, W.; Chen, Z. Interlayer temperature analysis of asphalt pavement based on grey correlation degree. *J. Munic. Technol.* **2022**, *40*, 55–61.
4. Bressette, T.; Zhou, H.; Stonex, A.; Hicks, R.G. Asphalt Rubber and Its Potential Use in China. In *Plan, Build, and Manage Transportation Infrastructures in China*; American Society of Civil Engineers: New York, NY, USA, 2008.
5. Huang, W.D.; Wang, W.; Huang, Y.; Li, Y. Influencing Factor Research on High-temperature Performance of Asphalt Rubber Mixture. *J. Tongji Univ. (Nat. Sci.)* **2010**, *38*, 1023–1028.
6. Ilyin, S.O.; Arinina, M.P.; Mamulat, Y.S.; Malkin, A.Y.; Kulichikhin, V.G. Rheological properties of road bitumens modified with polymer and solid nanosized additives. *Colloid J.* **2014**, *76*, 425–434. [CrossRef]
7. Xiao, F.; Amirkhanian, S.N.; Shen, J.; Putman, B. Influences of crumb rubber size and type on reclaimed asphalt pavement (RAP) mixtures. *Constr. Build. Mater.* **2009**, *23*, 1028–1034. [CrossRef]
8. Mashaan, N.S.; Ali, A.H.; Koting, S.; Karim, M.R. Performance Evaluation of Crumb Rubber Modified Stone Mastic Asphalt Pavement in Malaysia. *Adv. Mater. Sci. Eng.* **2013**, *2013*, 304676. [CrossRef]
9. Mull, M.A.; Stuart, K.; Ychia, A. Fracture resistance characterization of chemically modified crumb rubber asphalt pavement. *J. Mater. Sci.* **2002**, *37*, 557–566. [CrossRef]
10. Chen, Y.Z.; Li, Z.X. Mesostructure of crumb rubber asphalt mixture based on discrete element method. *J. Harbin Inst. Technol.* **2013**, *45*, 116–121.
11. Tan, Y.Q.; Xu, L.T.; Xin, X.; Zhou, C.X. A Study on Compacting Technologies for Rubber-modified asphalt Concrete in Marshall Process. *Highway* **2006**, *2*, 137–139.
12. Huang, W.D.; Gao, C.; Li, K. Research on Influence Factors of AR Mixture Fatigue Performanc. *J. Tongji Univ. (Nat. Sci.)* **2009**, *37*, 1608–1614.
13. Cao, G.S.; Xu, Z.Z.; Wang, X.C. Laboratory experimental research on road performance of rubber particles asphalt mixture. *Eng. J. Wuhan Univ.* **2014**, *47*, 230–233.
14. Yu, M.; Wu, G.X. Research on the Mix Design of Dty Process Crumb Rubber-modified asphalt Mixture. *J. Build. Mater.* **2014**, *17*, 100–105.
15. Wang, Z.C. A Study on Dynamic Creep Tests and Viscoelastic Model of Dry-Blending Crumb Rubber-modified asphalt Mixtures. *J. Inn. Mong. Univ. (Nat. Sci. Ed.)* **2013**, *44*, 647–653.
16. Tang, N.P.; Huang, W.D.; Zheng, M. Influence of Mineral Fillers on Performance of Asphalt Rubber Mixtures. *J. Chongqing Jiaotong Univ. (Nat. Sci.)* **2014**, *33*, 49–52+114.
17. Su, J.; Li, P.; Wei, X.; Zhu, L.; Gao, J. Interface Transformation Behavior of Bonding/Lubrication of Aggregate-Asphalt System. *J. Mater. Civ. Eng.* **2020**, *32*, 04020380. [CrossRef]

18. Guo, F.C.; Zhang, J.P.; Pei, J.Z.; Ma, W.S. Evaluation of the compatibility between rubber and asphalt based on molecular dynamics simulation. *Front. Struct. Civ. Eng.* **2020**, *14*, 435–445. [CrossRef]
19. Guo, F.C.; Zhang, J.P.; Pei, J.Z.; Ma, W.S. Investigating the interaction behavior between asphalt binder and rubber in rubber asphalt by molecular dynamics simulation. *Constr. Build. Mater.* **2020**, *252*, 118956. [CrossRef]
20. Liseane, P.T.; Jorge, C.P.; Paulo, A.A.P.; Glicério, T.; Sara, R.A. Assessment of the digestion time of asphalt rubber binder based on microscopy analysis. *Constr. Build. Mater.* **2013**, *47*, 431–440.
21. Dong, D.W.; Huang, X.X.; Li, X.L.; Zhang, L.Q. Swelling process of rubber in asphalt and its effect on the structure and properties of rubber and asphalt. *Constr. Build. Mater.* **2012**, *29*, 316–322. [CrossRef]
22. Frantzis, P. Crumb Rubber-Bitumen Interactions: Diffusion of Bitumen into Rubber. *J. Mater. Civ. Eng.* **2004**, *16*, 387–390. [CrossRef]
23. Ni, Y.; Liao, M.Y.; Li, X. The Preparation Process and Properties of Crumb Rubber-modified asphalt. *Acta Sci. Nat. Univ. Sunyatseni* **2005**, *44*, 33–35.
24. Bradley, J.; Putman, A.M.; Serji, N.; Amirkhanian, M. Characterization of the Interaction Effect of Crumb Rubber Modified Binders Using HP-GPC. *J. Mater. Civ. Eng.* **2010**, *22*, 153–159.
25. Wang, L.; Xing, Y.; Chang, C. Experimental Studies on Microstructure and Technical Performance of Multiphase Compound Crumb Rubber-modified asphalt. In *ICCTP 2010: Integrated Transportation Systems, Proceedings of the 10th International Conference of Chinese Transportation Professionals, Beijing, China, 4–8 August 2010*; ASCE: New York, NY, USA, 2010; pp. 2920–2926.
26. Guillamot, F.; Goujard, L.; Simard, D.; Boulangé, L. Influence of interface energy on compatibility between ground tyre rubber and bitumen used in asphalt pavement. *Road Mater. Pavement Des.* **2013**, *14*, 372–383. [CrossRef]
27. Li, P.; Ding, Z.; Zou, P.; Sun, A. Analysis of physico-chemical properties for crumb rubber in process of asphalt modification. *Constr. Build. Mater.* **2017**, *138*, 418–426. [CrossRef]
28. *JTGE20-2011*; Standard Test Methods of Bitumen and Bituminous Mixtures for Highway Engineering. Occupation Standard of the People's Republic of China: Beijing, China, 2011.
29. *JT/J 797-2011*; Ground Vulcanized Rubber of Scrap Tires for Highway Engineering. Ministry of Transport of the People's Republic of China: Beijing, China, 2011.
30. Ding, Z.; Li, P.; Zhang, J.; Bing, H.; Yue, X. Analysis of Viscosity Test Conditions for Crumb-Rubber-Modified Asphalt. *Constr. Build. Mater.* **2020**, *245*, 118454. [CrossRef]
31. Li, P.L.; Ding, Z.; Chen, C. Analysis of Swelling and Degradation Characteristics of Crumb Rubber in Hot Asphalt. *J. South China Univ. Technol. (Nat. Sci. Ed.)* **2016**, *44*, 97–103.
32. Chen, C. Study on Viscosity Characteristics and Its Composing Mechanism of Asphalt Rubber. Master's Thesis, Chang'an University, Xi'an, China, 2016.
33. Su, J.; Li, P.; Sun, S.; Dong, C.; Zhu, L.; Akhtar, J. Evaluation on Contact Characteristics of Particle System Based on Meso-structure. *J. Mater. Civ. Eng.* **2020**, *32*, 04020391. [CrossRef]
34. Cai, X.L.; Wang, X.Y.; Wang, M.; Guan, B.W.; Xiong, R. Temperature Attenuation Characteristics of Rubber Asphalt Mixture Construction in Alpine Regions. *Highway* **2023**, *68*, 52–58.
35. Li, Y.G.; Li, J.Z.; Li, W. Micro-mechanism study and Road Engineering Application of Rubber Asphalt. *J. Highw. Transp. Res. Dev.* **2011**, *28*, 25–30. [CrossRef]

buildings

Article

Implementing All-Weather Photocatalysis of Exhaust Fumes Based on the g-C_3N_4/TiO_2/$SrAl_2O_4$: Eu^{2+}, Dy^{3+} Ternary Composite Coating

Bochao Zhou [1], Hailong Li [1], Ao Cui [1], Di Wang [2], Fucheng Guo [3] and Chao Wang [1,*]

1 Department of Road & Railway Engineering, Beijing University of Technology, Beijing 100124, China; bczhou@bjut.edu.cn (B.Z.); lihailong@emails.bjut.edu.cn (H.L.); cuiao@emails.bjut.edu.cn (A.C.)
2 Department of Civil Engineering, University of Ottawa, Ottawa, ON K1N 6N5, Canada; dwang6@uottawa.ca
3 School of Civil Engineering, Lanzhou Jiaotong University, Lanzhou 730070, China; fcguo@lzjtu.edu.cn
* Correspondence: wangchao@bjut.edu.cn; Tel.: +86-15801062996

Abstract: This study examines the use of $SrAl_2O_4$: Eu^{2+}, Dy^{3+} long-afterglow materials doped into g-C_3N_4/TiO_2 coatings for photodegradation. The prepared sample was tested for the purification of automotive exhaust fumes, with the optimal mass ratio of g-C_3N_4/TiO_2 and $SrAl_2O_4$: Eu^{2+}, Dy^{3+} determined to be 1:1. Characterization tests, including XRD, FT-IR, XPS, and TG-DSC, were conducted to evaluate the microstructure and properties of the samples. Under poor lighting conditions, g-C_3N_4/TiO_2 reduced CH and NO_x by 59 ppm and 13 ppm within 4 h, respectively, while g-C_3N_4/TiO_2/$SrAl_2O_4$: Eu^{2+}, Dy^{3+} decreased CH and NO_x by 98ppm and 34ppm, respectively, resulting in a significant improvement in degradation efficiency. The addition of long-afterglow materials significantly improves the efficiency of photocatalysts in purifying exhaust fumes in low-light environments, providing potential value for all-weather exhaust treatment in the future.

Keywords: photocatalysis; vehicle exhaust degradation; TiO_2; g-C_3N_4; $SrAl_2O_4$: Eu^{2+}, Dy^{3+}

1. Introduction

With the development of modern industry and transportation, vehicle exhaust pollution is becoming an increasingly serious concern. Exhaust fumes contain a large amount of harmful substances, such as nitrogen oxides (NO_x), carbon monoxide (CO), and hydrocarbons (CH) [1], which threaten human health and exacerbate air pollution. Therefore, the treatment of vehicle exhaust fumes is an urgent matter.

Numerous studies have shown that photocatalytic technology has significant potential in the field of exhaust degradation and purification [2–4]. It offers an attractive opportunity to address environmental remediation effectively and sustainably [5,6]. As automotive exhaust fumes first come into contact with the road, the application of photocatalytic materials to purify harmful gases on the road surface can improve the air quality in cities. As one of the typical semiconductor materials, titanium dioxide (TiO_2) is widely used in the field of photocatalysis due to its low cost, positive effect on environmental protection, stable chemical properties, and excellent photocatalytic performance compared to other materials [7–9]. However, its wide band gap of 3.28 eV can only absorb ultraviolet light, and its photogenerated electrons and holes are prone to recombination, reducing its photocatalytic efficiency and limiting its widespread use [10]. In order to improve the photocatalytic efficiency of TiO_2, there are studies on TiO_2 integrated with metal materials, non-metallic materials, and various semiconductors. Researchers in recent years have attempted to narrow the band gap and improve the photocatalytic efficiency of titanium dioxide by adding metals (Pt, Zn, Ag, Cu, Au) [11–15]. Nevertheless, the metal-doped TiO_2 exhibits instability and is prone to corrosion issues. This will lead to a gradual decline in the photocatalytic activity of the materials [16]. Zhang et al. introduced a new material made of N-doped TiO_2 into the road field, and when simulating exhaust fumes with CO and NO, it was found that its degradation efficiency was significantly improved [17]. Heffner et al. demonstrated that doping TiO_2 with C resulted in a significant improvement in optical performance, achieving a band gap reduction of 0.3 eV [18]. However, the non-metallic doping process involves high-temperature heat treatment or long-term hydrothermal treatment, which will consume a large amount of energy [19]. The composite of MoS_2 and TiO_2 successfully constructed a heterojunction, which can increase the separation rate of photogenerated carriers and significantly enhance photocatalytic activity [20]. In addition, the heterojunction will also provide a large surface area, promote the adsorption of reactants on the catalyst surface and accelerate photocatalysis [21]. The previous research has put forward innovative ideas to enhance the efficiency of TiO_2.

Graphite carbon nitride (g-C_3N_4) is widely used in the field of photocatalysis due to its advantages such as low cost, non-toxicity, good photochemical stability, and environmental friendliness [22]. However, its low specific surface area and easy recombination of photogenerated carriers hinder its practical application. Constructing a heterojunction to produce electron hole pairs with a high separation efficiency and wide light response range is an effective way to improve the photocatalytic ability of materials [23,24]. Therefore, the composite of g-C_3N_4 and TiO_2 offers a potential way to develop a new effective photocatalysis. Tan et al. developed a green route to prepare TiO_2 and g-C_3N_4 composite materials with two types of heterojunctions. The material has a large interface surface contact, which can accelerate electron migration and efficiently degrade indoor formaldehyde [25]. After successful recombination of TiO_2 and g-C_3N_4, the heterostructure between the two can effectively promote the separation of electrons and holes, significantly improving the photocatalytic reduction efficiency of carbon dioxide (CO_2) [26].

As infrastructure is booming, monofunctional building materials can no longer meet the needs of development. Instead, it is desirable for one material to have multiple functions, so the field of composites has grown rapidly [27]. Photocatalysts are unlikely to undergo photocatalytic reactions in low-light environments. In order to ensure that TiO_2/g-C_3N_4 continues to undergo photocatalytic degradation reactions in the absence of light, it is necessary to find an all-weather photocatalytic material to solve the above problem [28]. Aluminum salt long-afterglow material ($SrAl_2O_4$: Eu^{2+}, Dy^{3+}) is a good type of photoluminescence material with a good light storage performance. Its application and research are also the most extensive [29]. It is non-toxic, inexpensive, and has good chemical stability [30]. When there is light, it can store energy; when there is no light, electrons can be released, resulting in luminescence [31,32]. There are studies demonstrating that long-afterglow materials can be regarded as an additional light source that can tsupport TiO_2 for photocatalysis in a dark environment [33]. This integration aims to achieve a synergistic effect, thereby enhancing the catalytic activity of the photocatalyst and facilitating the degradation of automotive exhaust fumes.

This study aimed to develop a ternary composite photocatalyst of g-C_3N_4/TiO_2/$SrAl_2O_4$: Eu^{2+}, Dy^{3+} for the degradation of road vehicle exhaust emissions. The ternary composite photocatalyst degradation coating was prepared, and comprehensive testing was conducted to assess its fundamental properties. A self-made reaction chamber was irradiated with different intensities of visible light, and the purification rate of automobile exhaust fumes was tested. On this basis, the successful formation of the composite was verified by X-ray diffraction (XRD), Fourier transform infrared spectroscopy (FT-IR), X-ray photoelectron spectroscopy (XPS), and thermogravimetric analysis differential scanning calorimetry (TG-DSC), and the mechanism of performance improvement was explained. This study proposes a composite material that can degrade automotive exhaust fumes in both low-light and no-light environments, as well as provides a useful reference for the photocatalytic degradation of automotive exhaust fumes in the road domain, broadening the application scenarios and raising the potential for timely application.

2. Materials and Methods

2.1. Materials

The dispersing agent (99%, Beijing Mairuida Technology Co., Ltd., Beijing, China), talcum powder (Beijing Mairuida Technology Co., Ltd.), titanium dioxide (TiO_2, 99%, Beijing Mairuida Technology Co., Ltd.), epoxy resin diluent (98%, Shanghai McLean Biochemical Technology Co., Ltd., Shanghai, China), calcium carbonate ($CaCO_3$, 98%, Shanghai McLean Biochemical Technology Co., Ltd.), coalescent ($C_{12}H_{24}O_3$, 98%, Shanghai McLean Biochemical Technology Co., Ltd.), $SrAl_2O_4$: Eu^{2+}, Dy^{3+}: $SrCO_3$ (55%), Al_2O_3 (42%), Eu_2O_3 (1%), Dy_2O_3 (1%), and H_3BO_3 (1%) (Chuangrong Chemical Industry Technology Co., Ltd., Guangzhou, China), melamine ($C_3H_6N_6$, 99%, Shanghai Dibai Biotechnology Co., Ltd., Shanghai, China), epoxy resin (Beijing Honghu United Chemical Products Co., Ltd., Beijing, China), antifoaming agent ($C_3H_6Cl_2$, Beijing Puxitang Biotechnology Co., Ltd., Beijing, China), silane coupling agent (98%, Sinopharm, Beijing, China), fumed silica (99%, Beijing Hanlongda Technology Development Co., Ltd., Beijing, China), and epoxy resin curing agent (98%, Beijing China Ocean Co., Ltd., Beijing, China) were used as received from the vendors without further processing.

2.2. Preparation of g-C_3N_4/TiO_2 Composite Photocatalyst

Melamine ($C_3H_6N_6$) is the most common precursor, chosen as the precursor for g-C_3N_4 due to its low synthesis cost and ease of operation [34]. $C_3H_6N_6$ and TiO_2 were weighted as a mass ratio of 1:1 [35], ground, and blended until a homogeneous mixture was achieved. The resulting powder sample should then be transferred carefully to a ceramic crucible, covered, and placed in a muffle furnace. For the calcination process, it is advisable to keep the crucible cover semi-closed and set the temperature control system of the muffle furnace to 550 °C. After two hours of calcination, stop heating and open the oven door. It is important to allow the crucible to cool naturally to room temperature before removing it from the furnace for further grinding.

2.3. Preparation of Long-Afterglow Vehicle Exhaust Degradation Coatings

The materials used to create a long-lasting coating for vehicle exhaust degradation included epoxy resin, epoxy resin diluent, defoamer, dispersant, silane coupling agent, talcum powder, calcium carbonate, fumed silica, $SrAl_2O_4$: Eu^{2+}, Dy^{3+}, g-C_3N_4/TiO_2 composite photocatalyst, film-forming assistant alcohol ester XII, and epoxy resin curing agent. Firstly, 20 g of epoxy resin, 10 g of epoxy resin diluent, 0.5 g of defoaming agent, 3 g of dispersant, and 6 g of silane coupling agent were weighed and stirred for 9 min at a speed of 110 r/min. Next, 2.5 g talcum powder and 2.5 g calcium carbonate were added and stirred for 12 min at a speed of 180 r/min. Then, 0.5 g of anti-settling agent fumed silica was added, with a total mass of 35 g, to the phosphor powder and composite photocatalyst with mass ratios of 1:1, 1:2, 1:3, 2:1, and 3:1, respectively, as well as 9 g of film-forming aid alcohol ester dodecyl and 10 g of epoxy resin curing agent. We commenced stirring at a speed of 105 r/min for 15 min. Finally, a long-afterglow automobile exhaust purification coating was obtained after adding 1 g of antifoaming agent and stirring for 5 min at a speed of 130 r/min.

2.4. Evaluation of Basic Properties of Coatings

For an accurate assessment of the coating's condition and appearance, it is recommended to conduct a visual inspection. The rutted plate specimen should be evenly coated with a g-C_3N_4/TiO_2/$SrAl_2O_4$: Eu^{2+}, Dy^{3+} solution of 2 kg/m^2 and then cured at room temperature. Once cured, the specimen should be placed outside to absorb light fully. At night, observe the specimen's luminous effect, from its initial stimulation to the minimum brightness visible to the human eye (0.32 mcd/m^2) until it is no longer visible (<0.32 mcd/m^2). Keep a record of the afterglow duration. The coating has a solid content of 45.5%, ensuring robust stability with no indication of caking or delamination after storage. Its drying time is relatively short, only about 30 min. When exposed to light, the coating's afterglow duration is about 8 h before the light complete fades away, gradually decreasing in intensity over time. Therefore, this coating can provide a potential and feasible solution for automotive exhaust photocatalytic degradation under low light or even in a no-light environment.

2.4.1. Anti-Slip Properties

The asphalt mixture underwent rigorous evaluation, focusing on its friction coefficient and surface texture depth. In accordance with the Field Test Methods of Highway Subgrade and Pavement [36], the friction coefficient was measured using a pendulum friction meter. Furthermore, the surface texture depth was precisely determined by adhering to the Standard Test Methods of Bitumen and Bituminous Mixtures for Highway Engineering [37]. This approach ensured that the asphalt mixture met the required standards for friction and texture depth.

2.4.2. Adhesion

The adhesion strength of the paint was assessed through the implementation of the scratch test for paint and varnish films [38]. The evaluation procedure involved applying the paint to a plate specimen and subsequently dividing it into a grid pattern using a cutter. The cutter was required to make cuts deep enough to penetrate the coating, resulting in a total of 6 horizontal and 6 vertical splits. The detached coating blocks that were scattered from the specimen were counted, and the percentage of detachment was used to determine the adhesion index. The index was then evaluated based on the percentage of detachment obtained.

2.4.3. Water Resistance

In compliance with the technical specifications outlined in Pavement Marking Coatings [39], the coating should endure a duration of 24 h of water soaking without exhibiting any irregularities. To determine the coating's resistance to water, an immersion test was conducted in accordance with determination of resistance to water of films [40]. This entailed the following procedures: (1) A Marshall specimen coated with a heat-reflective coating was placed face down in a container. (2) The container was filled with deionized water until the Marshall specimen was submerged by two-thirds. (3) Following the 24 h soaking period, the test piece was removed and any residual moisture was absorbed using filter paper. The coated specimen was then inspected for any signs of discoloration, blistering, peeling, folding, rust, loss of reflectivity, or any other abnormalities.

2.4.4. Abrasion Resistance

The determination of coating wear resistance is achieved via the Rotary Wear Resistance Test, adhering strictly to the protocol outlined in the national standard [41]. This procedure involves employing a specifically designed hard rubber friction wheel, embedded with emery abrasive, to abrade the coating surface under precise testing conditions, including the maintenance of a constant load on the pressurizing arm. The wear resistance is then determined by calculating the mean value of the coating's mass loss after undergoing a specified number of grinding rotations.

2.5. *Evaluation of Exhaust Degradation Effect*

We weighed 40 g of a ternary composite photocatalytic coating, which was evenly applied to the rutted board. The vehicle exhaust was introduced to the reaction chamber, surrounded by tinfoil paper to achieve a good shading effect. The rutted plate specimen was placed in the closed reaction chamber, with the light source placed inside the reaction chamber for standby use. The gas outlet of the reaction chamber was connected to a vehicle exhaust analyzer, and the recorded values were monitored until the concentration stabilized, as shown in Figure 1.

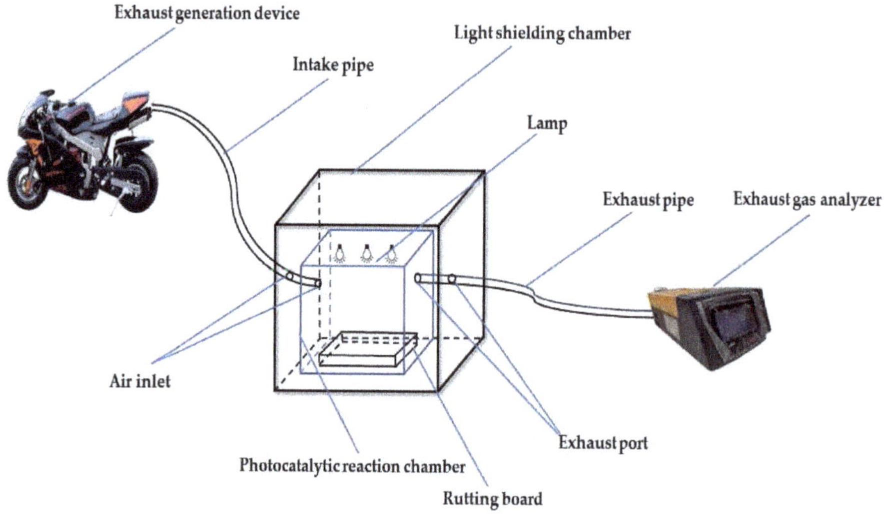

Figure 1. Schematic diagram of photocatalytic degradation for vehicle exhaust.

The calculation formula is as follows:

$$\begin{aligned} \mu_1 &= (A_1 - A_2)/A_1 \times 100\% \\ \mu_2 &= (B_1 - B_2)/B_1 \times 100\% \\ \mu_3 &= (C_1 - C_2)/C_1 \times 100\% \end{aligned} \quad (1)$$

The CO concentration value A_1, the NO concentration value B_1, the CH concentration value C_1, and time were recorded. The light source was turned on, and three different light intensities were set, with the strong light being 800 W/m^2, the weak light being 400 W/m^2, and no light being 0 W/m^2. After the degradation process began, when the exhaust concentration in the reaction chamber decreased and once again stabilized, the CO concentration value A_2, the NO concentration value B_2, and the CH concentration value C_2 were recorded. The degradation efficiencies μ_1, μ_2, and μ_3 of the coating under different light conditions were calculated from A_1, B_1, C_1 and A_2, B_2, C_2.

2.6. Characterization and Analysis of Microscopic Experimental Materials

XRD is a valuable technique that is utilized to analyze the elemental composition and morphology of a sample. FT-IR is a highly efficient method for obtaining molecular structure and chemical bond information on samples by measuring and analyzing their infrared spectra. XPS is employed to deeply determine the elemental composition and chemical bonding state of materials, contributing to a comprehensive understanding of their properties. TG-DSC is a crucial tool to study the thermal stability of materials by measuring the mass changes and thermal effects of the test sample. These microscopic experiments have been used to analyze the structure and morphology of ternary photodegradable composite coatings.

3. Results and Discussion

3.1. Evaluation of Basic Performance of Coating

After comprehensive testing, the long-lasting vehicle exhaust degradation coating displayed remarkable properties. It remained free from clumping in the container, ensuring smooth mixing. Following a period of static storage, there was no observable clumping or stratification, maintaining a consistent and uniform distribution, thereby indicating its reliable storage stability. We analyzed the adhesion index in correlation with the degree of detachment. The testing revealed that 11% of the entire coating area consisted of scattered coating blocks, which corresponded to an evaluation level of 2, indicating a favorable condition. This coating's resistance to water was also thoroughly tested, confirming its durability without experiencing any discoloration, bubbling, peeling, wrinkling, rusting, loss of gloss, or any other phenomena outlined in Section 2.4.3.

In accordance with the procedure outlined in Section 2.4.4, the wear resistance of the coating was subjected to rigorous testing, and the findings are presented in Figure 2. Wear resistance is an important consideration factor for coatings in practical applications [42]. The data indicate that the coating exhibited a favorable relative wear resistance and a moderate level of wear.

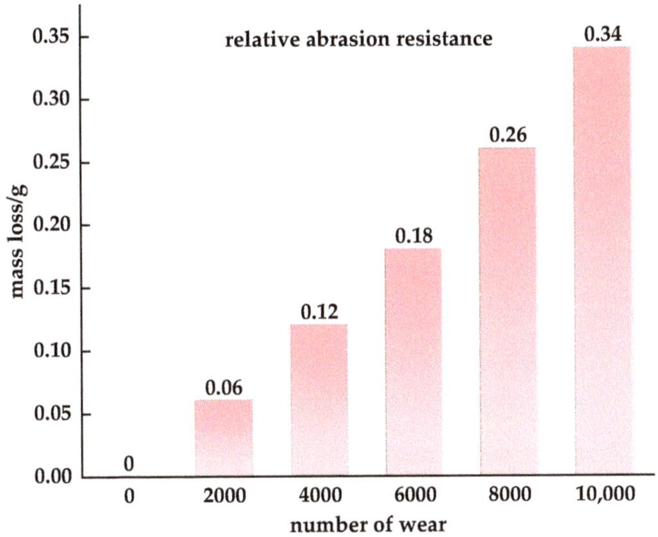

Figure 2. Abrasion test results of g-C_3N_4/TiO_2/$SrAl_2O_4$ coating.

3.2. Evaluation and Comparative Analysis of Exhaust Degradation Performance

The degradation efficiency of vehicle exhaust fumes under no-light conditions with varying mass ratios of the composite photocatalyst g-C_3N_4/TiO_2, and the extended afterglow materials was shown in Figure 3. As the mass ratio of the composite photocatalyst g-C_3N_4/TiO_2 to the long-afterglow materials was gradually reduced, the visible photocatalytic efficiency of the three main components in the exhaust fumes showed a tendency to increase at first and then decrease. From the change curve of the mass ratio from 1:1 to 1:3, we can surmise that when the content of g-C_3N_4/TiO_2 composite photocatalyst was high, the low degradation efficiency of the g-C_3N_4/TiO_2 composite photocatalyst under the no-light condition affected the degradation effect of the whole long-afterglow vehicle exhaust degradation coating, and when the content of the long-afterglow materials was high, the degradation performance of the overall long-afterglow vehicle exhaust degradation coating was not possible due to its low content of g-C_3N_4/TiO_2 photocatalyst. Therefore, the optimum mass ratio of g-C_3N_4/TiO_2 composite photocatalyst to the long-afterglow materials is 1:1.

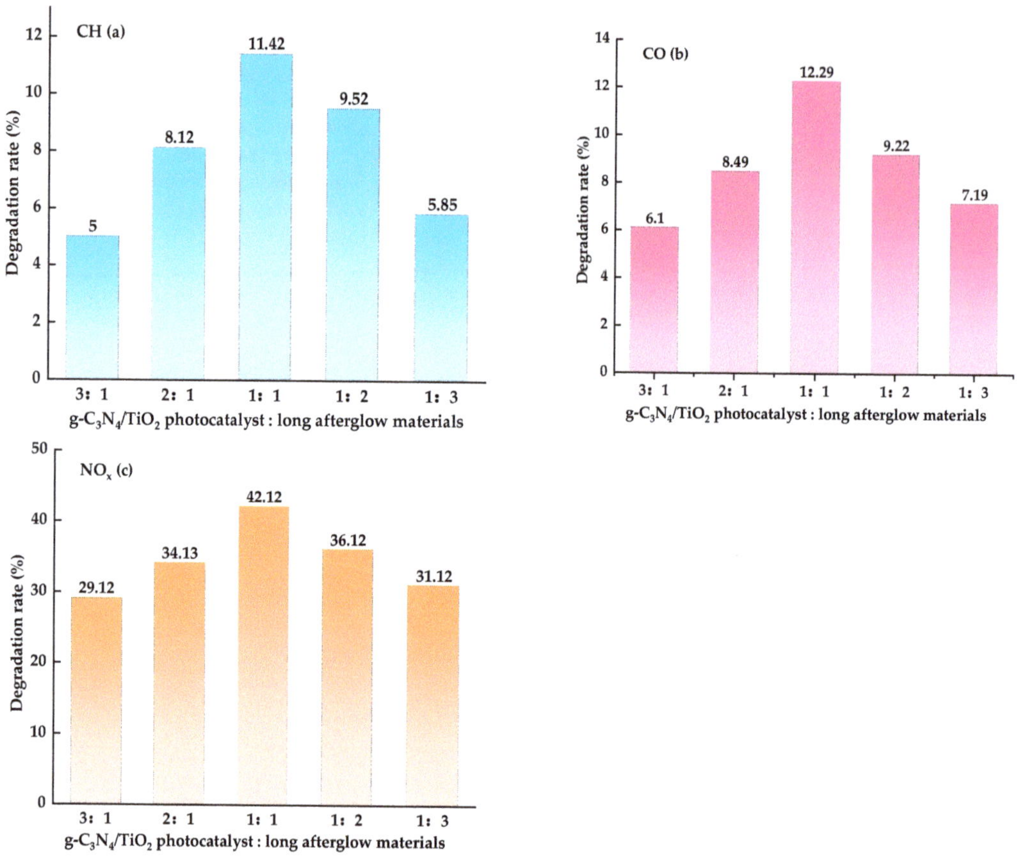

Figure 3. Degradation efficiency of $g\text{-}C_3N_4/TiO_2$ and long-afterglow materials with different mass ratios on exhaust fumes' different gases: (**a**) CH, (**b**) CO, (**c**) NO_x.

Figures 4–6 show the efficiency of coatings with or without long-afterglow materials $SrAl_2O_4$: Eu^{2+}, Dy^{3+} in degrading NO_x, CH, and CO under different lighting conditions. For all coatings with long-afterglow materials, in these experiments, the mass ratio of the $g\text{-}C_3N_4/TiO_2$ composite photocatalyst to the long-afterglow materials was 1:1. From the strong light decomposition curve, it is clear that the catalytic reaction of the $g\text{-}C_3N_4/TiO_2/SrAl_2O_4$: Eu^{2+}, Dy^{3+} vehicle exhaust decomposition coating and the $g\text{-}C_3N_4/TiO_2$ composite photocatalytic converter is relatively rapid when the light source is switched on. During the photocatalytic process, each product shows mainly a concave curve, which indicates that after a period of degradation, the degradation products produced by each gas will hinder the contact between the polluted gas and the photocatalytic materials to a certain extent. This leads to a gradual decrease in the visible photocatalytic performance of the sample, which is in line with the experimental results mentioned above. Under a strong-light environment (800 W/m^2), the degradation effect of the $g\text{-}C_3N_4/TiO_2$ composite photocatalyst is higher than that of the $g\text{-}C_3N_4/TiO_2/SrAl_2O_4$: Eu^{2+}, Dy^{3+} vehicle exhaust degradation coating. This is because the formation of the coating will have a slight negative impact on the original degradation effect. Moreover, the presence of $SrAl_2O_4$: Eu^{2+}, Dy^{3+} will to some extent hinder the full contact between exhaust fumes and $g\text{-}C_3N_4/TiO_2$, leaving it unable to demonstrate its advantages in strong-light environments.

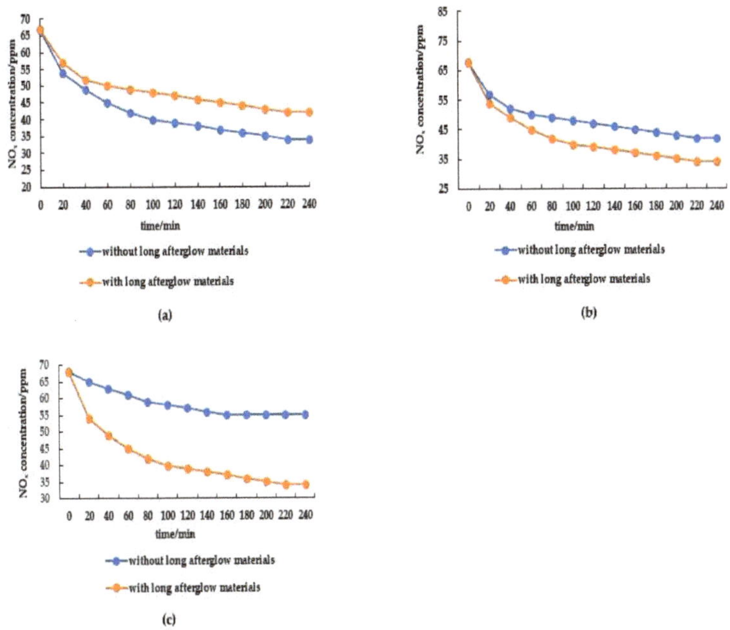

Figure 4. Degradation efficiency results of NO_x with and without long-afterglow materials $SrAl_2O_4$: Eu^{2+}, Dy^{3+} under different light intensities: (**a**) 800 W/m^2, (**b**) 400 W/m^2, (**c**) 0 W/m^2.

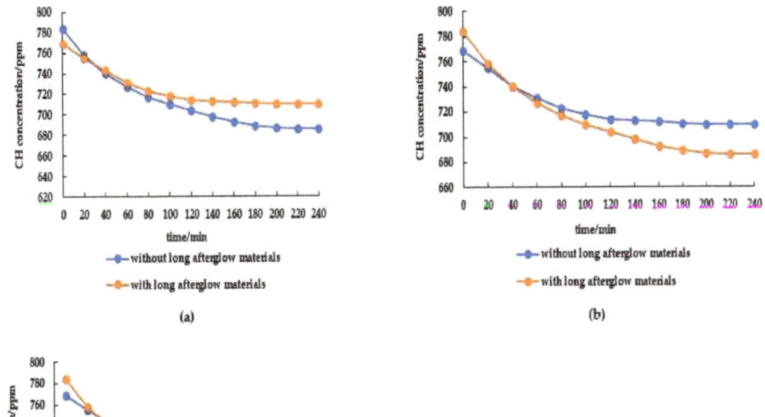

Figure 5. Degradation efficiency results of CH with and without long-afterglow materials under different light intensities: (**a**) 800 W/m^2, (**b**) 400 W/m^2, (**c**) 0 W/m^2.

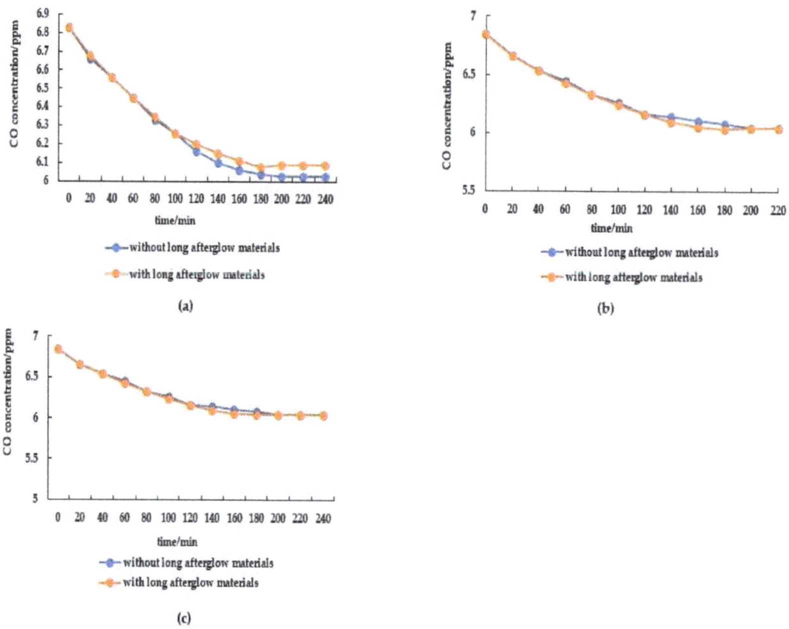

Figure 6. Degradation efficiency results of CO with and without long-afterglow materials $SrAl_2O_4$: Eu^{2+}, Dy^{3+} under different light intensities: (**a**) 800 W/m^2, (**b**) 400 W/m^2, (**c**) 0 W/m^2.

In a weak-light environment (400 W/m^2), the g-C_3N_4/TiO_2 photocatalyst demonstrated significant reductions in pollutant concentrations after four hours. Specifically, it achieved a decrease in NO_x concentration of 26 ppm and a reduction in CH concentration of 59 ppm. However, with the introduction of long-afterglow materials, these improvements were further enhanced. Within the same four-hour period, the concentration of NO_x decreased by an additional 34 ppm, while the concentration of CH decreased by a remarkable 98 ppm. The degradation effect of the long-afterglow automotive exhaust emission control coatings was higher than that of the g-C_3N_4/TiO_2 composite photocatalyst. This is due to the fact that the light intensity is an important factor that affects the degradation effect of the g-C_3N_4/TiO_2 photocatalyst. The lower the light intensity, the worse the degradation efficiency, but the long-afterglow materials can compensate for the light intensity, leading to a slightly greater degradation efficiency. However, the difference between the two is not significant, and the weak-light condition is not sufficient to prove the advantages of the long-afterglow coatings for vehicle exhaust degradation.

In a dark environment (0 W/m^2), the g-C_3N_4/TiO_2 photocatalyst showed a decrease in NO_x degradation from 68 ppm to 55 ppm within 4 h, with a concentration difference of 13 ppm. The degradation of CH decreased from 769 ppm to 710 ppm, with a concentration difference of 59 ppm. However, after the introduction of long-afterglow materials, the photocatalyst showed significant photocatalytic activity within the same time period, reducing the degradation of NO_x from 68 ppm to 34 ppm, with a concentration difference of 34 ppm. The degradation of CH correspondingly decreased from 784 ppm to 686 ppm, with a concentration difference of 98 ppm. That is, the g-C_3N_4/TiO_2 composite photocatalyst showed a weak degradation effect, while the long-afterglow automotive exhaust degradation coating still maintained a good degradation efficiency.

The advantages of g-$C_3N_4/TiO_2/SrAl_2O_4$: Eu^{2+}, Dy^{3+} ternary composite photocatalysts in the degradation of NO_x, CH, and CO could not be reflected under strong-light conditions of 800 W/m^2. However, in low-light conditions of 400 W/m^2, the ternary composite coating showed a 30.8% increase in NO_x degradation efficiency and a 66.1% increase in CH degradation efficiency compared to g-C_3N_4/TiO_2. Under no-light conditions of 0 W/m^2, the improvement was more significant; we noted that the incorporation of long-afterglow materials led to a significant increase in the degradation rate of NO_x, which was 2.6 times higher compared to g-C_3N_4/TiO_2, and likewise for CH, for which its rate was 1.7 times greater than that of g-C_3N_4/TiO_2.

As shown in Figure 6, the results indicated an overall decrease in CO concentration. However, compared to g-C_3N_4/TiO_2, under the low-light and no-light conditions, the improvement of g-$C_3N_4/TiO_2/SrAl_2O_4$: Eu^{2+}, Dy^{3+} in CO degradation rate was too small to show regular changes. The reason for this phenomenon may be that a reversible reaction occurs between CO and CO_2 when the catalyst is irradiated with light [43]. The two are both reactants and products, and after a certain period of time, their concentrations reach a relative equilibrium state. However, due to the influence of physical adsorption, the concentration of CO will still show a slight downward trend. When we considering all three lighting conditions, it was apparent that the photocatalyst's degradation efficiency exceeded that of g-C_3N_4/TiO_2 when long-afterglow materials $SrAl_2O_4$: Eu^{2+}, Dy^{3+} were integrated.

3.3. Microscopic Experimental Results

3.3.1. XRD

Figure 7 shows two distinct peaks, with one at around 27.5°and the other at approximately 13°. The g-C_3N_4 sample exhibits two standard peaks at 12.87° and 27.45°, which correspond to the planes of (100) and (002), respectively (JCPDS 87–1526). The peak at 27.45° is attributed to the conjugated aromatic structure of g-C_3N_4, while the faint peak at 12.87° is a result of the in-planar structure. It is noteworthy that the peak at 27.45° is relatively more intense than the peak at 12.87°, which highlights the importance of the conjugated aromatic structure in the sample [44]. The above two peaks indicate that g-C_3N_4 was successfully prepared [45]. The peaks at around 2θ = 25.3°, 37.8°, 53.9°, 55.1°, 62.7°, and 68.8° are attributed to the (101), (004), (105), (211), (204), and (116) crystal planes of rutile TiO_2 [46]. Meanwhile, when the long-afterglow materials were added, the typical characteristic peaks of TiO_2 and g-C_3N_4 could still be observed, which further confirmed that the addition of the long-afterglow materials did not damage the structure of the g-C_3N_4/TiO_2 composite catalyst.

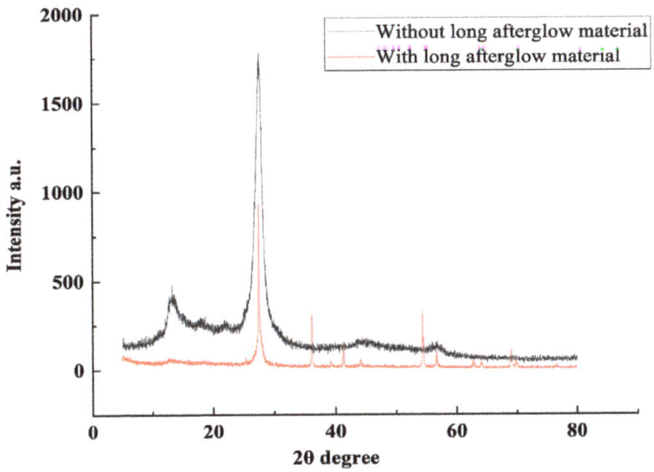

Figure 7. Comparison of XRD patterns of g-C_3N_4/TiO_2 doped with or without long-afterglow materials.

3.3.2. FT-IR

The presence of characteristic peaks in the vibrational spectrum of a material can provide useful information about the properties of the materials. Figure 8 exhibits a clearly defined peak at 808 cm^{-1}, which corresponds to the breathing vibration of the g-C_3N_4 triazine structure [47]. The region between 1240 cm^{-1} and 1640 cm^{-1} shows several characteristic peaks. These include peaks at 1240 cm^{-1}, 1329 cm^{-1}, 1430 cm^{-1}, and 1635 cm^{-1}, which are attributed to the C-N and C=N stretching vibrations of g-C_3N_4 [6]. Finally, the broad absorption peak in the range of 3000–3300 cm^{-1} is attributed to the presence of N-H and N=H bonds remaining from incomplete condensation of melamine during the heating process [48]. In a comparison between scenarios with and without long-afterglow materials added, the shapes of the two plots are largely consistent, indicating that the addition of the long-afterglow materials did not destroy the structure of the original g-C_3N_4/TiO_2, and also reflecting the successful composite of the three after the addition of the long-afterglow materials. These observations are expected to facilitate its use in a range of applications, including photocatalysis, sensing, and energy conversion, as they provide useful insights into the structural and functional properties of the materials.

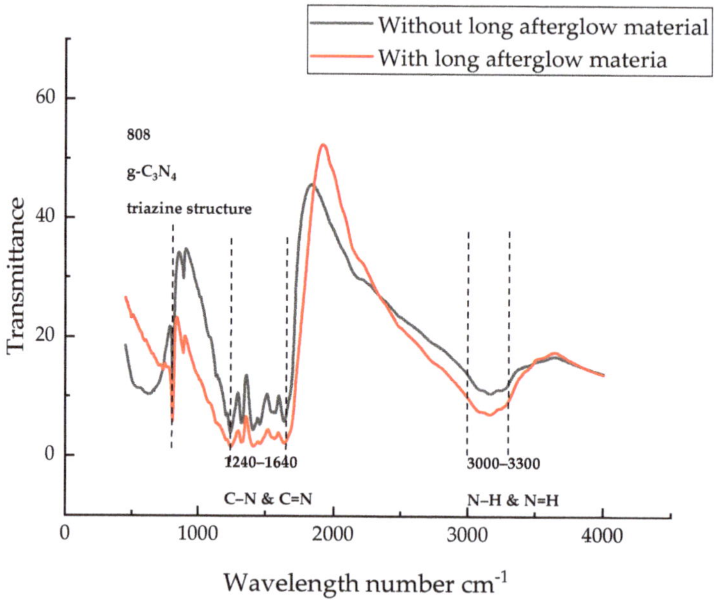

Figure 8. Infrared spectrogram with and without long-afterglow materials, evaluating the stability of the structure of the original g-C_3N_4/TiO_2.

3.3.3. XPS

The chemical binding states of each element in g-C_3N_4/TiO_2/$SrAl_2O_4$: Eu^{2+}, Dy^{3+} and g-C_3N_4/TiO_2 were characterized through XPS analysis. As we can see from Figure 9a, all elements including C, N, O, and Ti were detected in the survey spectra of g-C_3N_4/TiO_2, and additional Sr, Al, Eu, and Dy were in g-C_3N_4/TiO_2/$SrAl_2O_4$: Eu^{2+}, Dy^{3+}. The C1s spectrum shows three peaks at 284.8 eV, 286.2 eV, and 288.3 eV. The peak at 284.8 eV corresponds to a surface unstable C-C group, originating from pure graphitic carbon from the XPS instrument itself; the peak at 286.2 eV corresponds to the C-O group of g-C_3N_4; and the peak at 288.3 eV corresponds to the C-N or N=C-N_2 group of g-C_3N_4 [49]. The N1s spectrum shows three peaks at 398.4 eV, 399.1 eV, and 400.2 eV, assigned to the C=N-C group, C≡C group, and N-H group [50]. The O1s spectrum shows two peaks at 530.9 eV and 532.9 eV depicted in g-C_3N_4/TiO_2, representing the Ti-O band and -OH group, respectively, and an additional peak at 531.5 eV in g-C_3N_4/TiO_2/$SrAl_2O_4$: Eu^{2+}, Dy^{3+} representing the Al-O band. Within the spectrum of g-C_3N_4/TiO_2/$SrAl_2O_4$: Eu^{2+}, Dy^{3+}, a slight shift of the -OH peak to a higher binding energy position is observed, potentially attributed to the introduction of long-afterglow materials. Both samples showed peaks at around 455.4 and 461.2 eV, which we identified as the 2p3/2 and 2p1/2 peaks, indicating the presence of Ti^{4+} in both samples. No other peaks related to Ti^{2+} or Ti^{3+} were found. In Figure 9f, there is only one fitted peak for the Al 2p oxidation state, originating from $SrAl_2O_4$ (at 73.8 eV). The high-resolution scan of Sr 3d reveals two distinct peaks corresponding to $Sr3d_{5/2}$ (133.0 eV) and $Sr3d_{3/2}$ (134.8 eV), which are associated with the Sr from the two sites in $SrAl_2O_4$ [51]. In addition, there are $Eu3d_{5/2}$ (1110.1~1179.9 eV) and $Dy3d_{5/2}$ (1280.1~1349.4 eV) in g-C_3N_4/TiO_2/$SrAl_2O_4$ from the XPS survey; however, their contents are relatively low and it is difficult to further analyze the characteristic peaks. The XPS results indicate that a combination of chemical bonds in the form of g-C_3N_4/TiO_2/$SrAl_2O_4$: Eu^{2+}, Dy^{3+} is preferable, instead of physically mixing separate g-C_3N_4, TiO_2, and $SrAl_2O_4$: Eu^{2+}, Dy^{3+}.

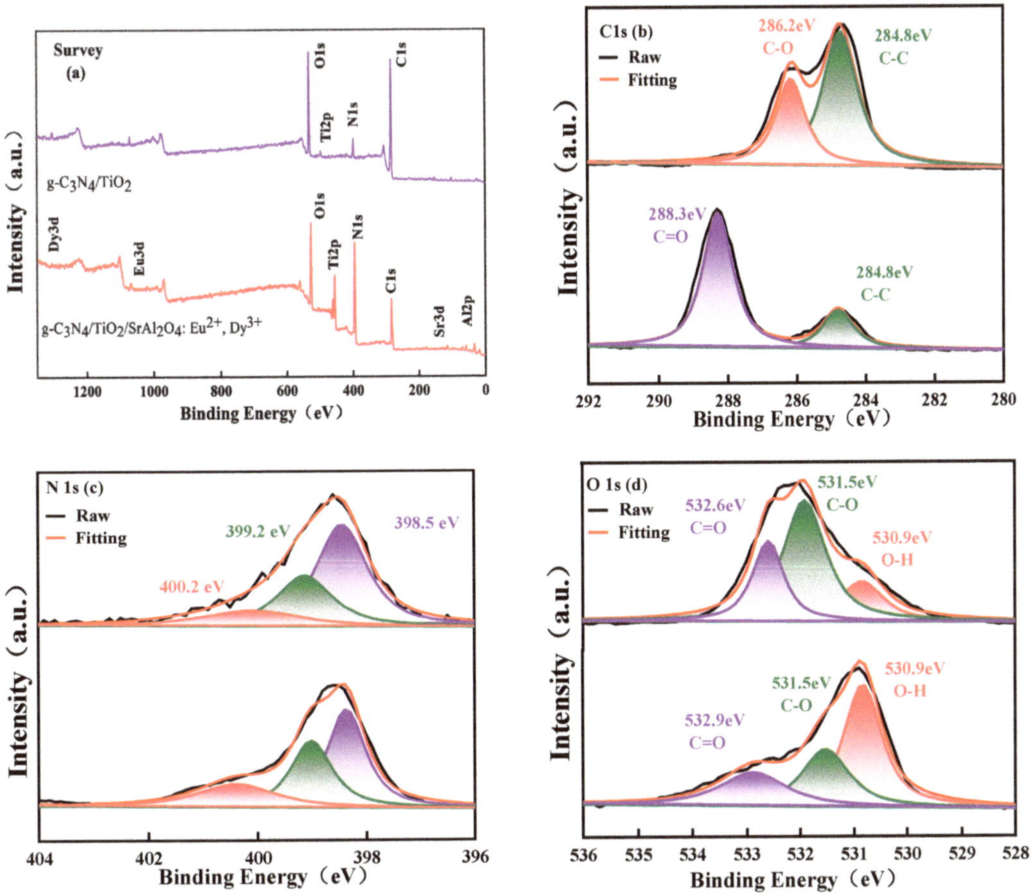

Figure 9. *Cont.*

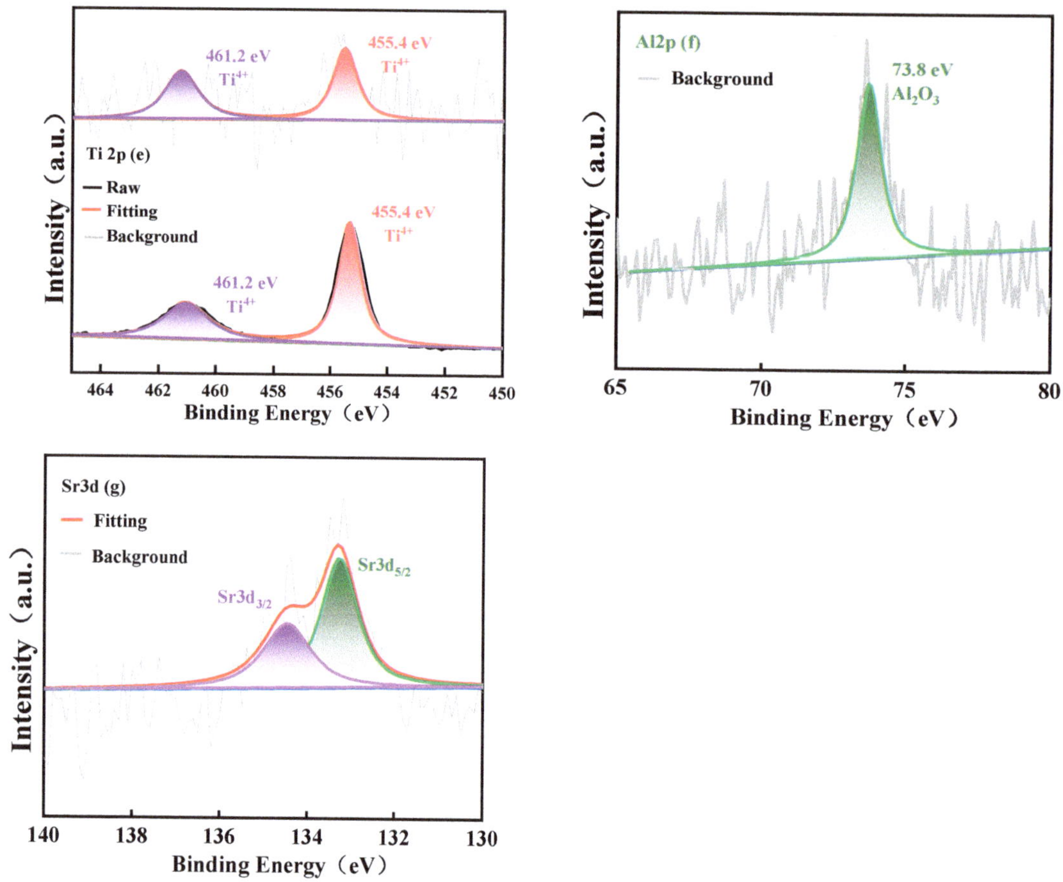

Figure 9. XPS spectra of $g-C_3N_4/TiO_2$ and $g-C_3N_4/TiO_2/SrAl_2O_4$: Eu^{2+}, Dy^{3+}: (**a**) survey and deconvoluted (**b**) C1s, (**c**) N1s, (**d**) O1s, (**e**) Ti 2p, (**f**) Al 2p, (**g**) Sr 3d.

3.3.4. TG-DSC

Figure 10 is a thermal analysis diagram of the ternary composite photocatalyst. In the TG curve, there is a 3.8% mass loss step between 100 and 250°, which can be attributed to the desorption of water adsorbed by the relevant materials [52]. Between 250 and 400 °C, the mass of the compound undergoes a 28.4% decline. This may correspond to the decomposition of $g-C_3N_4$ in composite materials. The decomposition temperature of the composite material is lower than that of pure $g-C_3N_4$, which may be attributed to the distribution of TiO_2 on the curved surface of $g-C_3N_4$, reducing the van der Waals force of $g-C_3N_4$ [53]. Organic solvents or additives are used in the preparation process of long-afterglow materials, which may decompose or evaporate in the temperature range of 250–400 °C and may also cause quality loss. The final mass loss between 400 and 600 °C can be attributed to the thermal decomposition of Sr_2CO_3 [54]. Based on the DSC curve depicted in Figure 10, the ternary composite material exhibits consistent heat absorption within the range of 30–600°, with no particularly prominent exothermic peaks.

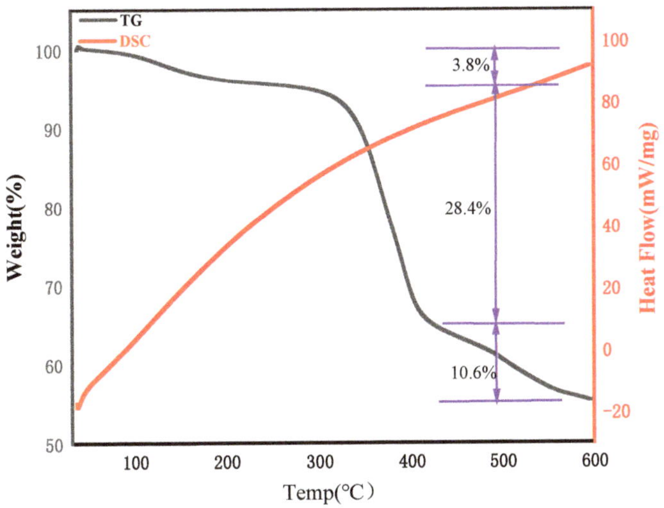

Figure 10. TG-DSC curves of thermal decomposition of g-$C_3N_4/TiO_2/SrAl_2O_4$: Eu^{2+}, Dy^{3+} under a nitrogen atmosphere.

4. Conclusions

This research study was conducted to develop and evaluate a composite photocatalyst coating composed of g-$C_3N_4/TiO_2/SrAl_2O_4$: Eu^{2+}, Dy^{3+} and compare its properties with those of g-C_3N_4/TiO_2. This study aimed to determine whether the addition of $SrAl_2O_4$: Eu^{2+}, Dy^{3+} could enhance the photocatalytic performance of the ternary composite photocatalyst under varying lighting conditions. The key findings of this research are as follows.

Firstly, we evaluated the anti-slip properties, adhesion, water resistance, and abrasion resistance of the ternary compound coating and found that it exhibited a good basic performance, and that the afterglow luminescence process lasted for more than 8 h in a dark environment. Secondly, this study demonstrated that the degradation efficiency of the ternary composite photocatalyst was slightly lower than that of g-C_3N_4/TiO_2 when exposed to intense light. However, it was significantly improved in the absence of light. Lastly, microscopic experiments revealed that the introduction of $SrAl_2O_4$: Eu^{2+}, Dy^{3+} cannot damage the original structure and stability of g-C_3N_4/TiO_2.

This study presents a composite photocatalyst, consisting of g-$C_3N_4/TiO_2/SrAl_2O_4$: Eu^{2+}, Dy^{3+}, which has demonstrated great promise in efficiently degrading pollutants in automotive exhaust fumes in all-weather conditions. This composite photocatalyst offers a potential solution to the growing environmental concern about automotive emissions, with the ability to be practically implemented in engineering applications. However, further research is required to fully understand the specific mechanisms behind the photocatalyst's remarkable ability to degrade pollutants in exhaust fumes. To gain a deeper understanding, future studies may employ theoretical calculations to reveal objective reaction laws from experimental phenomena, which will aid in the development of more directional and forward-looking materials. Overall, this study offers hope for the development of practical and effective solutions to mitigate the harmful effects of automotive emissions on the environment.

Author Contributions: B.Z.: Writing—Review and Editing, Supervision, Funding Acquisition, Project Administration; H.L.: Investigation, Data Curation, Writing—Original Draft; A.C.: Methodology, Resources, Validation; D.W., Writing—Review and Editing; F.G.: Writing—Review and Editing; C.W.: Conceptualization, Supervision. All authors have read and agreed to the published version of the manuscript.

Funding: This research was funded by the "Beijing Municipal Education Commission", grant number KM202310005002, and the "Transportation Science and Technology Research Project of Hebei Province", grant number 202206.

Data Availability Statement: The raw data supporting the conclusions of this article will be made available by the authors on request.

Conflicts of Interest: The funders had no role in the design of this study; in the collection, analyses, or interpretation of the data; in the writing of the manuscript; or in the decision to publish the results.

Abbreviations

NO_x	nitrogen oxides
CO	carbon monoxide
CH	hydrocarbons
TiO_2	titanium dioxide
g-C_3N_4	graphite carbon nitride
CO_2	carbon dioxide
$SrAl_2O_4$: Eu^{2+}, Dy^{3+}	a kind of long-afterglow material
XRD	X-ray diffraction
FT-IR	Fourier transform infrared spectroscopy
XPS	X-ray photoelectron spectroscopy
TG-DSC	thermogravimetric and differential scanning calorimetry
$C_3H_6N_6$	melamine
μ	degradation efficiency
A	CO concentration value
B	NO concentration value
C	CH concentration value

References

1. Twigg, M.V. Progress and future challenges in controlling automotive exhaust gas emissions. *Appl. Catal. B Environ.* **2007**, *70*, 2–15. [CrossRef]
2. Li, X.; Wan, F.; You, L.; Wu, S.; Yang, C.; Zhang, L.; Barbieri, D.M. A review on photocatalytic asphalt pavement designed for degradation of vehicle exhausts. *Transp. Res. Part D Transp. Environ.* **2023**, *115*, 103605. [CrossRef]
3. Wu, J.; Luo, Y.; Qin, Z. Composite-modified nano-TiO_2 for the degradation of automobile exhaust in tunnels. *Constr. Build. Mater.* **2023**, *408*, 133805. [CrossRef]
4. Zhao, J.; Sun, J.; Meng, X.; Li, Z. Recent advances in vehicle exhaust treatment with photocatalytic technology. *Catalysts* **2022**, *12*, 1051. [CrossRef]
5. Song, X.; Hu, Y.; Zheng, M.; Wei, C. Solvent-free in situ synthesis of g-C_3N_4/{0 0 1} TiO_2 composite with enhanced UV-and visible-light photocatalytic activity for NO oxidation. *Appl. Catal. B Environ.* **2016**, *182*, 587–597. [CrossRef]
6. Zhang, M.; Jiang, W.; Liu, D.; Wang, J.; Liu, Y.; Zhu, Y.; Zhu, Y. Photodegradation of phenol via C_3N_4-agar hybrid hydrogel 3D photocatalysts with free separation. *Appl. Catal. B Environ.* **2016**, *183*, 263–268.
7. Hou, Z.; Chu, J.; Liu, C.; Wang, J.; Li, A.; Lin, T.; François-Xavier, C.P. High efficient photocatalytic reduction of nitrate to N_2 by Core-shell Ag/SiO_2@ cTiO_2 with synergistic effect of light scattering and surface plasmon resonance. *Chem. Eng. J.* **2021**, *415*, 128863. [CrossRef]
8. Hsu, C.Y.; Mahmoud, Z.H.; Abdullaev, S.; Ali, F.K.; Naeem, Y.A.; Mizher, R.M.; Karim, M.M.; Abdulwahid, A.S.; Ahmadi, Z.; Habibzadeh, S.; et al. Nano titanium oxide (nano-TiO_2): A review of synthesis methods, properties, and applications. *Case Stud. Chem. Environ. Eng.* **2024**, *9*, 100626. [CrossRef]
9. Ng, K.H.; Yuan, L.S.; Cheng, C.K.; Chen, K.; Fang, C. TiO_2 and ZnO photocatalytic treatment of palm oil mill effluent (POME) and feasibility of renewable energy generation: A short review. *J. Clean. Prod.* **2019**, *233*, 209–225. [CrossRef]
10. Bindhu, M.R.; Willington, T.D.; Hatshan, M.R.; Chen, S.-M.; Chen, T.-W. Environmental photochemistry with Sn/F simultaneously doped TiO_2 nanoparticles: UV and visible light induced degradation of thiazine dye. *Environ. Res.* **2022**, *207*, 112108. [CrossRef]
11. Bibi, S.; Shah, S.S.; Muhammad, F.; Siddiq, M.; Kiran, L.; Aldossari, S.A.; Mushab, M.S.S.; Sarwar, S. Cu-doped mesoporous TiO_2 photocatalyst for efficient degradation of organic dye via visible light photocatalysis. *Chemosphere* **2023**, *339*, 139583. [CrossRef] [PubMed]
12. Esrafili, A.; Salimi, M.; Sobhi, H.R.; Gholami, M.; Kalantary, R.R. Pt-based TiO_2 photocatalytic systems: A systematic review. *J. Mol. Liq.* **2022**, *352*, 118685. [CrossRef]
13. Ferreira VR, A.; Pereira, C.M.; Silva, A.F.; Azenha, M.A. Ag-doped hollow TiO_2 microspheres for the selective photo-degradation of bilirubin. *Appl. Surf. Sci.* **2023**, *641*, 158457. [CrossRef]

14. Karuppasamy, P.; Nisha NR, N.; Pugazhendhi, A.; Kandasamy, S.; Pitchaimuthu, S. An investigation of transition metal doped TiO_2 photocatalysts for the enhanced photocatalytic decoloration of methylene blue dye under visible light irradiation. *J. Environ. Chem. Eng.* **2021**, *9*, 105254. [CrossRef]
15. Veziroglu, S.; Ullrich, M.; Hussain, M.; Drewes, J.; Shondo, J.; Strunskus, T.; Adam, J.; Faupel, F.; Aktas, O.C. Plasmonic and non-plasmonic contributions on photocatalytic activity of Au-TiO_2 thin film under mixed UV–visible light. *Surf. Coat. Technol.* **2020**, *389*, 125613. [CrossRef]
16. Kuvarega, A.T.; Mamba, B.B. TiO_2-based photocatalysis: Toward visible light-responsive photocatalysts through doping and fabrication of carbon-based nanocomposites. *Crit. Rev. Solid State Mater. Sci.* **2017**, *42*, 295–346. [CrossRef]
17. Zhang, L.; Lu, Q.; Shan, R.; Zhang, F.; Muhammad, Y.; Huang, K. Photocatalytic degradation of vehicular exhaust by nitrogen-doped titanium dioxide modified pavement material. *Transp. Res. Part D Transp. Environ.* **2021**, *91*, 102690. [CrossRef]
18. Heffner, H.; Faccio, R.; López-Corral, I. C–doped TiO_2 (B): A density functional theory characterization. *Appl. Surf. Sci.* **2021**, *551*, 149479. [CrossRef]
19. Khan, H.; Shah, M.U.H. Modification strategies of TiO_2 based photocatalysts for enhanced visible light activity and energy storage ability: A review. *J. Environ. Chem. Eng.* **2023**, *11*, 111532. [CrossRef]
20. Sun, Y.; Yang, Y.L.; Chen, H.J.; Liu, J.; Shi, X.L.; Suo, G.; Hou, X.; Ye, X.; Zhang, L.; Lu, S.; et al. Flexible, recoverable, and efficient photocatalysts: MoS_2/TiO_2 heterojunctions grown on amorphous carbon-coated carbon textiles. *J. Colloid Interface Sci.* **2023**, *651*, 284–295. [CrossRef]
21. Acharya, R.; Parida, K. A review on TiO_2/g-C_3N_4 visible-light-responsive photocatalysts for sustainable energy generation and environmental remediation. *J. Environ. Chem. Eng.* **2020**, *8*, 103896. [CrossRef]
22. Patnaik, S.; Sahoo, D.P.; Parida, K. Recent advances in anion doped g-C_3N_4 photocatalysts: A review. *Carbon* **2021**, *172*, 682–711. [CrossRef]
23. Li, D.; Li, R.; Zeng, F.; Yan, W.; Deng, M.; Cai, S. The photoexcited electron transfer and photocatalytic mechanism of g-C_3N_4/TiO_2 heterojunctions: Time-domain ab initio analysis. *Appl. Surf. Sci.* **2023**, *614*, 156104. [CrossRef]
24. Ma, X.; Yang, H.; Wang, B.; Wu, L.; Ma, X.; Li, Y.; Deng, H. Mechano-synthesized TiO_2/g-C_3N_4 composites for rapid photocatalytic removal of perrhenate. *J. Environ. Chem. Eng.* **2023**, *11*, 109423. [CrossRef]
25. Tan, Y.; Shu, R.; Xu, H.; Song, L.; Zhang, R.; Ouyang, C.; Xia, M.; Hou, J.; Zhang, X.; Yuan, Y.; et al. Supercritical carbon dioxide-assisted TiO_2/g-C_3N_4 heterostructures tuning for efficient interfacial charge transfer and formaldehyde photo-degradation. *J. Environ. Chem. Eng.* **2023**, *11*, 110992. [CrossRef]
26. Wang, H.; LHi Chen, Z.; Li, J.; Li, X.; Huo, P.; Wang, Q. TiO_2 modified g-C_3N_4 with enhanced photocatalytic CO_2 reduction performance. *Solid State Sci.* **2020**, *100*, 106099. [CrossRef]
27. Ahmad, M.R.; Pan, Y.; Chen, B. Physical and mechanical properties of sustainable vegetal concrete exposed to extreme weather conditions. *Constr. Build. Mater.* **2021**, *287*, 123024. [CrossRef]
28. Hai, O.; Pei, M.; Yang, E.; Ren, Q.; Wu, X.; Zhu, J.; Zhao, Y.; Du, L. Exploration of long afterglow luminescence materials work as round-the-clock photocatalysts. *J. Alloys Compd.* **2021**, *866*, 158752. [CrossRef]
29. Rojas-Hernandez, R.E.; Rubio-Marcos, F.; Rodriguez, M.Á.; Fernandez, J.F. Long lasting phosphors: $SrAl_2O_4$: Eu, Dy as the most studied material. *Renew. Sustain. Energy Rev.* **2018**, *81*, 2759–2770. [CrossRef]
30. Wu, Y.; Gan, J.; Wu, X. Study on the silica-polymer hybrid coated $SrAl_2O_4$: Eu^{2+}, Dy^{3+} phosphor as a photoluminescence pigment in a waterborne UV acrylic coating. *J. Mater. Res. Technol.* **2021**, *13*, 1230–1242. [CrossRef]
31. Chen, M.L.; Li, S.S.; Wen, L.; Xu, Z.; Li, H.H.; Ding, L.; Cheng, Y.H. Exploration of double Z-type ternary composite long-afterglow/graphitic carbon nitride@ metal–organic framework for photocatalytic degradation of methylene blue. *J. Colloid Interface Sci.* **2023**, *629*, 409–421. [CrossRef] [PubMed]
32. Li, S.S.; Liu, M.; Wen, L.; Xu, Z.; Cheng, Y.H.; Chen, M.L. Exploration of long afterglow luminescent materials composited with graphitized carbon nitride for photocatalytic degradation of basic fuchsin. *Environ. Sci. Pollut. Res.* **2023**, *30*, 322–336. [CrossRef] [PubMed]
33. Yang, X.; Waterhouse GI, N.; Lu, S.; Yu, J. Recent advances in the design of afterglow materials: Mechanisms, structural regulation strategies and applications. *Chem. Soc. Rev.* **2023**, *52*, 8005–8058. [CrossRef] [PubMed]
34. Papamichail, P.; Nannou, C.; Giannakoudakis, D.A.; Bikiaris, N.D.; Papoulia, C.; Pavlidou, E.; Lambropoulou, D.; Samanidou, V.; Deliyanni, E. Maximization of the photocatalytic degradation of diclofenac using polymeric g-C_3N_4 by tuning the precursor and the synthetic protocol. *Catal. Today* **2023**, *418*, 114075. [CrossRef]
35. Zhou, B.; Cui, A.; Wu, H.; Han, W.; Wang, C. Composition design and performance evaluation of g-C_3N_4/TiO_2 composite photocatalyst for automobile exhaust. *Munic. Technol.* **2024**, *42*, 186–192 + 236.
36. *JTG 3450-2019*; Field Test Methods of Highway Subgrade and Pavement. Ministry of Transport of the People's Republic of China: Beijing, China, 2019.
37. *JTG E20-2011*; Standard Test Methods of Bitumen and Bituminous Mixtures for Highway Engineering. Ministry of Transport of the People's Republic of China: Beijing, China, 2011.
38. *GB/T 9286-2021*; Paintsand Varnishes—Cros-Cuttest. State Administration for Market Regulation, China National Standardization Administration: Beijing, China, 2021.
39. *JT/T 280-2022*; Pavement Marking Paint. Ministry of Transport of the People's Republic of China: Beijing, China, 2022.

40. *GB/T 1733-1993;* Determination of Resistance to Water of Films. China National Bureau of Technical Supervision: Beijing, China, 1993.
41. *GB/T 1768-1979;* Method of Test for Abrasion Resistance of Paint Films. China National Standardization Administration: Beijing, China, 1979.
42. Cai, P.; Xu, M.; Wei, M.; Zhang, T.; Yao, F. Preparation and characterization of SiO_2–TiO_2 superhydrophilic coatings with photocatalytic activity induced by low temperature. *Colloids Surf. A Physicochem. Eng. Asp.* **2024**, *686*, 133264. [CrossRef]
43. Zhao, J.; Yang, X. Photocatalytic oxidation for indoor air purification: A literature review. *Build. Environ.* **2003**, *38*, 645–654. [CrossRef]
44. Firoozabadi, S.R.; Khosravi-Nikou, M.R.; Shariati, A. Kinetic study and optimization of hierarchical TiO_2 flower-like/exfoliated g-C_3N_4 composite for improved carbon dioxide photoconversion to methane and methanol under visible light. *J. Clean. Prod.* **2024**, *451*, 142019. [CrossRef]
45. Ding, P.; Ji, H.; Li, P.; Liu, Q.; Wu, Y.; Guo, M.; Zhou, Z.; Gao, S.; Xu, W.; Liu, W.; et al. Visible-light degradation of antibiotics catalyzed by titania/zirconia/graphitic carbon nitride ternary nanocomposites: A combined experimental and theoretical study. *Appl. Catal. B Environ.* **2022**, *300*, 120633. [CrossRef]
46. Zhang, H.; Bian, H.; Wang, F.; Zhu, L.; Zhang, S.; Xia, D. Enhanced photocatalytic reduction of CO_2 over pg-C_3N_4-supported TiO_2 nanoparticles with Ag modification. *Colloids Surf. A Physicochem. Eng. Asp.* **2023**, *674*, 131989. [CrossRef]
47. Wang, J.; Wang, G.; Cheng, B.; Yu, J.; Fan, J. Sulfur-doped g-C_3N_4/TiO_2 S-scheme heterojunction photocatalyst for Congo Red photodegradation. *Chin. J. Catal.* **2021**, *42*, 56–68. [CrossRef]
48. Chi, X.; Liu, F.; Gao, Y.; Song, J.; Guan, R.; Yuan, H. An efficient B/Na co-doped porous g-C_3N_4 nanosheets photocatalyst with enhanced photocatalytic hydrogen evolution and degradation of tetracycline under visible light. *Appl. Surf. Sci.* **2022**, *576*, 151837. [CrossRef]
49. Qu, J.; Wang, P.; Wang, Y.; Li, Z.; Yang, F.; Han, C.; Wang, L.; Yu, D. Determination of phospholipids in soybean oil using a phospholipase-choline oxidase biosensor based on g-C_3N_4-TiO_2 nanocomposite material. *J. Food Compos. Anal.* **2023**, *124*, 105717. [CrossRef]
50. Nasir, M.S.; Yang, G.; Ayub, I.; Wang, S.; Yan, W. In situ decoration of g-C_3N_4 quantum dots on 1D branched TiO_2 loaded with plasmonic Au nanoparticles and improved the photocatalytic hydrogen evolution activity. *Appl. Surf. Sci.* **2020**, *519*, 146208. [CrossRef]
51. Gültekin, S.; Yıldırım, S.; Yılmaz, O.; Keskin, İ.Ç.; Katı, M.İ.; Çelik, E. Structural and optical properties of $SrAl_2O_4$: Eu^{2+}/Dy^{3+} phosphors synthesized by flame spray pyrolysis technique. *J. Lumin.* **2019**, *206*, 59–69. [CrossRef]
52. Jiang, G.; Yang, X.; Wu, Y.; Li, Z.; Han, Y.; Shen, X. A study of spherical TiO_2/g-C_3N_4 photocatalyst: Morphology, chemical composition and photocatalytic performance in visible light. *Mol. Catal.* **2017**, *432*, 232–241. [CrossRef]
53. Moradi, S.; Isari, A.A.; Hayati, F.; Kalantary, R.R.; Kakavandi, B. Co-implanting of TiO_2 and liquid-phase-delaminated g-C_3N_4 on multi-functional graphene nanobridges for enhancing photocatalytic degradation of acetaminophen. *Chem. Eng. J.* **2021**, *414*, 128618. [CrossRef]
54. Lu, Y.; Li, Y.; Xiong, Y.; Wang, D.; Yin, Q. $SrAl_2O_4$: Eu^{2+}, Dy^{3+} phosphors derived from a new sol–gel route. *Microelectron. J.* **2004**, *35*, 379–382. [CrossRef]

buildings

MDPI

Article

Research on Laboratory Testing Method of Fatigue Performance of Semi-Rigid Base Considering Spatial Stress State

Longfei Wang [1], Zhizhong Zhao [1], Hao Liang [2], Yilong He [3], Xianzhang Kang [1] and Meng Xu [1,*]

[1] School of Transportation and Civil Engineering, Shandong Jiaotong University, Jinan 250357, China
[2] Shandong Luzhong Highway Construction Co., Ltd., Zibo 255030, China
[3] Shandong Expressway Environmental Protection Building Materials Co., Ltd., Jinan 250098, China
* Correspondence: xumeng@sdjtu.edu.cn

Abstract: In order to accurately assess the fatigue performance of semi-rigid base layers, this paper proposes a novel fatigue testing method for semi-rigid base layers that takes into account the spatial stress state. Based on this method, the fatigue performance of three types of reinforced semi-rigid base-layer materials (steel wire mesh, plastic geogrid, and fiberglass) was tested and compared with unreinforced materials. The fatigue strain evolution patterns of these materials were analyzed, and a fatigue strain characteristic value at the limit state was proposed as an evaluation index for the fatigue failure of base layer materials. The results showed that in terms of fatigue performance, plastic geogrid > steel wire mesh > fiberglass > unreinforced specimens. The development of tensile strain can be approximately classified into a three-stage growth pattern, consisting of "curve + straight line + curve". For the unreinforced specimens, the three stages of bending strain accounted for 10%, 70%, and 20% of the total fatigue life, respectively. The fatigue failure stages of the three types of reinforced materials had similar proportional ranges, representing 5%, 75%, and 20% of the total fatigue life, respectively. The fatigue strain characteristic values for plastic geogrid, steel wire mesh, fiberglass, and unreinforced specimens were 371 με, 280 με, 280 με, and 195 με, respectively. In summary, the use of reinforced materials within semi-rigid base layers enhances their fatigue performance, providing new insights and methods for extending the service life of road surfaces and offering scientific guidance for the practical application of reinforced materials in semi-rigid base layer road surfaces' fatigue performance.

Keywords: base layer; reinforcement materials; MTS; fatigue performance; spatial stress state

1. Introduction

Semi-rigid base refers to the base that is paved with inorganic binder-stabilized soil, can form a plate body and has a certain bending strength. Among various pavement structures, semi-rigid base layers have gained widespread application in various construction environments due to their combination of the advantages of both flexible and rigid base layers [1–4]. However, with the continuous increase in traffic volume and load-bearing capacity, many semi-rigid base layer road surfaces still exhibit structural damage, such as fatigue cracking, before reaching their intended service life. This is highly detrimental to the long-term development of pavement structure [5–8]. Therefore, enhancing the fatigue performance of semi-rigid base layers using advanced technology holds significant practical importance [9,10]. The researchers have investigated the effects of structure design [11,12], aggregate types [13,14], cement properties [15–18], and aggregate properties [10,19–21] on the fatigue performance of semi-rigid base layers. The idea that adding reinforcement materials in semi-rigid base layers draws inspiration from the concept of adding reinforcements in bridge structures to improve their flexural tensile performance [22,23]. By enhancing the bond between the reinforcement materials and the base layer materials, the flexural tensile strength and fatigue performance of semi-rigid base layers are improved.

Simultaneously, the occurrence and propagation of cracks in semi-rigid base layer road surfaces are mitigated or suppressed, offering a novel approach to extending road surface service life.

In order to determinate the fatigue failure of semi-rigid base layers accurately, the researchers focused on the influence of environmental conditions [8,24–26], stress intensity [27–29], and vehicle loads [30–33], which provided guidance for the high-performance development of semi-rigid base layers. But these studies also have some shortcomings because they mostly used four-point bending beam specimens to evaluate the fatigue performance of semi-rigid base layers; however, this method only considered unidirectional stress states in small beam specimens under load, while actual road surfaces experience bidirectional stress states under the action of wheel loads. Without accounting for the actual stress states and environmental conditions of road surfaces, the accuracy and reliability of test results are significantly limited. Complex adjustments are necessary in practical design. Therefore, the urgent challenge at hand is to design a new testing method that can simulate the actual spatial stress state in semi-rigid base layers accurately.

To address the issues mentioned above and rectify the considerable disparity between small beam specimen four-point bending fatigue testing methods and the actual stress states and environmental conditions on road surfaces, this paper leverages the Material Test System (MTS) to propose a novel fatigue testing approach that accounts for bidirectional stress states in semi-rigid base-layer road surfaces. Firstly, starting with a mixed design, this paper determines the optimal aggregate grading and prepares reinforced cement-stabilized crushed stone slab specimens. Next, a novel fatigue testing approach is designed to collect fatigue strain data from the specimens, ensuring the reliability and authenticity of the testing process. Finally, through an analysis of specimen fatigue strain, this paper investigates the fatigue failure patterns of semi-rigid base layers and explores the impact of reinforcement factors on the fatigue life, fatigue cracking, and fatigue strain of semi-rigid base layer road surfaces. The results indicate that the indoor testing method for assessing the fatigue performance of semi-rigid base layers, which takes into account spatial stress states, is practical, and that the test data align with theoretical foundations and exhibit a high level of reliability.

2. Raw Materials and Mix Design

2.1. Raw Materials

2.1.1. Cement

Ordinary Portland cement (Type P·O 42.5) was selected as the cement material. The selection was based on the standards and test procedures outlined in "Common portland cement" (GB 175-2007) [34], "Test Methods of Cement and Concrete for Highway Engineering" (JTG E30-2005) [35], and "Technical Guidelines for Construction of Highway Roadbases" (JTG/T F20-2015) [36]. Relevant tests were conducted to evaluate the performance of the cement material. The test results and performance indicators are presented in Table 1.

Table 1. Test Results for Various Parameters of Cement (P·O 42.5).

Test Project		Unit	Index Requirement	Test Result	Test Method
Stability		mm	≤ 5	2.5	T 0505—2005
Setting time	Initial setting time	min	≥ 180	330	
	Final setting time		360~600	400	
Water requirement for standard consistency		%	30.0	28.4	
Specific surface area		m^2/kg	300~450	380	T 0504—2005
Fineness (80 μm)		%	≤ 10.0	3.6	T 0502—2005
Density		kg/m^3	≈ 3100	3120	T 0503—2005
28 d dry shrinkage		%	≤ 0.10	0.06	T 0511—2005
Abrasion resistance		kg/m^2	≤ 3.0	2.1	T 0510—2005

Table 1. *Cont.*

Test Project		Unit	Index Requirement	Test Result	Test Method
Flexural strength	3 d	MPa	≥3.5	4.5	T 0506—2005
	28 d		≥6.5	9.7	
Compressive strength	3 d	MPa	≥16.0	21.5	
	28 d		≥42.5	48.6	

From Table 1, it can be observed that the technical specifications of the Ordinary Portland cement (Type P·O 42.5) comply with the required standards. Therefore, it is suitable for use in fatigue performance testing for reinforced semi-rigid base layer road surfaces.

2.1.2. Aggregates

Limestone was used as the aggregate material. In accordance with the "Test Methods of Aggregate for Highway Engineering" (JTG E42-2005) [37] and "Technical Guidelines for Construction of Highway Roadbases" (JTG/T F20-2015) [36], various tests were conducted to evaluate the performance of the aggregates. The test results are presented in Table 2. It is evident that the technical specifications for both coarse and fine aggregates meet the requirements stipulated in the standards.

Table 2. Test Results for Various Parameters of Aggregates.

	Test Project	Unit	Index Requirement	Test Result	Test Method
Coarse aggregate	Crush value	%	≤26	12.7	T 0316—2005
	Needle flake particle content	%	≤18	8.2	T 0312—2005
	Dust content below 0.075 mm	%	≤2	0.7	T 0310—2005
	Soft stone content	%	≤5	1.3	T 0320—2005
Fine aggregate	Particle analysis	—	Satisfying gradation	Satisfying gradation	T 0327—2005
	Plasticity index	—	≤17	8.2	T 0118—2007
	Organic matter content	%	<2	0.7	T 0336—1994
	Sulfate content	%	≤0.25	0.12	T 0341—1994

2.1.3. Reinforcement Materials

By referring to the literature, three reinforcing materials, namely selected steel wire mesh, geogrid and glass fiber, which have been mentioned many times, were selected, and then, combined with factors such as material acquisition difficulty and raw material price, galvanized steel wire mesh, biaxial plastic geogrid and glass fiber geogrid were finally selected as reinforcing materials in this paper.

(1) Galvanized Steel Wire Mesh

As shown in Figure 1, galvanized steel wire mesh with 3 cm openings and a 0.7 mm wire diameter was chosen as the reinforcement material. Performance tests of the galvanized steel wire mesh were conducted based on "Welded wire fabric coated with zine" (GB/T 33281-2016) [38], "Test method for gravimetric determination of the mass per unit area of galvanized coatings on steel products" (GB/T 1839-2008) [39], and "Metallic materials—Tensile testing—Part 1: Method of test at room temperature" (GB/T 228.1-2021) [40]. The test results and performance indicators are presented in Table 3.

(2) Biaxial Plastic Geogrid

As shown in Figure 2, polypropylene biaxial plastic geogrid with a designation of TGSG50-50 was selected as the reinforcement material. Performance tests of the biaxial plastic geogrid were carried out based on "Geosynthetics—Plastic geogrids" (GB/T 17689-2008) [41]. The test results and performance indicators can be found in Table 3.

Figure 1. Galvanized Steel Wire Mesh.

Table 3. Test Results of Various Parameters for Reinforcement Materials.

Reinforcement Materials	Test Project		Unit	Index Requirement	Test Result	Test Method
Galvanized Steel Wire Mesh	Diameter		mm	0.70 ± 0.04	0.69	GB/T 33281-2016 [38] GB/T 1839-2008 [39] GB/T 228.1-2021 [40]
	Warp mesh size (clear)		mm	30.00 ± 0.75	30.22	
	Weft mesh size (clear)		mm	30.00 ± 0.30	30.15	
	Solder joint tensile force		N	>40	118	
	Zinc layer quality		g/m^2	>122	161	
Bidirectional Plastic Geogrid	Ultimate tensile strength	Portrait	kN/m	≥50.0	51.9	GB/T 17689-2008 [41]
		Landscape orientation	kN/m	≥50.0	51.6	
	Elongation at nominal strength	Portrait	%	≤15.0	13.7	
		Landscape orientation	%	≤13.0	12.9	
	Tensile strength at 2% elongation	Portrait	kN/m	≥17.5	18.4	
		Landscape orientation	kN/m	≥17.5	18.6	
	Tensile strength at 5% elongation	Portrait	kN/m	≥35.0	35.3	
		Landscape orientation	kN/m	≥35.0	35.7	
Glass Fiber Geogrid	Meridional fracture strength		kN/m	≥50.0	51.7	GB/T 21825-2008 [42]
	Zonal breaking strength		kN/m	≥50.0	51.2	
	Elongation at break		%	≤4.0	2.5	
	Elongation at zonal break		%	≤4.0	2.6	

Figure 2. Biaxial Plastic Geogrid.

(3) Glass Fiber Geogrid

As shown in Figure 3, a glass fiber geogrid with a designation of EGA50-50 was chosen as the reinforcement material. Performance tests of the glass fiber geogrid were conducted based on "Glass fibre geogrid" (GB/T 21825-2008) [42]. The test results and performance indicators are presented in Table 3.

Figure 3. Glass Fiber Geogrid.

From Table 3, it is evident that the technical specifications of the galvanized steel wire mesh, biaxial plastic geogrid, and glass fiber geogrid all meet the required standards. Therefore, they can be used in fatigue performance testing for reinforced semi-rigid base layer road surfaces.

2.2. Mix Design

2.2.1. Aggregate Grading Design

Various aggregates were subjected to sieve analysis in accordance with the "Test Code for Highway Engineering Aggregates" (JTG E42-2005). The mix design for aggregates was selected based on the "Technical Specifications for Road Base Construction" (JTG/T F20-2015). For the semi-rigid base layer material, cement-stabilized crushed stone was chosen, and the mix design selected followed the recommended C-B-1 type as per the specifications. The aggregate grading curve is shown in Figure 4.

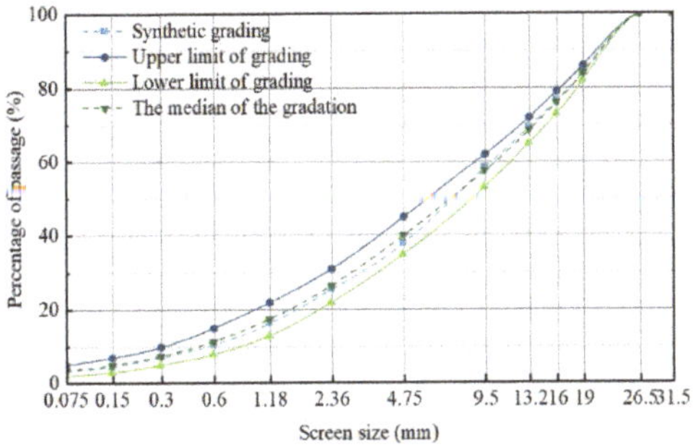

Figure 4. Gradation Curve of Aggregate.

Taking into consideration the recommended aggregate grading range and practical experience in aggregate design, after repeated calculations and comparisons, the optimal aggregate ratio for various particle sizes was determined to be 20 mm~30 mm:10 mm~20 mm:5 mm~10 mm:0 mm~5 mm = 18:34:17:31.

2.2.2. Optimal Moisture Content

For cement-stabilized materials, cement dosages of 3%, 4%, 5%, 6%, and 7% were considered, and external water was used. Moisture content levels of 4%, 4.5%, 5%, 5.5%, 6%, and 6.5% were evaluated. A heavy-impact compaction test was conducted on the cement-stabilized crushed stone mixture prepared in the determined proportions to establish the optimal moisture content. Additionally, the unconfined compressive strength at 7 days was measured. The results are illustrated in Figure 5.

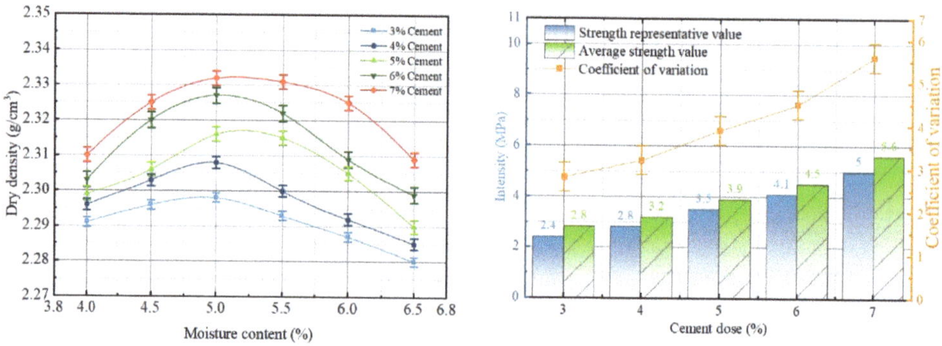

(**A**) Optimal Moisture Content (**B**) Unconfined Compressive Strength

Figure 5. Optimal Moisture Content and Unconfined Compressive Strength.

The representative value of unconfined compressive strength should meet or exceed the strength standard values. Considering the relatively high market price of cement, and while prioritizing cost-effective solutions that ensure project quality in practical engineering, a cement dosage of 5% was selected along with a moisture content of 5.2% and a maximum dry density of 2.318 g/cm^3. These parameters were chosen based on both design and construction experience. The combination of these values was deemed suitable for the project. To validate the technical performance of the synthetic graded mixture, it was confirmed that the strength requirements were met, and the mixture exhibited good resistance to erosion and cracking.

2.3. Specimen Preparation

Specimen preparation methods drew from the techniques used for inorganic binder-stabilized material beam specimens. The mold for specimen molding was inspired by the rutting plates used in asphalt mixture rutting tests, with enhancements such as increased height and thickness. The dimensions are as shown in Figure 6 (inner dimensions designed as 300 mm × 300 mm × 100 mm). The molding process is illustrated in Figure 6.

Figure 6. Specimen Preparation Flowchart.

Due to the large number of specimens required for the experiments, a small mixer, as shown in Figure 6, was chosen for mixing. Unreinforced specimens were selected as the reference group. Different specimens were molded after the placement of reinforcement materials and were finally statically compacted. According to the testing protocol, specimens were cured after demolding, with a 90-day curing period and water immersion treatment on the final day. Given the significant time span of specimen preparation, large temperature variations, and substantial material requirements, it was deemed necessary to increase the number of specimens to ensure the accuracy of the experiments. Additionally, the selection of the stress ratio directly affects the results of fatigue testing. After considering various factors, three stress ratios (0.5, 0.6, 0.7) were chosen as the test loads, and the number of specimens prepared is outlined in Table 4.

Table 4. Number of Specimens Prepared.

Type of Reinforcement	Stress Ratio	Test Quantity	Reserve Quantity
Unreinforced	0.5, 0.6, 0.7	3 × 10 = 30	3 × 2 = 6
Wire mesh	0.5, 0.6, 0.7	3 × 5 = 15	3 × 1 = 3
Biaxial plastic geogrid	0.5, 0.6, 0.7	3 × 5 = 15	3 × 1 = 3
Glass fiber grille	0.5, 0.6, 0.7	3 × 5 = 15	3 × 1 = 3

3. Test Method

Currently, there are two main types of test methods for studying the fatigue performance of road surfaces. One type of method involves subjecting real vehicle loads to actual road surfaces for fatigue testing or using large-scale full-scale testing to simulate fatigue experiments [43–45]. These methods are advantageous in that they can better reflect the real fatigue performance of road surfaces. However, they are costly and time-consuming, and the test results are significantly influenced by the environment and road structure. Therefore, these methods are less commonly used and are not suitable for current research on the fatigue performance of semi-rigid road surfaces. The other type of method primarily involves indoor small-scale fatigue testing. Currently, the four-point bending beam fatigue test method is the most commonly used. However, this method results in a uniaxial stress state in the beam specimen under load, whereas actual road surfaces experience complex spatial stress states under the action of vehicle wheels. As a result, the accuracy and reliability of test results are greatly limited, as it does not consider the actual stress states and environmental effects on the road surface. To address the issue of the mismatch between the four-point bending fatigue test method and the actual stress state, this paper presents a two-way stress-bearing fatigue test method for semi-rigid base-layer road surfaces based on the Material Test System (MTS) material testing system.

3.1. Fatigue Test Loading Device

The fatigue loading test utilizes the MTS 810 material testing system. The system comprises a load frame with a loading head connected to it. A hydraulic oil source is connected to the load frame to provide pressure, and a control system is linked to the load frame to control the loading mode on the specimen.

The design of the loading head was inspired by the testing wheel of the rutting tester. The testing wheel has an outer diameter of 200 mm and a width of 50 mm, with the tire made of high-strength rubber. The loading head features multiple bolt holes, and the actuator head is securely fixed to the loading head using several bolts and bolt holes. The loading head's structure, which includes the testing wheel, can be used to apply load to the specimens, effectively simulating the load conditions that road surfaces experience when vehicles are driven over them.

At the beginning of the test, turn on the hydraulic oil source and compressor of the MTS material test device to check whether the mechanical equipment and power supply of the test system are normal. The driving rod on the beam is adjusted to the appropriate length, and the acting head is installed on the driving rod through a large bolt, and adjusted to the appropriate position above the specimen. The loading head is installed on the acting head by small bolts, and the upper surface of the loading head is horizontal. The connection lines of the stress sensor and displacement sensor on the computer are connected to the test device to set the loading mode. The beam is controlled by the load frame control unit to make the loading head just contact with the upper surface of the specimen.

Before the test, the corresponding operation is carried out through the control windows on the software, such as assigning sensors, setting limits, clearing sensors, setting error limits, adjusting parameters, etc., and the relevant test program is compiled. After repeated debugging to reliable and stable, can be carried out formal test.

The fatigue test was conducted under stress control mode, the loading waveform was half sine wave, and the load frequency was 10 Hz. The test time was relatively short under the stress control mode, the failure was clearly defined, and the complete fracture was the failure, which was more consistent with the actual stress condition of the structure layer.

In order to protect the test equipment, the procedure shall specify the relevant protection procedure for the termination of the test. In the process of test loading, the output data of the sensor is recorded. When the number of loading times is reached or the surface structure of the specimen is damaged, the loading is stopped. By lifting control of the drive rod, the loading head is lifted upward, the damaged specimen is removed, and the other specimen is placed to continue the above test. After a batch of samples is completed or all tests are completed for the day, the obtained data will be stored.

3.2. *Fatigue Failure Ultimate Load*

The flexural ultimate load test not only determines the failure ultimate load of cement-stabilized crushed stone materials, but also provides the basic parameters for load application standards. The data acquisition process is illustrated in Figure 7A. The flexural ultimate load test on unreinforced slab specimens was conducted using a load applied at the center point of the testing wheel, and the maximum load at which the center point of the specimen failed is defined as the ultimate failure load. The center point of the specimen failed, as shown in Figure 7B. The flexural ultimate load of the cement-stabilized crushed stone slab was found to be 12.31 kN.

(**A**) The Data Acquisition Process (**B**) The Center Point of The Specimen Failed

Figure 7. Specimen Failure States.

3.3. *Supportive Fatigue Test Method*

In conventional indoor bending fatigue tests, beam specimens are subjected to loading in a simply supported state without any support. Typically, the specimens fracture instantly, making it difficult for observers to carefully observe the crack formation before the complete failure. In actual road surface structures, there is often a supportive layer like a subbase or subgrade beneath the semi-rigid base layer. The presence of this layer significantly delays crack propagation at the bottom of the base layer, allowing the base layer to maintain its integrity and function normally for an extended period.

3.3.1. Supportive Material

To better simulate the real stress conditions of road surface structures, suitable supportive materials were selected and placed at the bottom of the cement-stabilized crushed stone slab specimens during the experiments. These materials were chosen to emulate the supporting layer in actual road surface structures, providing resistance to compression, distributing loads, and reducing deformation. Four materials, as shown in Figure 8, were considered, with the elastic modulus as a crucial parameter for the supportive material. Based on the experimentally determined elastic modulus of typical supportive layers in semi-rigid base layers, which generally falls within the range of 300 MPa to 500 MPa, and considering the technical specifications of different support materials, rubber sheets were found to best meet the experimental requirements for supportive material.

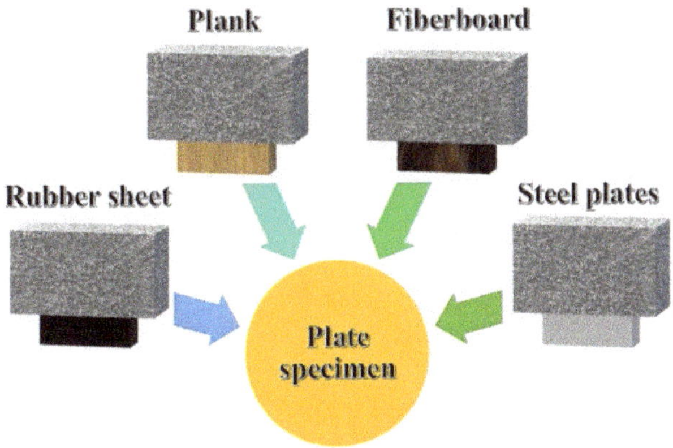

Figure 8. Plate Support Material.

3.3.2. Support Method for Specimens

The experimental support method involves central contact, ensuring that the slab specimen is in close contact with the rubber sheet while also making direct contact with the lower fixture base to facilitate bending fatigue testing under support conditions.

The dimensions of the rubber sheet are 200 mm × 200 mm × 50 mm, and its compressive elastic modulus meets the requirements. The total height of the supportive material is adjusted by adding or removing 1 mm thick rubber sheets to ensure close contact between the supportive material and the slab specimen.

3.3.3. Fatigue Strain Testing of Specimens

To visually summarize the development of fatigue strain in the reinforced semi-rigid base layers, strain gauges were used to measure the stress, load-carrying capacity, and fatigue performance of the reinforced cement-stabilized crushed stone slab under loading conditions. The strain gauge was attached to the center of the lower surface of the cement-stabilized crushed stone plate specimen. When a load was applied to the specimen, it deformed, causing the strain gauges to deform in tandem, resulting in changes in their resistance values. The resistance strain measurement device was capable of measuring these changes, converting them into strain values, or outputting analog signals proportional to strain. After data collection and processing with a computer, the required strain values were obtained. The strain measurement device and resistance strain gauges used in this study are shown in Figure 9.

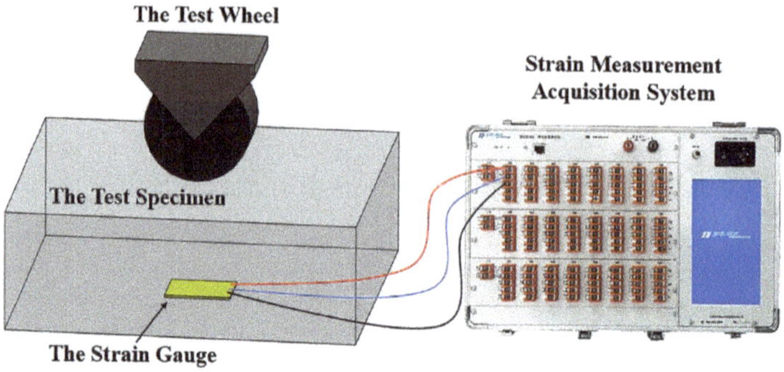

Figure 9. Strain Measurement Device and Resistance Strain Gauge.

Prior to the start of the experiment, strain gauges were attached to the central position on the lower surface of the cement-stabilized crushed stone slab specimens. Once the experimental setup was complete, the loading head was lowered until it made contact with the specimen. The MTS fatigue test system and strain measurement acquisition system were then initiated. The fatigue test was conducted in stress control mode, utilizing a half-sine wave waveform with a load frequency of 10 Hz, which closely matched the stress conditions of actual structural layers. As illustrated in Figure 10, fatigue strain data were collected during the test.

Figure 10. Fatigue Strain Test Data Acquisition System.

4. Analysis of Fatigue Cracking Characteristics of Stiffened Water Stabilizer Plate

The fatigue cracking of semi-rigid base pavement mostly starts from the fatigue cracking of base pavement. Under the continuous action of load, the cracks extend to the asphalt surface, causing fatigue cracking of the whole pavement and fatigue failure of the pavement. When the bearing fatigue test method is used to test the cement stable broken slate specimen, because there are supports at the bottom of the specimen, the crack of the specimen will not break quickly in a short time after the crack appears, and the process from crack appearance to complete fracture lasts a long time, so the crack cracking condition of the specimen can be closely observed.

Most cracks appear in the bottom of the specimen with pores, segregation and other defects, and continue to expand along the long axis of the initial crack. Among them, the two-directional tensile stress at the center of the plate specimen is basically the same. When the two-directional stress $\sigma_x = \sigma_y$, axial symmetry within a certain range is formed locally, and the maximum tensile stress in each direction is the same. At this time, cracking may occur in any direction.

As the number of fatigue loads continues to increase, multiple cracks in different directions begin to connect, forming cracks with a certain length, and the crack length gradually increases. When the crack length approaches the length of the specimen, one or more main cracks will be formed through the center of the bottom of the specimen. After the formation of the main crack, the width of the crack gradually increases, and it can be clearly observed from the side of the specimen that the main crack develops from bottom to top, the deflection of the specimen changes, the stiffness of the specimen decreases sharply, and finally the specimen is completely cracked and destroyed.

The fatigue cracking of cement-stabilized broken slate specimens has the following characteristics: (1) The crack is mainly caused by the original defects at the bottom of the specimen, and the crack distribution in the center of the specimen is multi-directional and finally forms different bottom cracks. (2) Before the formation of the main crack of the specimen, the length and width of the crack of the specimen develop slowly, and after the formation of the main crack of the specimen, the length and width of the crack of the specimen develop more rapidly. (3) There will be a number of cracks at the bottom of the specimen; the depth of the crack surface at the center is the largest, when the crack expands from the bottom to the bottom, the length of the crack at the bottom of the specimen develops at the same time; and when a crack develops to the surface of the specimen, the specimen will be completely cracked, and the width of the crack at the surface position of the specimen is less than the bottom position of the specimen.

5. Test Results and Analysis

The fatigue strain of semi-rigid base layer materials is primarily composed of elastic strain and plastic strain. The degree of fatigue damage can be represented by the fatigue strain peak value, which corresponds to the maximum strain value under cyclic loading. Under the continuous action of cyclic loading, the material experiences elastic degradation and plastic accumulation failure. The maximum stress is constant at all stress ratios, so the effect of fatigue strain on the specimen is only caused by cyclic loading. When the stress level is high, both elastic and plastic strains develop rapidly, leading to a quick increase in elastic degradation and plastic accumulation, resulting in a shorter fatigue life. When the stress level is low, most of the fatigue strain is elastic, with a lesser contribution from plastic strain, which significantly extends the fatigue life.

5.1. Analysis of Fatigue Strain Growth Patterns of Specimens with Different Reinforcement Materials

Taking the fatigue life and strain values under a stress ratio of 0.5 from Figure 11 as an example, the fatigue failure of the specimens can be mainly divided into three stages, corresponding to the three magnified parts A, B and C in the Figure 11. Combining the fatigue failure mechanism of semi-rigid base layers and actual observations [46–48], the characteristics of these three stages are as follows:

First Stage: The MTS testing machine head closely contacts the specimen, and at the beginning of the test, the strain at the bottom of the specimen increases rapidly and is unstable. After a short period of fluctuation, the tensile strain values start to increase more slowly and stabilize.

Second Stage: Most of the strain in the specimen is elastic strain. The strain peak value at the bottom of the specimen increases relatively slowly, and the rate of increase approaches a certain value, approximately in linear relation to the number of fatigue cycles.

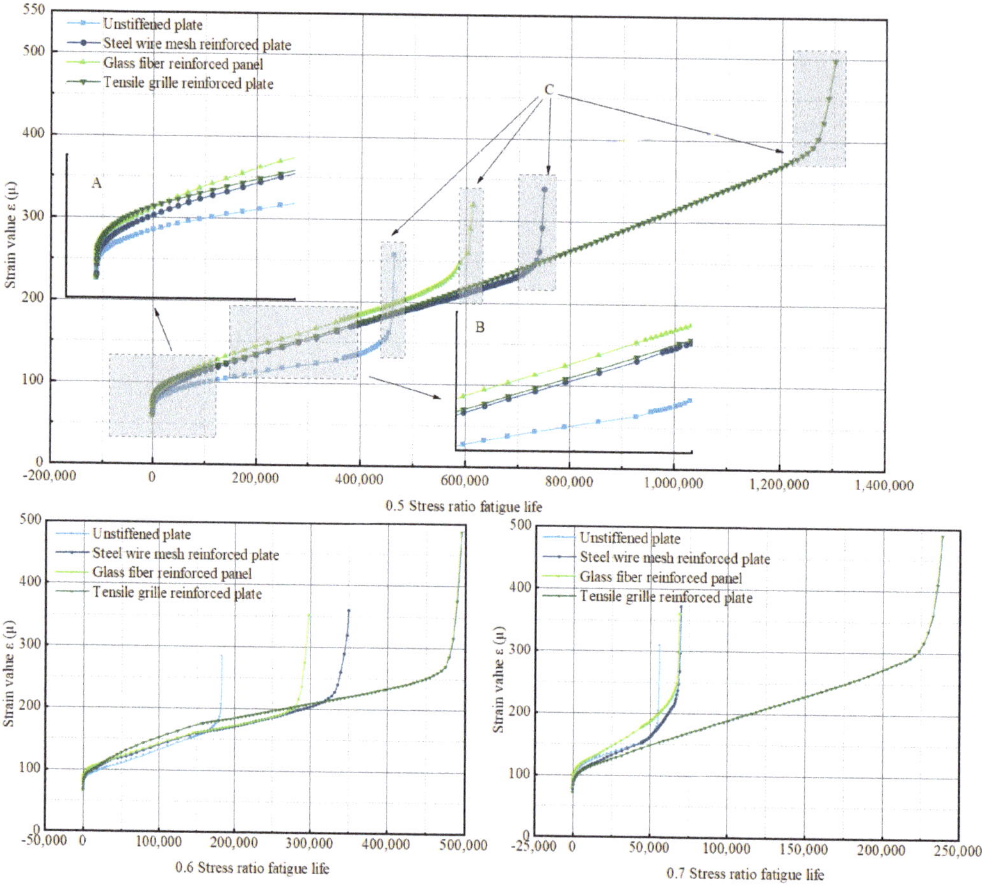

Figure 11. Strain Development Behavior of Specimens.

Third Stage: With an increasing number of cyclic loading cycles, plastic strain starts to increase. The strain peak value in the specimen increases rapidly and reaches the limit tensile strain, resulting in fatigue fracture and failure within a relatively short time.

A comprehensive analysis of Figure 11 reveals that reinforced materials exhibit better fatigue resistance compared to unreinforced materials. Among the reinforced materials, the geogrid reinforcement shows relatively good performance. Across different stress ratios, the fatigue life of specimens consistently follows this order: pull-plastic grid > steel wire mesh > fiberglass > unreinforced. The main reason is that the technical performance of the three kinds of reinforced materials is different. From the technical parameters and actual use of the three kinds of reinforced materials, the tensile strength, hardness and elongation of the pull-plastic grid is superior to those of the steel wire mesh and the glass fiber grid in many aspects, and the tensile grid has better bonding performance with the cement stabilized gravel. The pull-plastic grid forms an effective bearing and diffusion chain system in the cement stabilized gravel, increases the bearing capacity of the pavement base, and extends its service life. The strength of the glass fiber grating is lower than that of the pull-plastic grid, the elongation is also lower, and the bonding performance of the cement stabilized gravel material is general, but the glass fiber grating has a better bonding performance with the asphalt mixture, and the resistance effect of high temperature rutting in the asphalt pavement is better. The strength of the steel wire mesh is between the pull-plastic grid and the glass fiber grid, the elongation is also low, and the bonding performance of the cement-stabilized gravel material is good, but the steel wire mesh is easy to rust in the wet environment, resulting in its tensile strength gradually reduced, and the fracture failure occurs under the continuous action of load. Analyzing the proportion of fatigue life in each stage based on the three-stage fatigue failure analysis, for unreinforced specimens, the first stage accounts for 10% of the total fatigue life, the second stage accounts for 70%, and the third stage accounts for 20%. Similarly, for the other reinforced specimens, the analysis indicates that the proportions of fatigue failure in the three stages are consistent, at 5%, 75%, and 20% of the total fatigue life. This suggests that the inclusion of reinforcement materials accelerates the entry of the specimens into the elastic failure stage, thereby extending their fatigue life.

5.2. Analysis of Fatigue Strain Growth Patterns for Specimens under Different Stress Ratios

Analysis of fatigue strain values for different specimens under different stress ratios, as shown in Figure 12, reveals that reinforced specimens consistently exhibit better fatigue resistance compared to unreinforced ones at varying stress ratios. Stress levels have a significant impact on strain development, with higher stress ratios leading to greater strain growth rates. In terms of initial strain values, it is evident that all specimens at high stress ratios have significantly higher initial values than those at low stress ratios. Analyzing the strain growth rates, it becomes apparent that in reinforced specimens, the strain growth rates during the second and third stages are notably higher at high stress ratios compared to low stress ratios. Strain development during the second and third stages of reinforced specimens shows a linear relationship with the number of loading cycles, with the strain growth rate during the third stage being higher than that during the second stage.

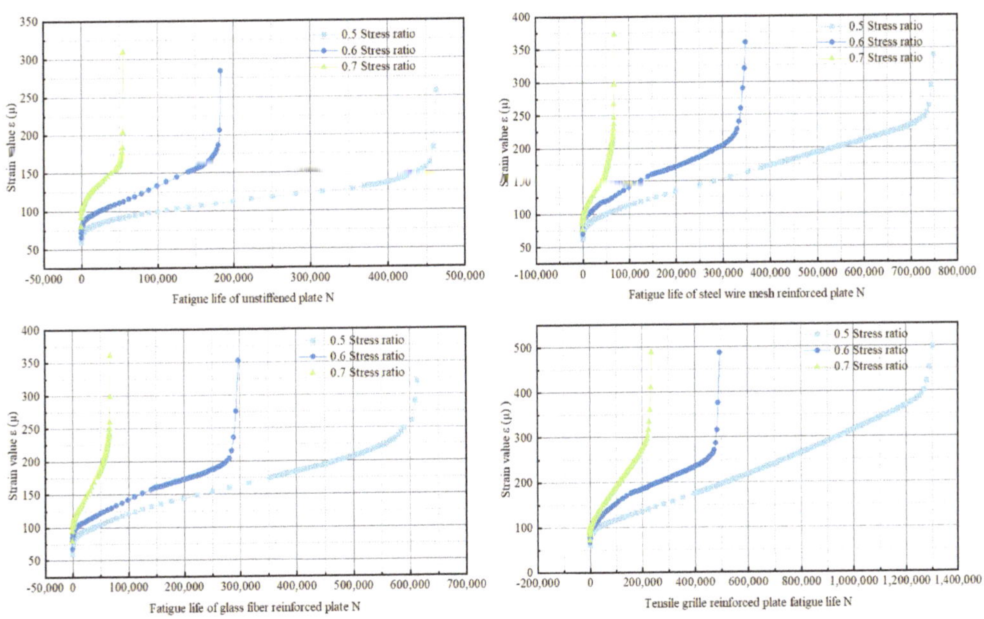

Figure 12. Changes in Strain Characteristics under Different Stress Levels for Specimens.

5.3. *Analysis of Fatigue Strain Characteristics*

In the context of the novel supporting fatigue test designed in this study, the use of characteristic values to represent critical points of fatigue strain-related to the various stages of fatigue failure was considered. Comparative analysis revealed variations in the initial tensile strain and tensile strain growth patterns of cement-stabilized crushed stone slabs reinforced with different materials and under different stress levels. By examining the fatigue strain development curves of reinforced specimens and the fatigue failure mechanism of semi-rigid base layers, a regression analysis was conducted to determine the initial strain value, strain inflection point, and ultimate strain value for different reinforced specimens.

Taking the example of the pull-plastic grid, which exhibits the best fatigue resistance, its tensile strain development pattern shows stage-wise changes. The tensile strain development in the first and third stages appears to follow a curve, while the second stage demonstrates linear strain growth. A linear regression analysis was performed on the tensile strain development during the second stage. The point at which the linear regression intersects with the curve in the third stage represents the tensile strain inflection point, and the point of abrupt change in the curve during the third stage indicates the ultimate tensile strain point.

With respect to the varying stress ratios, the analysis of strain characteristic values in Figure 13 reveals the following: at a stress ratio of 0.5, the reinforced panel with pull-plastic grid exhibits an initial strain value of approximately 79 με, a tensile strain inflection point of approximately 228 με, and an ultimate tensile strain value of about 453 με. At a stress ratio of 0.6, the initial strain value is approximately 82 με, the tensile strain inflection point is about 196 με, and the ultimate tensile strain value is roughly 316 με. At a stress ratio of 0.7, the initial strain value is around 72 με, the tensile strain inflection point is approximately 209 με, and the ultimate tensile strain value is about 360 με. In summary, the strain characteristic values align with the theoretical foundations of fatigue failure, effectively representing the three stages of semi-rigid fatigue failure mechanisms, and they offer a representative assessment of the extent of fatigue failure for the specimens. Additional characteristic values for other specimens are provided in Table 5.

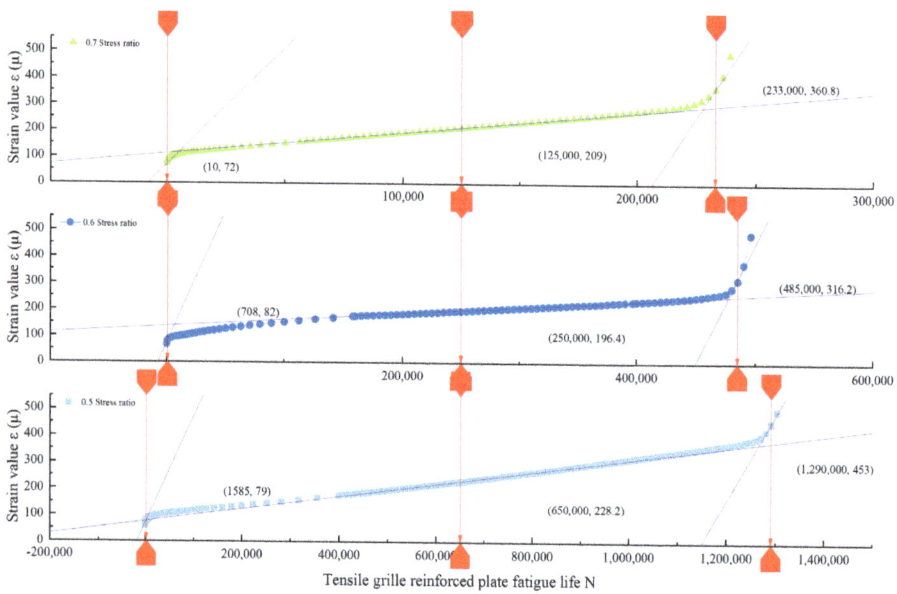

Figure 13. Changes in Strain Characteristics of Pull Plastic Geogrid Specimens under Different Stress Ratios.

Table 5. Strain Characteristics under Different Stress Ratios for Various Reinforced Plates.

Material Type	Stress Ratio	Strain Value (με) Initial Strain Value	Turning Point of Strain	Ultimate Tensile Strain Value
Unstiffened plate	0.5	72	153	182
	0.6	83	176	196
	0.7	101	171	203
Steel wire mesh reinforced plate	0.5	77	170	272
	0.6	90	161	290
	0.7	98	155	277
Tensile grille reinforced plate	0.5	79	228	453
	0.6	82	196	316
	0.7	72	209	360
Glass fiber grid reinforced panel	0.5	81	177	265
	0.6	92	160	276
	0.7	108	175	280

6. Conclusions

This study proposed a novel fatigue testing method for semi-rigid base layer road surfaces based on MTS, which simulates bidirectional stress conditions. By preparing reinforced cement-stabilized crushed stone slab specimens and conducting indoor fatigue strain tests, the study explored the impact of reinforcement factors on fatigue strain in semi-rigid base layers. It provides an experimental foundation for the practical application of reinforcing materials in semi-rigid base layer road surfaces. Compared with the current fatigue test method, which mainly adopts the four-point bending beam fatigue test method, the unidirectional stress state of the beam specimen under load is changed to the bidirectional stress state of the actual road surface under the action of the wheel, the authenticity of the test results is improved, the accuracy and reliability are guaranteed, and the complicated correction in the actual design is simplified. The main conclusions are as follows:

a. The strain growth rate in semi-rigid base layers is directly proportional to the stress ratio, and reinforced specimens consistently exhibit better fatigue resistance than unreinforced specimens. When comparing different reinforcing materials for their fatigue resistance, it is evident that they perform as follows: pull-plastic grid > steel wire mesh > glass fiber > unreinforced specimens.
b. The failure of semi-rigid base layers can be primarily divided into three stages, with corresponding strain changes approximated as a combination of curved, linear, and curved changes. For unreinforced specimens, the proportions of fatigue life in each stage are 10%, 70%, and 20%. The addition of reinforcement materials prolongs the elastic failure stage of semi-rigid base layers and enhances their fatigue life, with proportions of fatigue life in each stage reaching 5%, 75%, and 20%.
c. Using characteristic strain values to represent the extent of fatigue failure in the novel support fatigue test satisfies the theoretical foundations of fatigue failure. It effectively represents the three stages of semi-rigid base layer fatigue failure mechanisms, providing a representative assessment of the extent of fatigue failure for the specimens.

Author Contributions: Conceptualization, L.W. and Z.Z.; methodology, L.W., H.L., Y.H. and X.K.; software, M.X.; formal analysis, X.K.; investigation, L.W. and Z.Z.; resources, X.K.; data curation, L.W., Z.Z., Y.H. and X.K.; writing—original draft preparation, Z.Z. and Y.H.; writing—review and editing, L.W. and M.X.; visualization, H.L. and M.X.; supervision, H.L., Y.H. and M.X.; project administration, Y.H. and M.X.; funding acquisition, M.X. All authors have read and agreed to the published version of the manuscript.

Funding: This study is supported by the National Key R&D Program of China (2022YFB2601900), National Natural Science Foundation of China (U2233210 and 51978034), the Beijing Scholars Foundation (No. 067), the Beijing Natural Science Foundation, the Beijing Municipal Education Commission (KZ202110016020) and the Shandong Jiaotong University Scientific Research Fund Project (23YK029).

Data Availability Statement: The raw data supporting the conclusions of this article will be made available by the authors on request.

Conflicts of Interest: Author Hao Liang was employed by the Shandong Luzhong Highway Construction Co., Ltd. Author Yilong He was employed by the Shandong Expressway Environmental Protection Building Materials Co., Ltd. The remaining authors declare that the research was conducted in the absence of any commercial or financial relationships that could be construed as a potential conflict of interest.

References

1. Cao, J.; Zhang, B. Study on long-term performance of four types of asphalt pavement with different base layers. *Highway* **2020**, *65*, 1–665.
2. Shen, W.; Zheng, X.; Li, H. Classification and service status of pavement base materials. *J. Wuhan Univ. Technol.* **2021**, *43*, 1–5.
3. Tang, Y.; Zhang, K. Analysis on the Fatigue Properties of Semi-Rigid Base Asphalt-Pavement Structures With Horizontal Interfacial Cracks. *J. Test. Eval.* **2017**, *45*, 20160091. [CrossRef]
4. Jia, W.; Hu, M.; Liu, Z.; Li, T.; Chen, S.; Gao, J. Experimental Study on Interlayer Mechanical Properties of Semi-Rigid Base Asphalt Pavement Using Interlayer Treatment Method. In *IOP Conference Series: Earth and Environmental Science*; IOP Publishing: Bristol, UK, 2021; Volume 719, pp. 32070–32075.
5. Guo, F. Analysis and Research on Fatigue Characteristics of Semi-Rigid Base Layer of Permanent Pavement. Master's Thesis, Chongqing Jiaotong University, Chongqing, China, 2023.
6. Wang, L. Numerical Simulation Analysis of Mechanical Effects on Long-Life Pavement under Semi-Rigid Base Mode. Master's Thesis, Jilin University, Changchun, China, 2022.
7. Jia, K. Research on Fatigue Characteristics of Semi-Rigid Base Materials. Master's Thesis, Chang'an University, Xi'an, China, 2008.
8. Wang, L.; Lei, X.; Liu, S. Investigation of immersion influence on dynamic properties of high-speed railway subgrade with semi-rigid waterproof functional layer through field-excitation testing. *Can. Geotech. J.* **2017**, *55*, 19–33. [CrossRef]
9. Song, Y.; Jin, M.; Qiu, H. Alkali excitation slag recycled aggregate semi-rigid base. *J. Mater. Sci. Eng.* **2023**, *9*, 455–461.
10. Lv, S.; Hu, L.; Peng, X.; Cabrera, M.B.; Li, J. Investigation of structural resistance for semi-rigid layers in an actual stress state. *Constr. Build. Mater.* **2021**, *273*, 122041. [CrossRef]
11. Wang, X.; Huang, X.; Bian, G. Reflection crack propagation path analysis of semi-rigid base asphalt pavement based on numerical simulation. *Highway* **2018**, *63*, 1–6.
12. Liu, J. Structural design analysis of semi-rigid asphalt pavement on highways. *Transp. Manag. World* **2023**, *21*, 34–36.
13. Zhang, Z. Research on semi-rigid base materials and structural fatigue damage of asphalt pavement. *Zhonghua Constr.* **2023**, *10*, 154–156.
14. Zhu, Y. Research on Design Control Index of Semi-Rigid Base Asphalt Pavement. Master's Thesis, Southeast University, Nanjing, China, 2019.
15. Liu, J. Study on Design Index and Road Performance of Chinese-French Asphalt Concrete and Semi-Rigid Base Material. Master's Thesis, Highway Science Research Institute of the Ministry of Transport, Beijing, China, 2019.
16. Zhao, G. Effect of high dosage of cement on performance of cement-stabilized gravel base. *China Munic. Eng.* **2019**, *06*, 92–95+109.
17. Busch, C.; Th Gersen, F.; Henrichsen, A. Development and Validation of a Mechanistic Recursive-Incremental Deterioration Model for Cement-Stabilized Base Courses. *Transp. Res. Rec.* **2018**, *1974*, 128–137. [CrossRef]
18. Zhang, B.; Zhang, X.; Zhong, Y.; Li, X.; Hao, M.; Sang, X.; Wang, X.; Liu, J. Research on Fatigue Model of Semi-Rigid Base Asphalt Pavement before and after Polymer Grouting. *Adv. Civ. Eng.* **2021**, *2021*, 6658943. [CrossRef]
19. Mohammadinia, A.; Disfani, M.M.; Narsilio, G.; Hemachandra, P.; Aye, L. Application of semi-rigid composite permeable pavements in road network. *Geotech. Innov. Chall. Transp. Infrastruct.* **2017**, 119–126.
20. Tian, X. Research on the Cracking Process and Influencing Factors of Regenerated Semi Rigid Base. Master's Thesis, Shaoxing University of Arts and Sciences, Shaoxing, China, 2023. [CrossRef]
21. Yu, L.; Pei, J.; Li, R. Research on Maintenance Strategies and Benefits of Semi rigid Base Crushed Petrochemical. In *Chinese Association for Science and Technology, Ministry of Transport, Chinese Academy of Engineering, Hubei Provincial People's Government. 2022 World Transportation Conference (WTC2022) Paper Collection (Highway Engineering)*; China Communication Press: Beijing, China, 2022; p. 6. [CrossRef]
22. Peng, T.; Gong, S.; Zhu, Y. Research on the Performance of Reinforced Soil Abutment Composite Structures under Static Load. In *Chinese Association for Science and Technology, Ministry of Transportation, Chinese Academy of Engineering, Hubei Provincial People's Government. Proceedings of the World Transportation Congress (WTC2022) (Bridge and Tunnel Engineering)*; China Communication Press: Beijing, China, 2022; p. 9.
23. González, M.; Villalobos, F.; Méndez, A.; Carrillo, P. Study of the static and seismic performance of a geogrid reinforced soil wall as bridge abutment in Concepción. *Obras Y Proy.* **2018**, *24*, 41–52. [CrossRef]
24. Zhang, H.; Zang, G.; Wu, B. Simulation analysis of dynamic strain response of semi-rigid base asphalt pavement. *Transp. Sci. Technol.* **2017**, *5*, 65–67.

25. Jin, F. Damage Characteristics and Service State Evaluation of Semi-Rigid Base. Master's Thesis, Harbin Institute of Technology, Harbin, China, 2020. [CrossRef]
26. Li, X. Stress analysis and treatment method of reflection crack in semi-rigid base road under temperature action. *Highway* **2017**, *62*, 54–60.
27. Li, S.; Sun, P.; Sun, T.; Nie, D.; Wang, D. Fatigue life analysis of semi-rigid base pavement based on three-dimensional scale variation of hidden cracks. *Highway* **2014**, *59*, 61–69.
28. Sun, R. Research on Reinforcement Effect and Application Technology of Geogrates in Widening Highway. Master's Thesis, Shandong University, Jinan, China, 2018.
29. Gao, Y. Theoretical Analysis of Reflective Cracking in Asphalt Pavement with Semi-rigid Base. *Iran. J. Sci. Technol. Trans. Civ. Eng.* **2019**, *43*, S149–S157. [CrossRef]
30. Huang, L.; Song, W.; Jiang, Z.; Yang, S. The influence of temperature and vehicle speed on the modulus field and fatigue life of semi-rigid base asphalt pavement. *J. Hunan Univ. (Nat. Sci. Ed.)* **2021**, *48*, 148–156. [CrossRef]
31. Yu, W.; Huang, Y.; Li, G. Fatigue test of semi-rigid base Materials. *Highway* **1991**, *9*, 31–34.
32. Tian, H.; Zheng, C.; Qin, Y.; Lv, D.; Zhang, T. Research on fatigue cracking control Factors of semi-rigid base asphalt concrete Pavement. *Highway* **2014**, *59*, 32–36.
33. Zhu, J.; Guo, F. Experimental study on anti-aging performance of polymer modified high toughness cement-based composite concrete pavement. *Chem. Bond.* **2024**, *46*, 65–70.
34. *GB 175-2007;* Common Portland Cement. Standards Press of China: Beijing, China, 2007.
35. *JTG E30-2005;* Test Methods of Cement and Concrete for Highway Engineering. China Communications Press: Beijing, China, 2005.
36. *JTG/T F20-2015;* Technical Guidelines for Construction of Highway Roadbases. China Communications Press: Beijing, China, 2015.
37. *JTG E42-2005;* Test Methods of Aggregate for Highway Engineering. China Communications Press: Beijing, China, 2016.
38. *GB/T 33281-2016;* Welded Wire Fabric Coated with Zine. Standards Press of China: Beijing, China, 2016.
39. *GB/T 1839-2008;* Test Method for Gravimetric Determination of the Mass per Unit Area of Galvanized Coatings on Steel Products. Standards Press of China: Beijing, China, 2008.
40. *GB/T 228.1-2021;* Metallic Materials—Tensile Testing—Part 1: Method of Test at Room Temperature. Standards Press of China: Beijing, China, 2021.
41. *GB/T 17689-2008;* Geosynthetics—Plastic Geogrids. Standards Press of China: Beijing, China, 2008.
42. *GB/T 21825-2008;* Glass Fibre Geogrid. Standards Press of China: Beijing, China, 2008.
43. Ye, H.; Tang, S.; Duan, Z. Orthotropic composite bridge of steel fiber reinforced concrete panel fatigue test research. *Highway* **2022**, *3*, 115–121. [CrossRef]
44. Wang, F.; Xie, S. Research on Mulative Fatigue Damage of Asphalt Pavement Based on Multi-level Amplitude Loading. *Highw. Eng.* **2017**, *42*, 267–272.
45. Kim, Y.K.; Park, J.Y.; Han, S.H.; Lee, S.W. Investigation of Fatigue Characteristics of Roller-Compacted Concrete Pavement based on Full-Scale Fatigue Test and Accelerated Pavement Test. In Proceedings of the 1st International Conference for Road Engineers, Jeju, Republic of Korea, 29 May–1 June 2018.
46. Cui, X. Research on the Development Law of Semi-Rigid Base Structure and Material Properties. Master's Thesis, Southeast University, Nanjing, China, 2020.
47. Guo, F. Permanent Pavement Semi-Rigid Base Fatigue Characteristics Analysis and Research. Master's Thesis, Chongqing Jiaotong University, Chongqing, China, 2023.
48. Zou, J.R.; Zhang, Z.Q.; Tao, L.I. Failure Mechanism and Control Measures of Common Trunk Highway Semi-rigid Base Asphalt Pavement. *J. Highw. Transp. Res. Dev.* **2018**, *35*, 1–7.

buildings

MDPI

Article

An Explainable Evaluation Model for Building Thermal Comfort in China

Haiyang Liu and Enlin Ma *

School of Highway, Chang'an University, Xi'an 710064, China; 2018021023@chd.edu.cn
* Correspondence: maenlin@chd.edu.cn

Abstract: The concentration of atmospheric greenhouse gases is being amplified by human activity. Building energy consumption, particularly for heating and cooling purposes, constitutes a significant proportion of overall energy demand. This research aims to establish a smart evaluation model to understand the thermal requirements of building occupants based on an open-access dataset. This model is beneficial for making reasonable adjustments to building thermal management, based on factors such as different regions and building user characteristics. Employing Bayesian-optimized LightGBM and SHAP (SHapley Additive exPlanations) methods, an explainable machine learning model was developed to evaluate the thermal comfort design of buildings in different areas and with different purpose. Our developed LightGBM model exhibited superior evaluation performance on the test set, outperforming other machine learning models, such as XGBoost and SVR (Support Vector Regression). The SHAP method further helps us to understand the interior evaluation mechanism of the model and the interactive effect among input features. An accurate thermal comfort design for buildings based on the evaluation model can benefit the carbon-neutral strategy.

Keywords: thermal comfort; building energy consumption; machine learning; open-access dataset

1. Introduction

The rapid increase in global temperature and its associated detrimental impacts have made climate change one of the most pressing challenges of the 21st century [1]. A central aspect of this escalation in global temperatures is the increasing concentration of atmospheric greenhouse gases, notably amplified by human activities [2]. As per recent studies, urban regions are major contributors to greenhouse gas emissions, predominantly due to activities such as transportation, industrial operations, energy production and consumption, waste management, and the functioning of residential and commercial buildings [3–7]. Consequently, addressing urban carbon emissions has become imperative in the fight against global climate change [8]. Among the various factors contributing to urban carbon emissions, building energy consumption, especially for heating and cooling purposes, plays a predominant role [9]. It is estimated that buildings account for nearly 40% of global energy consumption [10], with a significant fraction of this energy being expended for maintaining thermal comfort [11,12]. Thermal comfort, a state of mind expressing satisfaction with the surrounding thermal environment, is crucial for ensuring the health, productivity, and well-being of building occupants [13–15]. Thermal comfort is a field of study that has garnered considerable attention, with research standards playing a pivotal role in establishing uniform testing protocols. Pioneering standards, such as ASHRAE Standard 55 [16] and ISO 7730 [17], provide comprehensive methodologies for assessing thermal comfort in various environments. These standards define the thermal environmental conditions for human occupancy and prescribe a range of factors, including temperature, humidity, airflow, and clothing insulation, which contribute to individual thermal satisfaction [18,19]. The quantification of comfort parameters has been further refined through the Predicted Mean Vote (PMV) and Predicted Percentage Dissatisfied (PPD) indices, which are now widely accepted benchmarks for evaluating thermal environments in relation to human satisfaction [18]. Such standards not only guide experimental design but also facilitate the comparison of findings across different studies, ensuring that assessments of thermal comfort are both reliable and replicable. However, achieving optimal thermal comfort in a manner that is both energy-efficient and aligned with occupants' preferences is a formidable challenge [20,21].

The challenge is compounded by the diversity in regional climates, building designs, and occupant preferences [22]. Different regions, influenced by their geographical positioning and topographical attributes, experience different temperature ranges and climatic conditions. Similarly, buildings, based on their design, materials used, and purpose (whether commercial, residential, or industrial), have varying energy needs and thermal characteristics [23]. Additionally, the preference for thermal comfort can differ significantly among occupants, influenced by factors such as age, health, clothing, and activities. This diversity necessitates a detailed, data-driven understanding of the thermal requirements of buildings and their occupants. With the advent of the digital age, vast amounts of data are being generated and made available through open-access datasets, providing an unparalleled opportunity to harness this information for understanding and addressing the thermal comfort needs of building occupants. Employing machine learning methodologies, researchers can now model complex relationships between multiple variables, offering insights that were previously elusive [24,25].

Recent advances in interpretability of machine learning models have emphasized the importance of explanation techniques that provide insight into model predictions. Global explanation methods, like permutation feature importance, offer an overall perspective on feature relevance across the entire dataset, but they do not account for the complex interactions between features within individual predictions. In contrast, local explanation techniques, such as LIME (Local Interpretable Model-agnostic Explanations) and SHAP (SHapley Additive exPlanations), provide granular insights into the contribution of each feature to individual predictions, reflecting the conditional interaction effects within the model [26]. SHAP, in particular, employs a game-theoretic approach to attribute the prediction output to its input features, thereby offering a cohesive and theoretically grounded method for local explanations [27]. The SHAP technique not only elucidates feature contributions but also enhances transparency and trust in complex models, a crucial aspect in fields like energy management where model decisions have significant impacts [28]. By incorporating local explanation methods, researchers can more effectively communicate model behavior, providing stakeholders with understandable and actionable insights into model predictions [29,30].

This paper delves into the above context, with the primary aim of establishing a smart evaluation model that leverages an open-access dataset [31] to understand the thermal requirements of building occupants. The emphasis is on creating a model that not only accurately predicts thermal comfort needs but also offers explanations for its predictions. The latter is particularly important as it offers architects, urban planners, and policy makers actionable insights into the factors influencing thermal comfort, facilitating informed decision-making. Furthermore, with China being the world's most populated country and undergoing rapid urbanization, the focus of this research on China offers timely insights. China's urban areas, characterized by their diverse climates ranging from the cold northeast to the hot and humid southeast, present a unique challenge and opportunity. An effective and efficient approach to ensuring thermal comfort in Chinese buildings can significantly contribute to the country's carbon-neutral strategy, echoing its commitments to global climate change mitigation efforts.

The organization of this paper is as follows. Section 2 delves into the data processing methodologies employed, from filtering the raw data to encoding categorical variables. In Section 3, the LightGBM model's establishment is detailed, along with insights into the hyperparameter optimization and training processes. Section 4 presents the results, critically analyzing the model's performance. Section 5 interprets the evaluation model's predictions mechanism through the SHAP (SHapley Additive exPlanations) method. Finally, Section 6 concludes the research, highlighting its contributions and implications.

In essence, this research sits at the nexus of urban development, thermal comfort, and sustainable energy consumption, providing a roadmap for future urban planning efforts aimed at achieving carbon neutrality while ensuring the well-being of occupants.

2. Data Preparation

2.1. Dataset Filtering

The open-access Chinese thermal comfort dataset [31], spearheaded by Xi'an University of Architecture and Technology in collaboration with seven other institutions, encompasses 41,977 data entries gathered from 49 cities spanning five climatic zones in China over the last two decades. Rigorous quality control measures were implemented on the raw data, involving systematic organization to guarantee its dependability. Each data entry encompasses environmental parameters, occupants' subjective feedback, building specifications, and individual details. In the raw dataset, certain non-essential features have a substantial amount of missing data. We first deleted these features, and subsequently removed samples with incomplete data to derive a filtered dataset. The features in the filtered dataset are shown in Table 1. A total of 11,899 samples are retained after dataset filtering.

Table 1. The features in the filtered dataset for building thermal comfort.

Categories	Features
General and building information	Seasons, city, climate zoom, building type, building function, thermal operation mode
Subject's personal information	Gender, age, height, weight
Subjective thermal comfort information	Thermal sensation voting scale (TSV), thermal comfort voting scale (TCV), thermal acceptability voting scale (TAV), clothing insulation, metabolic rate
Indoor physical parameters	Measured from different heights above the floor (0.1 m, 0.6 m, 1.1 m): Air temperature, relative humidity, air velocity
Outdoor environment parameters	Mean daily outdoor temperature

The details of the subjective thermal comfort indicators are delineated below:

- The thermal sensation scale is based on the ASHRAE Standard 55-2020 [32] 7-point voting scale, from -3 (cold) to 3 (hot). The thermal comfort scale is a 6-point scale from 0 (very comfortable) to 5 (very uncomfortable). The thermal acceptability scale is a 4-point scale from -1 (unacceptable) to 1 (acceptable).
- Clothing Insulation: Respondents were prompted to select the clothing type that matched their attire at the time of taking the survey. In instances where their specific clothing type was not listed, they were guided to choose the closest alternative. The insulation value for individual clothing items was determined based on ASHRAE 55-2020 [32]. For outfits composed of multiple garments, the total insulation value was computed by aggregating the insulation values of each individual piece.
- Metabolic Rate: The dataset features metabolic rate values for the Chinese population across various activity states. These values were ascertained in [33] using indirect calorimetry. The participants' activities at the time of completing the questionnaire were documented and subsequently translated into metabolic rate values. The corresponding values are sitting (0.9 met), sitting while typing (1.0 met), sitting with document filing (1.2 met), standing in an office setting (1.1 met), standing with document filing (1.3 met), and walking at a pace of 2 km/h (2.1 met).

2.2. Feature Selection

Feature selection is a critical step in the development of a robust and efficient model. Properly selecting the right features not only enhances the model's performance but also provides insights into the underlying processes governing the system. With the growing dimensions of data, especially in the age of big data, pruning irrelevant or redundant features becomes an imperative to prevent models from becoming overly complex and to reduce the computational overhead associated with training. In this study, we employ a two-pronged criterion for feature selection, aiming to streamline the input dataset while retaining the most informative predictors.

- Exclusion of Irrelevant Features: The primary objective of any modelling endeavor is to capture the underlying patterns in the data that are pertinent to the prediction or classification task at hand. Hence, the first step in our feature selection process is to remove any feature that does not have a direct or meaningful relationship with the evaluation indicators. Features that do not contribute significant information or might introduce noise into the system are systematically identified and excluded. This ensures that our model remains focused on pertinent information and is not swayed by irrelevant data.
- Addressing Feature Collinearity: The presence of highly correlated or collinear features can introduce instability in certain models and can also make the model's interpretations more challenging. When two or more features convey similar information, they are, in essence, redundant, and the inclusion of all these features does not necessarily improve the predictive power of the model but certainly increases the computational burden. In our methodology, if a set of features exhibit high collinearity (i.e., they are highly related), we adopt a conservative approach by retaining only a few representative features from that set and discarding the rest. This approach ensures that our model remains efficient without a compromise in its predictive capability.

Upon examining the data, we observe that indoor physical parameters have been gauged at three distinct heights above the floor: 0.1 m, 0.6 m, and 1.1 m. Figure 1 illustrates the significant correlation between these parameters across the three levels, as evidenced by their Pearson correlated coefficients. Guided by the principle of "Addressing Feature Collinearity", it is judicious to select a single set of indoor physical parameters from one specific height, given the strong interrelation between measurements from different heights. We have chosen the parameters measured at 0.6 m above the floor, as this height consistently exhibits the most robust correlation with the other two levels. Subsequently, the Spearman correlated coefficients (*SCC*) for the remaining features are shown in Figure 2. *SCC* is a rank correlation coefficient, and its calculation is based on the ranking of sample values of two variables in the data. *SCC* is agnostic to the numerical type and distribution of variables, thus exhibiting a broad scope of applicability. The formula for *SCC* is expressed as follows:

$$SCC(x,y) = 1 - \frac{6\sum_{i=1}^{n}((R(x_i) - R(y_i))^2}{n(n-1)^2} \tag{1}$$

where x and y are the variables to be studied, $R(x_i)$ is rank of sample x_i, $R(y_i)$ is rank of sample y_i, and n is the amount of all samples. The value of *SCC* ranges from −1 to +1, and the greater absolute value indicates stronger correlation between the two studied variables.

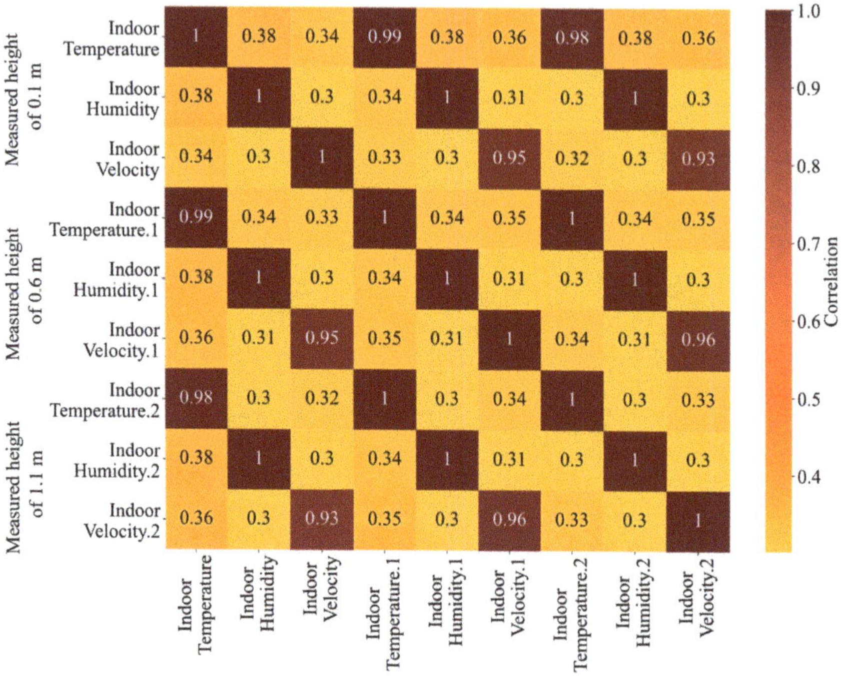

Figure 1. Pearson correlated coefficients between indoor physical parameters with different measured heights.

In the analysis presented within Figure 2, we focus solely on the absolute value of the *SCC*, emphasizing the strength of correlations between variables. Given the intricate internal dynamics observed in large sample sets, a mere reliance on significance might lead to misconceptions. Thus, the magnitude of the *SCC* holds primary importance in our approach. For the scope of this study, we designate thermal sensation (TSV), thermal comfort (TCV), and thermal acceptability (TAV) as evaluation outputs. Adhering to the principle of "Exclusion of Irrelevant Features", any feature demonstrating an *SCC* below 0.1 with these evaluation criteria is excluded from the modelling process. As tree-based learning models inherently yield a single feature output, separate models are necessitated for TSV, TCV, and TAV. Consequently, each model autonomously selects its most pertinent input features. The feature selection results are:

- For the TSV evaluation model, the related input features are building type, building function, thermal operation mode, clothing insulation, metabolic rate, and indoor air temperature.
- For the TCV evaluation model, the related input features are seasons, city, building type, building function, thermal operation mode, clothing insulation, metabolic rate, indoor air temperature, indoor relative humidity, and indoor air velocity.
- For the TAV evaluation model, the related input features are city, climate zoom, weight, clothing insulation, metabolic rate, indoor air temperature, and indoor air velocity.

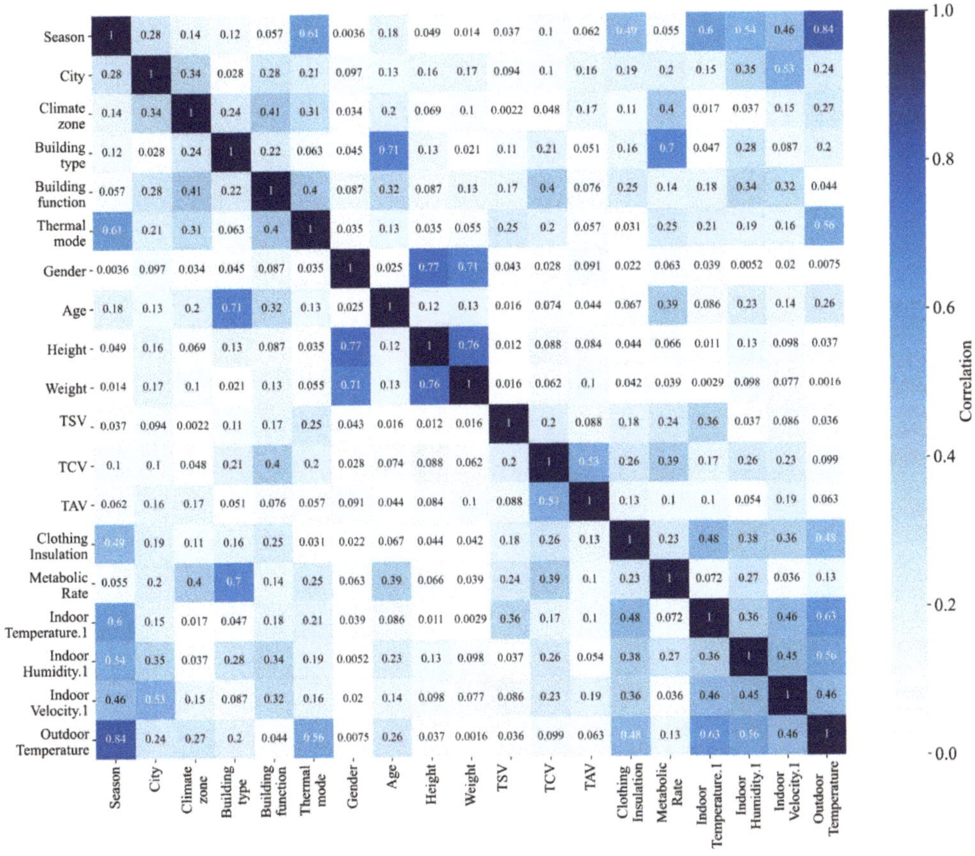

Figure 2. Spearman correlation coefficients for all features.

In pursuit of a broader applicability, this research seeks to formulate a versatile evaluation model tailored for cities not encompassed within the current dataset. As such, the "city" variable is excluded from the previously identified factors. All the selected features are shown in Table 2.

Table 2. The selected features in the dataset for the evaluation model of building thermal comfort.

Categories	Features
General and building information	Seasons, climate zoom, building type, building function, thermal operation mode
Subject's personal information	Weight
Subjective thermal comfort information	Clothing insulation, metabolic rate
Indoor physical parameters	Measured from 0.6 m above the floor: Air temperature, relative humidity, air velocity

In conclusion, the feature selection process adopted in this study is rigorous and is designed to produce a streamlined, informative, and non-redundant set of predictors. This not only facilitates efficient model training but also aids in deriving meaningful and interpretable results from the model. After feature selection, the data distribution of each feature (including the evaluation results TSV, TCV, and TAV) is shown in Figure 3.

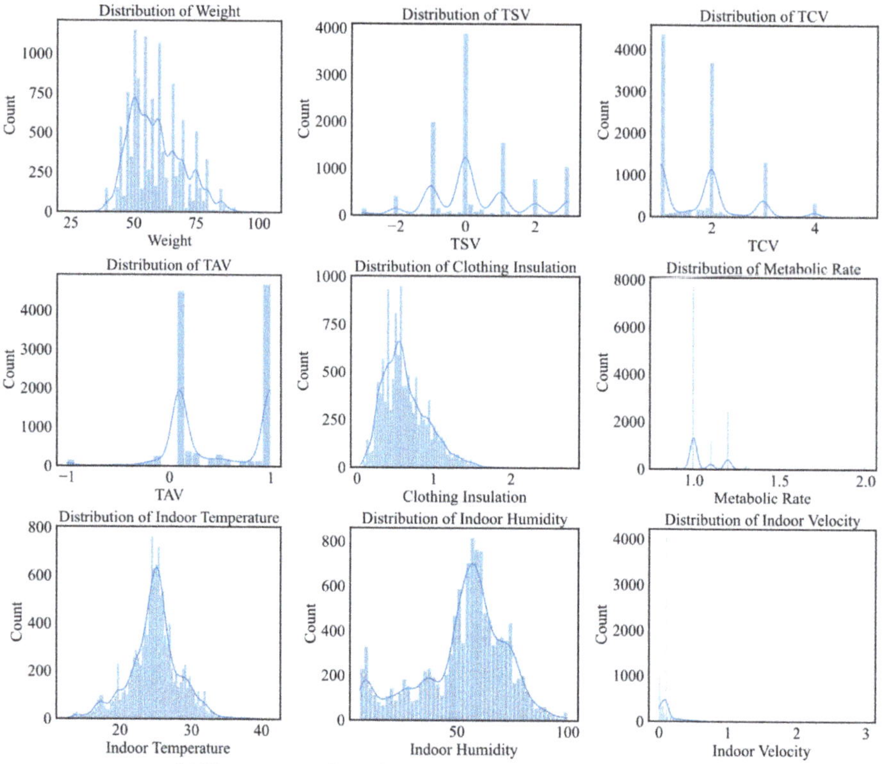

(**a**) Histograms to show the distribution of continuous variables

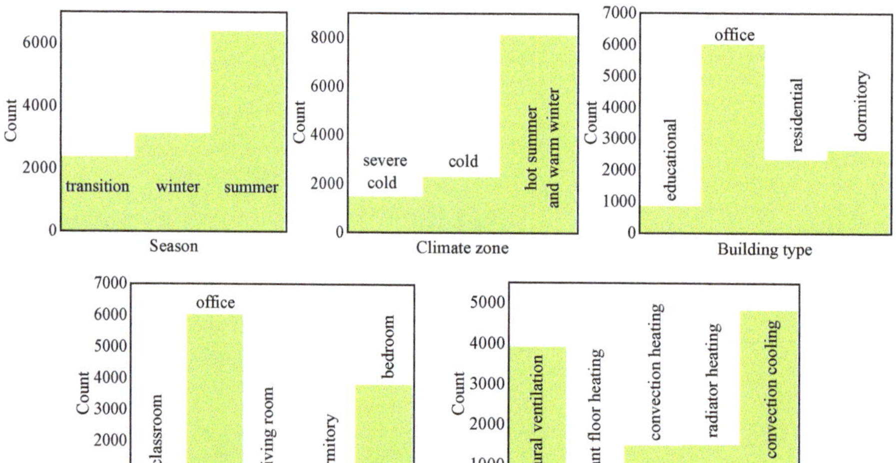

(**b**) Bar charts to show the frequency of each category

Figure 3. The distribution of each feature after feature selection.

It should be noted that, in the context of many traditional machine learning algorithms, preprocessing steps, like data normalization for numerical features and one-hot encoding for categorical variables, are essential to ensure optimal model performance. However, when working with the LightGBM model, such transformations are not required. This is due to the inherent design and mechanism of LightGBM, which can naturally handle different scales of numeric data and internally manages categorical variables through its histogram-based algorithm. Specifically, LightGBM applies a binning process to sort numerical values into discrete bins and utilizes a special algorithmic approach for categorical attributes, negating the necessity for manual one-hot encoding. This not only simplifies the preprocessing pipeline but also often results in faster training times and reduced memory usage without compromising model accuracy. However, in this study, the categorical variables are unordered, and thus it is preferable to employ one-hot encoding rather than label encoding. The encoding function can be expressed as follows:

$$\text{one-hot}\ (X) = [e_1, e_2, \ldots, e_i, \ldots, e_n] \tag{2}$$

$$\begin{cases} e_i = 0,\ 1 \\ \sum\limits_{i=1}^{n} e_i = 1 \end{cases} \tag{3}$$

where n is denoted as the total number of the categories of variable X, and e_i is the element of the one-hot vector whose value equals to 1 only in the corresponding categorical position that variable X indicates and equals 0 in the rest of the positions.

3. Model Establishment

3.1. LightGBM Model

LightGBM serves as an enhancement of the XGBoost and Gradient Boosting Decision Tree (GBDT) models [34]. It integrates the Exclusive Feature Bundling (EFB) and Gradient-based One-Side Sampling (GOSS) algorithms, positioning LightGBM as a leading model for tabular data prediction, boasting rapid training speeds and elevated prediction accuracy [35,36]. Typically, tabular data, characterized by rows representing samples and columns denoting features, often contain sparse categorical features abundant in zero elements, particularly when subjected to the one-hot encoding method. Such feature sparsity can detrimentally impede the efficacy of machine learning models. Addressing this, LightGBM leverages the EFB algorithm to amalgamate specific sparse features. Given that many sparse features frequently display mutual exclusivity, preventing them from being concurrently non-zero, the EFB algorithm consolidates these features into a singular new feature, thereby curtailing the feature dimension [34]. This approach efficiently mitigates training complexities while retaining commendable accuracy. Moreover, as an ensemble model of the Classification and Regression Tree (CART), LightGBM encapsulates the decision manifold inherent in the Decision Tree (DT), ensuring it remains impervious to discrepancies in value-type and distribution. Consequently, LightGBM emerges as an apt choice for evaluating building thermal comfort based on the selected tabular features.

3.2. Bayesian-Optimized Hyperparameters

A total of 20% of the entire dataset was randomly allocated as the test set, providing a basis for evaluating the performance of the model. The remaining 80% of the data was designated for hyperparameter tuning and model training processes. Hyperparameter optimization was undertaken using 5-fold cross-validation, i.e., the dataset was divided into five equal parts, with each part used as a validation set while the remaining four parts were combined to form a training set, in a rotational manner to ensure comprehensive evaluation. Subsequently, for model training, the remaining 80% of the data was further partitioned into a training set and a validation set in a 4:1 ratio, facilitating the iterative refinement of the model parameters. It is imperative that the test set remains completely separate from and uninvolved in the model establishment process, encompassing both hyperparameter optimization and model training phases, to preclude any potential for data leakage and ensure the integrity of the model's evaluation.

Bayesian optimization is a probabilistic model-based approach for global optimization of black-box functions that are expensive to evaluate. It operates by constructing a posterior distribution over the objective function and then subsequently selects points to evaluate by balancing exploration and exploitation. The method is particularly well suited for optimization of hyperparameters in machine learning algorithms. In this research, Bayesian optimization is employed to fine-tune hyperparameters of a LightGBM regression model. Here, $f(x)$ represents the cross-validated root mean squared error (RMSE) of the model predictions, with x denoting the vector of hyperparameters: $x = [x_1, x_2, x_3, x_4, x_5, x_6, x_7, x_8]$, where each x_i represents a hyperparameter in LightGBM (shown in Table 3). The goal is to find the hyperparameter vector x^* that minimizes $f(x)$, which in this scenario translates to the optimal model performance. The Gaussian process is often used to model the distribution over functions $p\ (f \mid D)$, where D represents the set of points $(x, f(x))$ already evaluated. Acquisition functions, such as Expected Improvement (EI) in this research, are then used to select the next query point by maximizing the expected utility. The Gaussian process posterior is updated with the new observations, and this process is repeated for a predefined number of iterations or until convergence criteria are met. This iterative process allows for the adaptive refinement of the search space, leading to more efficient optimization when compared to traditional grid or random search methods. This research provided initiates the optimization with 50 starting points and continues for an additional 500 iterations, progressively refining the model's hyperparameters towards the optimal configuration. The boosting type was set as "GBDT", and all of the other parameters of LightGBM, such as learning rate, were set as default. Since we need to build three different evaluation models for TSV, TCV, and TAV, respectively, the above optimization process will be conducted independently for each model. The Searching space and optimal value of each hyperparameter are shown in Table 3.

Table 3. Hyperparameter optimization for LightGBM.

Hyperparameters	Searching Space	The Optimal Value for TSV Evaluation	The Optimal Value for TCV Evaluation	The Optimal Value for TAV Evaluation
num_leaves	[24, 45]	28	25	43
feature_fraction	[0.1, 0.9]	0.86	0.59	0.57
bagging_fraction	[0.8, 1.0]	0.84	0.83	0.83
max_depth	[5, 15]	14	13	14
lambda_l1	[0, 1]	0.69	0.12	0.42
lambda_l2	[0, 1]	0.22	0.85	0.71
min_split_gain	[0.001, 0.1]	0.07	0.06	0.001
n_estimators	[100, 1000]	329	466	339

3.3. Model Training

The training progression of our LightGBM-based evaluation models is captured in Figure 4, which delineates the RMSE as the chosen loss metric over successive iterations for both the training and validation datasets. The loss curves of the Thermal Sensation Vote (TSV), Thermal Comfort Vote (TCV), and Thermal Acceptability Vote (TAV) models, as shown in Figure 4a–c, respectively, demonstrate a sharp decline in training RMSE. This illustrates the models' rapid learning curve and their ability to quickly assimilate the patterns within the training data. Concomitantly, the validation RMSE for each model converges to a low, indicating an effective generalization to the validation data which is pivotal in preventing overfitting—a phenomenon where a model exhibits high accuracy on training data yet fails to predict accurately on unseen data.

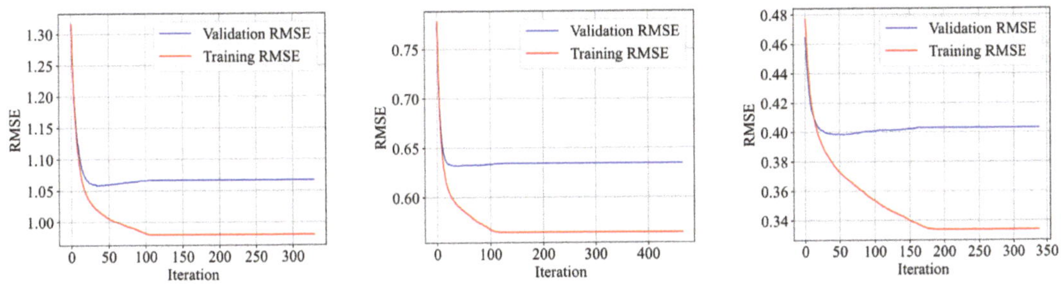

(**a**) Evaluation model for TSV (**b**) Evaluation model for TCV (**c**) Evaluation model for TAV

Figure 4. Loss curves in the training process.

Remarkably, the models achieve their best validation performance within the first 50 iterations, suggesting a swift convergence indicative of the efficiency of the LightGBM algorithm. The ongoing reduction in training RMSE post-convergence points to the potential for additional fine-tuning, should it be necessary. The depicted validation loss curves reinforce the balance attained by the models, which encapsulates sufficient complexity to learn from the training data while maintaining the ability to generalize to new datasets. This balance is vital, affirming the models' robustness and ensuring their applicability to a broader range of data, consistent with the objectives of the validation phase.

4. Model Performance

To quantitatively evaluate the accuracy of our developed model for evaluating building thermal comfort, we employed three widely accepted evaluation metrics: Mean Absolute Error (*MAE*), Root Mean Square Error (*RMSE*), and *Error*. Generally, for these metrics, lower values signify superior model performance. The definitions for these evaluation metrics are presented as follows:

$$MAE = \frac{1}{n}\sum_{i=1}^{n}|y_i - \hat{y}_i| \tag{4}$$

$$RMSE = \sqrt{\frac{1}{n}\sum_{i=1}^{n}(y_i - \hat{y}_i)^2} \tag{5}$$

$$Error = \hat{y}_i - y_i \tag{6}$$

where y_i and $\hat{y}_i$ denote as the true value and the predicted value, and $\overline{y}$ and $\overline{\hat{y}}$ denote as the averages of true value and predicted value.

Figure 5 provides a visual representation of the performance metrics for the evaluation models—TSV, TCV, and TAV—when applied to the testing set. The depicted box charts summarize the error distributions for each model, with the interquartile range (IQR) capturing the middle 50% of the data, delineated by the box's extent from the 25th to the 75th percentile. The central tendency of the models' errors is indicated by the median line and the mean symbol within the boxes, offering a dual perspective on the models' predictive accuracy. Notably, the span of the whiskers, extending to 1.5 times the IQR, illustrates the variability within the majority of the predictions, with the outliers marked as diamonds highlighting instances of significant deviation from the typical error range. Such graphical analysis aids in the comparative evaluation of model robustness and error consistency. The TSV model exhibits a slightly wider interquartile range, suggesting more variability in predictions compared to the TCV and TAV models. The latter models demonstrate a more compressed IQR, indicative of a tighter clustering of errors and, potentially, a more consistent predictive performance.

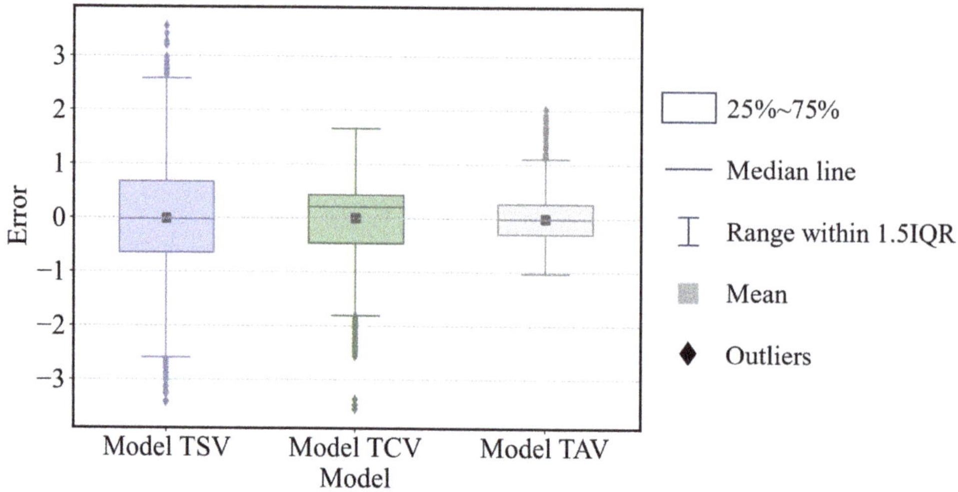

Figure 5. Performance of evaluation models on testing set.

In the TSV evaluation model, outliers are symmetrically distributed beyond the whiskers, whereas for the TCV model outliers are predominantly found below the lower whisker, and for the TAV model outliers are primarily above the upper whisker. This indicates that the TCV and TAV models tend to produce anomalously low and high results, respectively. It is imperative to consider the range span of TSV, TCV, and TAV, as a broader span implies a more challenging prediction task. Specifically, the spans for TSV, TCV, and TAV are 6, 5, and 2, respectively. In this context, as illustrated in Figure 6, the TCV model outperforms the others in terms of prediction across its range, which is also deemed the most crucial metric for assessing thermal comfort in buildings.

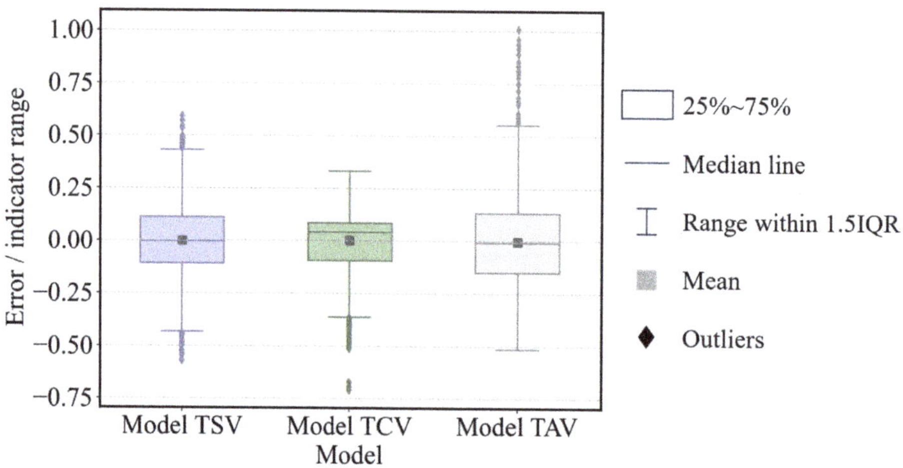

Figure 6. Performance of evaluation models on testing set considering the indicator ranges.

In Figure 7, we present a comparative analysis of the LightGBM-based evaluation models against a suite of established machine learning algorithms, namely KNN (k-Nearest Neighbor), RF (Random Forest), XGBoost, GBDT (Gradient Boosting Decision Tree), and SVR (Support Vector Regression). The KNN, RF, GBDT, and SVR models were constructed using the Scikit-learn library, while XGBoost was implemented via its dedicated library. All models, including LightGBM, were established with default parameter settings to exclude the impact of hyperparameter optimization. The comparative outcomes suggest that RF, XGBoost, and GBDT exhibit comparable levels of accuracy, likely attributable to their shared foundation in tree-based methodologies. Conversely, SVR and KNN appear less adept at managing the tabular dataset's large-scale nonlinearity, as evidenced by their respective error metrics. Although LightGBM demonstrates a marginal superiority in assessing TSV and TAV, it is distinctly more proficient in evaluating TCV. The consistent performance across multiple evaluation metrics shows the robustness of LightGBM, confirming its potential as a reliable tool for thermal comfort evaluation. Instead of the development of a novel model attuned to extensive thermal datasets, a core objective of this study is to elucidate the relative impact weights, marginal effects, and interplay among all pertinent factors, which we discuss comprehensively in Section 5.

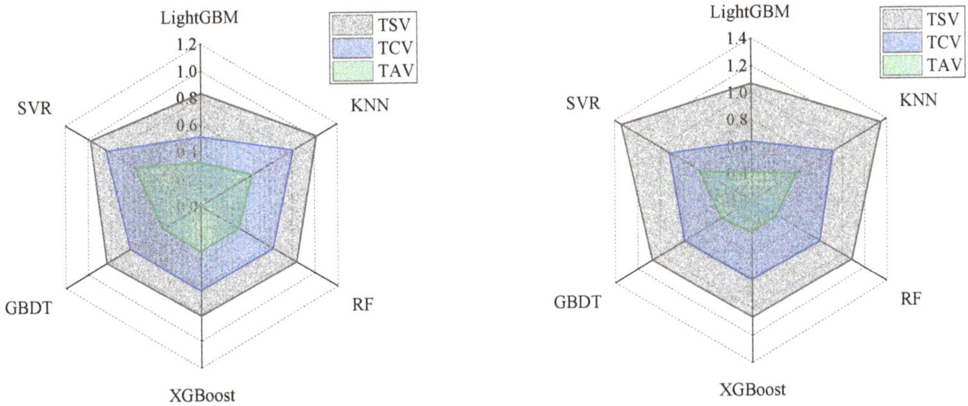

(**a**) MAE (**b**) RMSE

Figure 7. Overall performance of all machine learning models on testing set.

5. Model Interpretation

The interpretive analysis of the established LightGBM model was conducted using the SHAP (SHapley Additive exPlanations) method, which is grounded in game theory and relies on conditional expectations to elucidate the model's decision-making process [26,37–42]. The SHAP approach delineates the marginal contribution of each input feature to the predictive outcomes and helps understand the model's operational tendencies when evaluating the thermal comfort. This interpretative process exclusively employed the test dataset to reveal the model's explanatory insights. Particular attention was devoted to the TCV (Thermal Comfort Vote) evaluation model, attributed to its exceptional predictive accuracy and its acknowledged importance in gauging thermal comfort within building environments. By scrutinizing the TCV model, we discerned the influence weights, marginal effects, and interactive mechanisms of its contributing factors. This detailed examination enables a deeper comprehension of the factors that predominantly affect thermal comfort evaluations, guiding both the design of intelligent thermal regulation systems and the formulation of strategies for enhancing occupants' comfort and well-being.

5.1. Influence Weights

Figure 8 delineates the influence weights of various factors on the Thermal Comfort Vote (TCV) model through mean absolute SHAP values. It should be noted that category features, including building type, building function, thermal operation mode, and season, have been one-hot encoded, therefore, each category feature's mean absolute SHAP value indicates the sum of its one-hot features' mean absolute SHAP values. SHAP (SHapley

Additive exPlanations) values provide a profound understanding of feature contributions by assigning each feature an importance value for a particular prediction. A higher mean absolute SHAP value signifies a greater impact of the feature on the model's output. The bar chart reveals that 'Indoor Temperature' possesses the most significant influence on TCV, as evidenced by its highest mean absolute SHAP value. This suggests that variations in indoor temperature are the most substantial predictor of thermal comfort levels perceived by occupants. 'Building Type' also demonstrates a notable impact, implying that the structural and architectural characteristics encapsulated by this factor are critical in determining thermal comfort. 'Metabolic Rate' and 'Building Function' follow closely, indicating their substantial roles in influencing the thermal comfort outcomes, likely due to their direct relationship with human thermal regulation and the activities conducted within the building space. Conversely, 'Clothing Insulation', 'Indoor Humidity', 'Indoor Velocity', and 'Thermal Mode' display comparatively lower influence weights. Nonetheless, their contributions are non-negligible, suggesting a complex interplay of environmental conditions and personal factors that collectively shape the thermal comfort experience. It is noteworthy that 'Season' is the factor with the lowest mean absolute SHAP value, playing an inconsequential role in the evaluation model. The presence of multiple factors with varied influence weights reinforces the multifaceted nature of thermal comfort, which cannot be attributed to a singular environmental or personal characteristic. Instead, it emerges as an aggregate outcome of multiple interacting variables. The quantification of influence weights via SHAP values facilitates a nuanced understanding of the TCV model, allowing practitioners to prioritize interventions based on the factors most predictive of thermal comfort. Such insights can drive informed decisions in the design and management of building environments, optimizing occupant comfort while potentially enhancing energy efficiency.

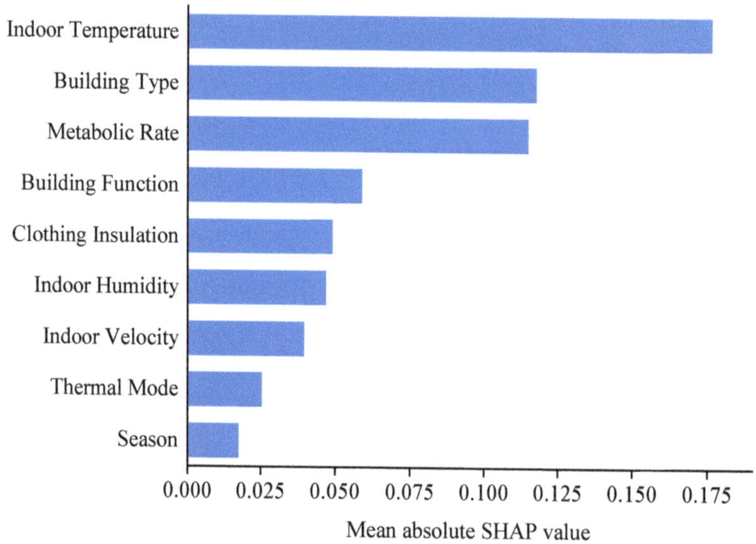

Figure 8. Mean absolute SHAP values for factors of TCV model.

5.2. Marginal Effects

Figure 9 illustrates the marginal impacts of various factors of the TCV model. Each dot within the figure symbolizes an independent data point. The hue of each dot corresponds to the specific factor's value for that data point. The SHAP value associated with each dot quantifies the marginal influence of the data point on the outcome, namely, the Thermal Comfort Voting (TCV) assessment. A positive SHAP value suggests that the respective feature value of the data point contributes to an increase in the output. The TCV scale ranges from 0, denoting 'very comfortable', to 5, indicating 'very uncomfortable'. Consequently, an elevated SHAP value denotes that the feature value of the data point adversely affects thermal comfort. The SHAP scatters depicted in the figure offers a granular view of the feature importance and impact on the predictive model.

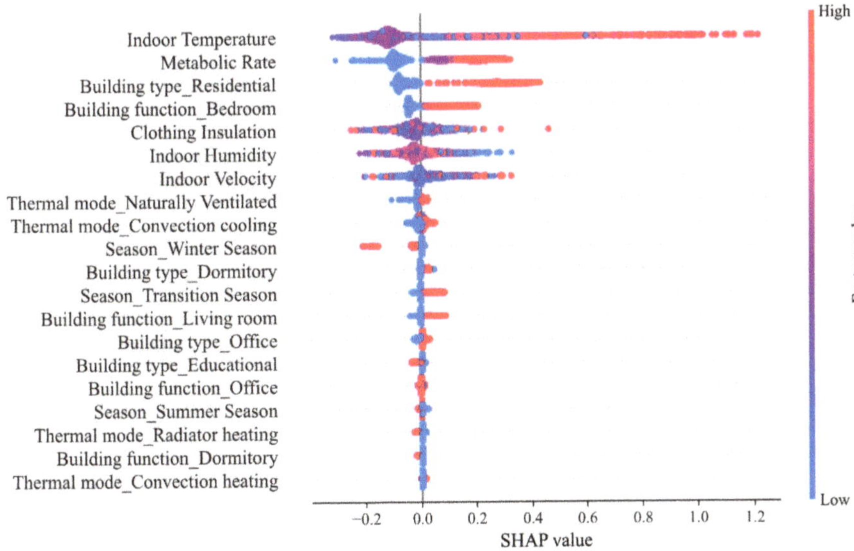

Figure 9. Variation trend of SHAP value with alteration to the value of each input variable of TCV model.

To help better understand the SHAP summary plot, we summarised five essential aspects of it: (a) Features list: The features are listed on the y-axis, ordered by the sum of SHAP value magnitudes across all samples. The feature at the top has the highest overall impact on model output; (b) SHAP value: The x-axis represents the SHAP value, which measures the impact of each feature on the model's prediction. Values to the right of the zero line indicate a positive impact on the model's prediction, while values to the left indicate a negative impact; (c) Color coding: The color indicates the feature value, with blue representing low and red representing high values. For instance, a high indoor temperature (red dot) is typically associated with a positive SHAP value, suggesting it increases the predicted outcome; (d) Data points: Each dot represents an individual data point. The spread of dots across the x-axis for a single feature shows the variability of the SHAP values for that feature across all the data points; (e) Impact direction: Features that have a mixture of colors across the zero line indicate that they can have both positive and negative impacts on the prediction, depending on their value for a given sample. The underscores signify categorical variables that have been one-hot encoded, with the subsequent text indicating the specific category within the variable. For instance, a red dot on the 'Building type_Residential' row suggests that the building type is residential for that data point, as indicated by a one-hot variable value of 1. Conversely, a blue dot denotes a non-residential building type, with the one-hot variable set to 0. It should be noted that two features ("Building function_Classroom" and "Thermal mode_Radiant floor heating") were omitted automatically due to their low influence.

Among the value-type features, the feature 'Indoor Temperature' prominently exhibits a concentration of red dots with high SHAP values, which, considering the TCV scale, suggests that higher indoor temperatures are generally associated with decreased thermal comfort. This is consistent with the physiological responses to thermal environments, where excessively high temperatures can lead to discomfort and even heat stress. The distribution of SHAP values for 'Indoor Temperature' underlines the critical balance required in maintaining temperatures within a range that maximizes comfort while minimizing energy consumption for cooling systems. Conversely, the 'Metabolic Rate' feature is characterized by a diverse spread of SHAP values, reflecting its complex relationship with thermal comfort. Notably, higher metabolic rates, indicated by red dots, contribute to a higher TCV, which in this context translates to a reduction in comfort levels. This is in line with the understanding that increased activity levels lead to higher internal heat production, which, if not offset by the thermal environment, can cause discomfort. This finding emphasizes the importance of designing building environments that are adaptable to the varying activity levels of occupants, suggesting that spaces should be versatile enough to accommodate different metabolic rates while still ensuring comfort. The insights derived from analyzing 'Indoor Temperature' and 'Metabolic Rate' highlight the interplay between environmental conditions and occupant activities in the context of thermal comfort. Effective thermal comfort design must therefore account for these factors, aiming to create an adaptive environment that can respond to both the dynamic nature of indoor temperatures and the diverse metabolic rates of occupants. This approach not only enhances occupant comfort but also promotes energy efficiency by aligning the building's climate control strategies with the actual needs of its users. The rest of the value-type features, such as clothing insulation and indoor humidity, did not show a clear mode in influencing the output, which might be revealed through interactive influence analysis.

Within the categorical variables assessed, particular attention is given to each category's relative impact on thermal comfort. For the variable 'Building Type', the category 'Residential' exhibits a pronounced detrimental influence on thermal comfort. In contrast, other categories, such as 'Dormitory', 'Office', and 'Educational', appear to exert negligible effects. This observation suggests that occupants may have less stringent thermal comfort expectations within public edifices, or that these structures may inherently possess superior thermal regulation capabilities compared to private dwellings. In addition, this disparity may be attributed to the economic aspects of thermal consumption costs and payment responsibility. Specifically, the cost of thermal energy in public spaces, which is not borne directly by individuals, potentially reduces thermal comfort concerns among users of these buildings. As for 'Building Function', spaces with public utility, including offices and dormitories, demonstrate no significant impact on thermal comfort, whereas private spaces such as 'Bedroom' and 'Living Room' are associated with the poorest thermal comfort levels. The influence patterns for 'Building Function' align with those observed in 'Building Type'. Regarding 'Thermal Mode', 'Radiator Heating' emerges as the most conducive to thermal comfort. Alternative modes, such as 'Natural Ventilation' and 'Convection Cooling', tend to negatively affect comfort margins. Seasonally, individuals report optimal thermal comfort in the winter, with the 'Transition Season' being the least comfortable period. Summer does not display a clear trend in thermal comfort preferences.

5.3. Interactive Mechanism

The interactive mechanism is to show the comprehensive effects of two features on building thermal comfort. In this part, we took the most relevant indicator "indoor temperature" as the basic index, and conducted four groups of interactive analysis, as shown in Figure 10. In the SHAP dependence graphs, the scales of color bars do not include the outliers. Each point in these graphs represents how the interaction of the two features at that specific data point influences the TCV score, offering insights into the complex interplay of environmental factors on thermal comfort. To help better understand the SHAP dependence graph, we summarised two essential aspects of it: (a) Feature value-prediction impact relationship: The horizontal axis usually represents the value of a specific feature (i.e., indoor temperature in Figure 10), while the vertical axis shows the SHAP value, indicating the impact of that feature value on the model's prediction (i.e., TCV value); (b) Color of data points: Data points can be colored to represent the values of other features (i.e., clothing insulation, metabolic rate, indoor humidity, and indoor velocity from Figure 10a–d, respectively), revealing the interaction effects between different features.

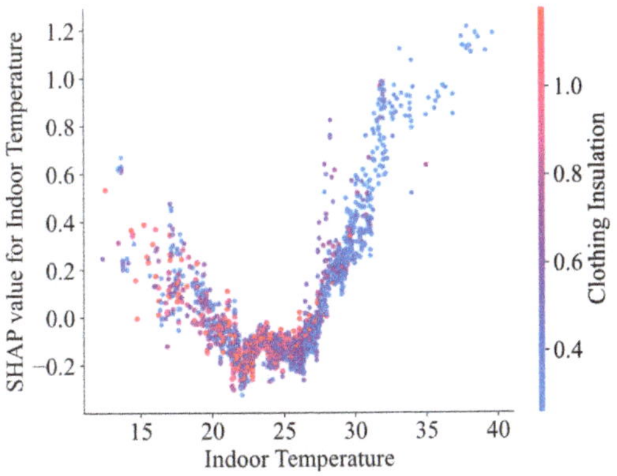

(**a**) Indoor temperature and clothing insulation

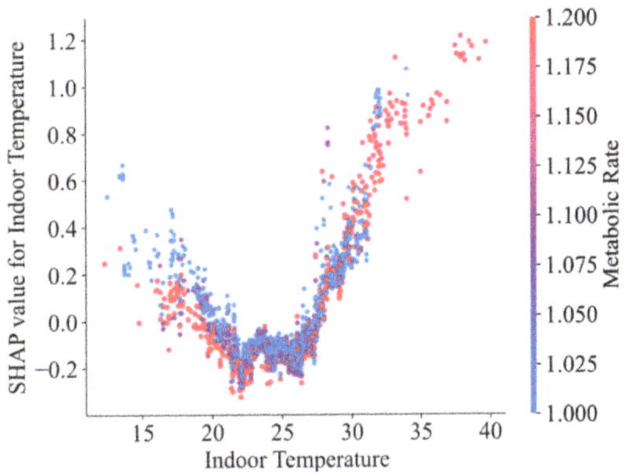

(**b**) Indoor temperature and metabolic rate

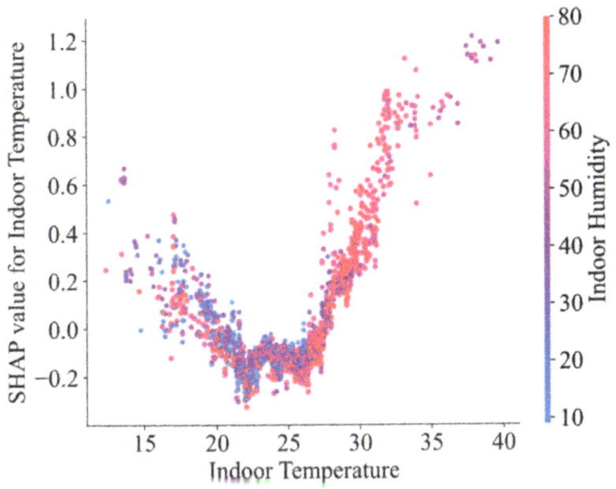

(**c**) Indoor temperature and indoor humidity

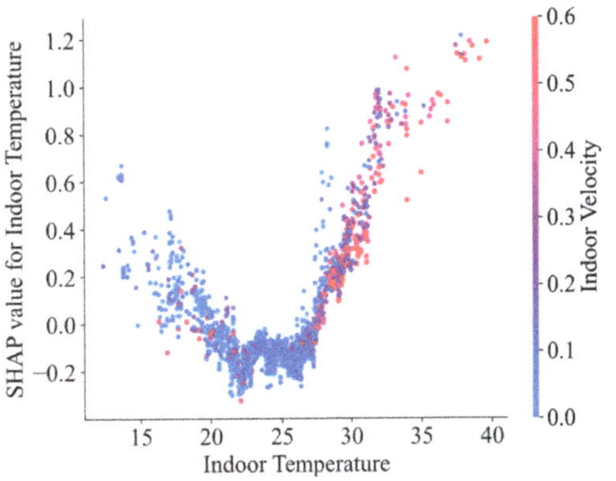

(**d**) Indoor temperature and indoor air velocity

Figure 10. SHAP dependence graph to show the interactive effects of different factors.

Figure 10a indicates a nonlinear relationship between indoor temperature and the SHAP values for this temperature, with a color gradient representing clothing insulation levels. As indoor temperature increases, SHAP values initially show a decline and then rise, suggesting an inverse U-shaped relationship. Lower SHAP values, indicating higher thermal comfort, are predominant at moderate temperatures, while extreme temperatures, both low and high, correspond to higher SHAP values, reflecting reduced thermal comfort. At lower temperatures, increased clothing insulation (as indicated by a gradient from blue to red) seems to mitigate the discomfort to some extent, as evidenced by the cluster of points with higher insulation levels associated with lower SHAP values. However, as the temperature rises beyond a certain threshold, even higher levels of clothing insulation cannot counteract the discomfort caused by high temperatures. In the mid-range of temperatures, there is a spread of SHAP values at varying levels of clothing insulation, implying a more complex interaction, where factors other than clothing and temperature may play a significant role in thermal comfort. This could include individual metabolic rates, the presence of direct sunlight, or other environmental factors not captured in this two-dimensional graph. At higher temperatures, the trend of increasing SHAP values regardless of clothing insulation suggests a limit to the compensatory role of clothing in managing thermal comfort. In these conditions, the physiological limits of heat dissipation might be reached and the discomfort becomes more pronounced, regardless of clothing insulation. Overall, the SHAP dependence graph reveals that, while clothing insulation can moderate the impact of indoor temperature on thermal comfort, this effect is bounded by the limits of physiological adaptation to temperature extremes. This suggests the importance of maintaining indoor temperatures within a moderate range to optimize thermal comfort, particularly in environments where the clothing insulation cannot be easily adjusted [43].

Figure 10b suggests that individuals with a higher metabolic rate (represented by red dots) tend to achieve thermal comfort more easily at lower indoor temperatures. This observation implies that the inherent heat generation from a higher metabolic rate may compensate for the lower ambient temperatures, thus aligning with the body's thermoregulatory needs to maintain a sensation of comfort. This phenomenon can be attributed to the body's endogenous thermal regulation system, where metabolic heat production plays a critical role. At lower temperatures, a higher metabolic rate can help maintain core body temperature, reducing the need for external heating sources and potentially leading to a more energy-efficient state of comfort. The concentration of red dots at the lower end of the indoor temperature spectrum on the SHAP graph indicates that, as the ambient temperature decreases, the thermal contribution of metabolic heat becomes increasingly significant. This aligns with thermoregulatory principles, where the human body's metabolic heat generation helps to offset the heat loss to the environment. The SHAP dependence graph indicates that the influence of metabolic rate on thermal comfort is attenuated at temperatures exceeding 27 °C, beyond which thermal comfort significantly declines with further increases in temperature, regardless of the metabolic rate. The implications of individual metabolic differences on thermal comfort are profound. They indicate that personalized comfort models could be beneficial in designing HVAC (Heating Ventilation and Air Conditioning) systems and in developing building energy management strategies that take into account the metabolic diversity of occupants. Adaptive thermal regulation systems that respond to individual metabolic rates can optimize energy consumption by reducing the reliance on artificial heating or cooling when the occupants' metabolic heat production is sufficient to achieve comfort.

Figure 10c delineates the interaction between indoor temperature and humidity, elucidating their collective effects on thermal comfort. At lower indoor temperatures, the contribution of humidity to Thermal Comfort Voting (TCV) appears to be ambiguous; conversely, at elevated indoor temperatures, humidity levels predominantly register as high, hinting at a homogeneity within the filtered dataset. This homogeneity notably underscores the dearth of observations from hot and arid climates [44], thereby limiting the model's capacity to accurately reflect the variations in comfort perceptions associated with such conditions. To foster the creation of comprehensive thermal comfort models, it is imperative to procure a dataset that is both diverse and representative, spanning the full gamut of climatic scenarios.

Figure 10d illustrates the relationship between indoor air velocity and thermal comfort across various temperature ranges. It can be observed that higher air velocities, which are predominantly prevalent during the summer months, correspond to lower TCV values, suggesting an increase in thermal comfort. This phenomenon is likely attributable to the prevalent cooling and ventilation strategies employed during these warmer periods. Conversely, during winter, instances of high indoor air velocity are comparatively scarce, thereby rendering the impact of air movement on thermal comfort less discernible. The lack of significant data points under cold conditions suggests that ventilation strategies may be less aggressive, possibly due to the heating requirements and the desire to minimize energy loss. This analysis underscores the importance of considering the seasonal context when evaluating the influence of air velocity on thermal comfort. Airflow, often a crucial factor in thermal comfort during hot conditions, might play a nuanced role in colder climates.

Such insights are vital for the design of HVAC systems that are responsive to the thermal needs of occupants while balancing energy efficiency across seasonal variations.

6. Conclusions

This research presents an innovative approach for evaluating building thermal comfort in China, utilizing a smart evaluation model underpinned by an open-access dataset. Through the integration of Bayesian-optimized LightGBM and SHAP methodologies, we have developed an explainable machine learning model that accurately predicts thermal comfort requirements across different regions and building types. The following key insights have been distilled from our study:

(1) Our model has demonstrated commendable accuracy in evaluating thermal comfort, with SHAP analysis providing granular insights into the model's internal workings. The ability of the model to generalize across the test set with high precision suggests its potential for widespread application in smart building management systems.
(2) The study underscores the paramount influence of indoor temperature on thermal comfort voting, reiterating the necessity for precise temperature control in the pursuit of occupant comfort. The notable impacts of building type and metabolic rate highlight the significance of architectural design and human physiological activity in thermal comfort perception.
(3) The insights gleaned from our analysis have significant policy implications. They can inform the development of energy-efficient thermal comfort standards and regulations that are sensitive to regional climatic diversity and personalized occupant needs. Accurate predictions of thermal comfort can aid substantially in the optimization of energy usage, aligning with the objectives of sustainable development and carbon neutrality. The model's ability to delineate the influence of distinct factors enables the design of energy-efficient and occupant-centric thermal environments.
(4) The research paves the way for future studies to incorporate additional variables, such as clothing adaptability, occupant behaviour, and building occupancy patterns. Such expansions could yield a holistic thermal comfort model that is both predictive and prescriptive, aiding stakeholders in creating energy-efficient, comfortable, and health-promoting built environments.

In conclusion, our study contributes a sophisticated, data-driven evaluation model to the field of building thermal comfort. This model not only serves as a tool for optimizing thermal comfort but also acts as a guide for sustainable building design and operation, ultimately supporting the global endeavor to mitigate climate change through improved energy stewardship in the building sector. With its capacity to elucidate complex relationships within large datasets, our research exemplifies the potential of machine learning to revolutionize building science and urban planning.

Author Contributions: H.L.: Methodology, Data visualization, Writing. E.M. Conceptualization, Supervision. All authors have read and agreed to the published version of the manuscript.

Funding: This research received no external funding.

Data Availability Statement: The raw dataset can be found at https://www.nature.com/articles/s41597-023-02568-3 (accessed on 1 November 2023). The filtered data used for this article will be uploaded to be publicly available on the GitHub repository.

Acknowledgments: The authors gratefully acknowledge Yang et al. [31] for their open-access Chinese thermal comfort dataset.

Conflicts of Interest: The authors declare that they have no known competing financial interests or personal relationships that could have appeared to influence the work reported in this paper.

References

1. Houghton, J. Global warming. *Rep. Prog. Phys.* **2005**, *68*, 1343. [CrossRef]
2. Quadros, B.; Mizgier, M. Urban green infrastructures to improve pedestrian thermal comfort: A systematic review. *Urban For. Urban Green.* **2023**, *88*, 128091. [CrossRef]
3. EU Directive. 31/EU of the European Parliament and of the Council of 19 May 2010 on the Energy Performance of Buildings. *Off. J. Eur. Union* **2010**, *153*, 13–35.
4. Yuan, D.; Jiang, W.; Sha, A.; Xiao, J.; Wu, W.; Wang, T. Technology method and functional characteristics of road thermoelectric generator system based on Seebeck effect. *Appl. Energy* **2023**, *331*, 120459. [CrossRef]
5. Jiang, W.; Yuan, D.; Shan, J.; Ye, W.; Lu, H.; Sha, A. Experimental study of the performance of porous ultra-thin asphalt overlay. *Int. J. Pavement Eng.* **2022**, *23*, 2049–2061. [CrossRef]
6. Jiang, W.; Li, P.; Sha, A.; Li, Y.; Yuan, D.; Xiao, J.; Xing, C. Research on Pavement Traffic Load State Perception Based on the Piezoelectric Effect. *IEEE Trans. Intell. Transp. Syst.* **2023**, *24*, 8264–8278. [CrossRef]
7. Vergés, R.; Gaspar, K.; Forcada, N. Assessment of the energy implications adopting adaptive thermal comfort models during the cooling season: A case study for Mediterranean nursing homes. *Energy Build.* **2023**, *299*, 113598. [CrossRef]
8. Yue, L.; Zhongqing, C. Seasonal thermal comfort and adaptive behaviours for the occupants of residential buildings: Shaoxing as a case study. *Energy Build.* **2023**, *292*, 113165. [CrossRef]
9. Sun, Y.; Zhang, C.; Zhao, Y.; Li, J.; Ma, Y.; Zhu, C. A systematic review on thermal environment and thermal comfort studies in Chinese residential buildings. *Energy Build.* **2023**, *291*, 113134. [CrossRef]
10. International Energy Agency (IEA). *Global Status Report for Buildings and Construction*; IEA: Paris, France, 2019.
11. Ma, Z.; Wang, J.; Ye, S.; Wang, R.; Dong, F.; Feng, Y. Real-time indoor thermal comfort prediction in campus buildings driven by deep learning algorithms. *J. Build. Eng.* **2023**, *78*, 107603. [CrossRef]
12. Uddin, M.N.; Lee, M.; Ni, M. The impact of socio-demographic factors on occupants' thermal comfort and sensation: An integrated approach using statistical analysis and agent-based modeling. *Build. Environ.* **2023**, *346*, 110974. [CrossRef]
13. Wu, Y.; Cao, B.; Zhu, Y. Development of an automatic personal comfort system (APCS) based on real-time thermal sensation prediction. *Build. Environ.* **2023**, *246*, 110958. [CrossRef]
14. Ma, S.; Deng, W.; Lu, J.; Zhou, T.; Liu, B. Investigation of thermal comfort and preferred temperatures for healthcare staff in hospitals in Ningbo, China. *J. Build. Eng.* **2023**, *80*, 108029. [CrossRef]
15. Lee, J.; Cha, S.H.; Hong, T.; Koo, C. Empirical investigation of occupant-centric thermal comfort in hotel guestrooms. *Renew. Sustain. Energy Rev.* **2024**, *189*, 114046. [CrossRef]
16. *ANSI/ASHRAE Standard 55-2017*; Thermal Environmental Conditions for Human Occupancy. ASHRAE: Peachtree Corners, GA, USA, 2017.
17. *ISO 7730:2005*; Ergonomics of the Thermal Environment—Analytical Determination and Interpretation of Thermal Comfort Using Calculation of the PMV and PPD Indices and Local Thermal Comfort Criteria. ISO: Geneva, Switzerland, 2005.
18. Fanger, P.O. *Thermal Comfort: Analysis and Applications in Environmental Engineering*; Danish Technical Press: Copenhagen, Denmark, 1970.
19. De Dear, R.J.; Brager, G.S. Thermal comfort in naturally ventilated buildings: Revisions to ASHRAE Standard 55. *Energy Build.* **2002**, *34*, 549–561. [CrossRef]
20. Kramer, T.; Garcia-Hansen, V.; Omrani, S.; Zhou, J.; Chen, D. Personal differences in thermal comfort perception: Observations from a field study in Brisbane, Australia. *Build. Environ.* **2023**, *245*, 110873. [CrossRef]
21. Su, X.; Yuan, Y.; Wang, Z.; Liu, W.; Lan, L.; Lian, Z. Human thermal comfort in non-uniform thermal environments: A review. *Energy Built Environ.* 2023, *in press*. [CrossRef]
22. Wu, Y.; Jiang, A.; Liu, H.; Li, B.; Wang, W.; Zheng, X. Effect of the mood of indoor activities on thermal comfort after winter commutes with different metabolic levels. *Energy Build.* 2023, *in press*. [CrossRef]
23. Zhang, J.; Lu, J.; Deng, W.; Beccarelli, P.; Lun, I.Y.F. Thermal comfort investigation of rural houses in China: A review. *Build. Environ.* **2023**, *235*, 110208. [CrossRef]
24. Deng, M.; Fu, B.; Menassa, C.C.; Kamat, V.R. Learning-Based personal models for joint optimization of thermal comfort and energy consumption in flexible workplaces. *Energy Build.* **2023**, *298*, 113438. [CrossRef]
25. Arowoiya, V.A.; Moehler, R.C.; Fang, Y. Digital Twin Technology for Thermal Comfort and Energy Efficiency in Buildings: A State-of-the-Art and future directions. *Energy Built Environ.* 2023, *in press*. [CrossRef]
26. Lundberg, S.M.; Lee, S.I. A unified approach to interpreting model predictions. *Adv. Neural Inf. Process. Syst.* **2017**, *30*, 4768–4777.
27. Strumbelj, E.; Kononenko, I. Explaining prediction models and individual predictions with feature contributions. *Knowl. Inf. Syst.* **2014**, *41*, 647–665. [CrossRef]
28. Ribeiro, M.T.; Singh, S.; Guestrin, C. "Why should I trust you?" Explaining the predictions of any classifier. In Proceedings of the 22nd ACM SIGKDD International Conference on Knowledge Discovery and Data Mining, San Francisco, CA, USA, 13–17 August 2016; pp. 1135–1144.
29. Yang, Y.; Yuan, Y.; Han, Z.; Liu, G. Interpretability analysis for thermal sensation machine learning models: An exploration based on the SHAP approach. *Indoor Air* **2022**, *32*, e12984. [CrossRef]
30. Iban, M.C. An explainable model for the mass appraisal of residences: The application of tree-based Machine Learning algorithms and interpretation of value determinants. *Habitat Int.* **2022**, *128*, 102660. [CrossRef]

31. Yang, L.; Zhao, S.; Zhai, Y.; Gao, S.; Wang, F.; Lian, Z.; Duanmu, L.; Zhang, Y.; Zhou, X.; Cao, B.; et al. The Chinese thermal comfort dataset. *Sci. Data* **2023**, *10*, 662. [CrossRef] [PubMed]
32. *ASHRAE Standard 55-2020*; Thermal Environmental Conditions for Human Occupancy. ASHRAE: Atlanta, GA, USA, 2020.
33. Zhai, Y.C.; Li, M.; Gao, S.; Yang, L.; Zhang, H.; Arens, E.; Gao, Y. Indirect calorimetry on the metabolic rate of sitting, standing and walking office activities. *Build. Environ.* **2018**, *145*, 77–84. [CrossRef]
34. Ke, G.; Meng, Q.; Finley, T.; Wang, T.; Chen, W.; Ma, W.; Ye, Q.; Liu, T.-Y. Lightgbm: A highly efficient gradient boosting decision tree. *Adv. Neural Inf. Process. Syst.* **2017**, *30*, 3149–3157.
35. Jin, D.; Lu, Y.; Qin, J.; Cheng, Z.; Mao, Z. SwiftIDS: Real-time intrusion detection system based on LightGBM and parallel intrusion detection mechanism. *Comput. Secur.* **2020**, *97*, 101984. [CrossRef]
36. Liang, W.; Luo, S.; Zhao, G.; Wu, H. Predicting hard rock pillar stability using GBDT, XGBoost, and LightGBM algorithms. *Mathematics* **2020**, *8*, 765. [CrossRef]
37. Mangalathu, S.; Hwang, S.H.; Jeon, J.S. Failure mode and effects analysis of RC members based on machine-learning-based SHapley Additive exPlanations (SHAP) approach. *Eng. Struct.* **2020**, *219*, 110927. [CrossRef]
38. Nor, A.K.M.; Pedapati, S.R.; Muhammad, M.; Leiva, V. Abnormality detection and failure prediction using explainable Bayesian deep learning: Methodology and case study with industrial data. *Mathematics* **2022**, *10*, 554. [CrossRef]
39. Parsa, A.B.; Movahedi, A.; Taghipour, H.; Derrible, S.; Mohammadian, A. Toward safer highways, application of XGBoost and SHAP for real-time accident detection and feature analysis. *Accid. Anal. Prev.* **2020**, *136*, 105405. [CrossRef] [PubMed]
40. Wen, X.; Xie, Y.; Wu, L.; Jiang, L. Quantifying and comparing the effects of key risk factors on various types of roadway segment crashes with LightGBM and SHAP. *Accid. Anal. Prev.* **2021**, *159*, 106261. [CrossRef] [PubMed]
41. Nguyen, H.D.; LaFave, J.M.; Lee, Y.J.; Shin, M. Rapid seismic damage-state assessment of steel moment frames using machine learning. *Eng. Struct.* **2022**, *252*, 113737. [CrossRef]
42. Li, Z.; Lei, H.; Ma, E.; Lai, J.; Qiu, J. Ensemble technique to predict post-earthquake damage of buildings integrating tree-based models and tabular neural networks. *Comput. Struct.* **2023**, *287*, 107114. [CrossRef]
43. Zafarmandi, S.; Matzarakis, A.; Norford, L. Effects of clothing's thermal insulation on outdoor thermal comfort and thermal sensation: A case study in Tehran, Iran. *Sustain. Cities Soc.* **2024**, *100*, 104988. [CrossRef]
44. Abdollahzadeh, S.M.; Heidari, S.; Einifar, A. Evaluating thermal comfort and neutral temperature in residential apartments in hot and dry climate: A case study in Shiraz, Iran. *J. Build. Eng.* **2023**, *76*, 107161. [CrossRef]

MDPI AG
Grosspeteranlage 5
4052 Basel
Switzerland
Tel.: +41 61 683 77 34

Buildings Editorial Office
E-mail: buildings@mdpi.com
www.mdpi.com/journal/buildings

www.ingramcontent.com/pod-product-compliance
Lightning Source LLC
Chambersburg PA
CBHW061444210926
51850CB00008B/1603
* 9 7 8 3 7 2 5 8 5 7 9 1 3 *